Uwe Hansmann, Lothar Merk
Martin S. Nicklous, Thomas Stober

Pervasive Computing

The Mobile World

Second Edition
Forewords by Pertti Korhonen,
Philippe Kahn and Nick Shelness
With 211 Figures

 Springer

Uwe Hansmann
Lothar Merk
Martin S. Nicklous
Thomas Stober

IBM Deutschland Entwicklung GmbH
Pervasive Computing
Schönaicher Straße 220
71032 Böblingen, Germany

Library of Congress Cataloging-in-Publication Data applied for

ISBN 3-540-00218-9 Springer-Verlag Berlin Heidelberg New York
ISBN 3-540-67122-6 (first edition) Springer-Verlag Berlin Heidelberg New York

Springer-Verlag Berlin Heidelberg New York,
a member of BertelsmannSpringer Science+Business Media GmbH
http://www.springer.de

© Springer-Verlag Berlin Heidelberg 2001, 2003
Printed in Germany

Cover design: KünkelLopka , Heidelberg
Typesetting: Data conversion by G & U e.Publishing Services, Flensburg
Printed on acid-free paper 33/3142 GF   543210

Foreword

Mobile communications has changed the way we communicate. By removing the restriction of place, people world-wide have found new and rewarding ways of connecting with others – both privately and for business. The possibility of anytime, anywhere communications brings unprecedented choice and freedom. By virtue of being the most cost-effective form of communication, mobile technologies have, in a mere decade, surpassed the number of users that it has taken the fixed network more than a century to reach.

Recent advances in mobile network technologies have brought about a significant increase in available bandwidth, providing a solid basis for the transition from voice-only mobile services to web-based content services. These new services will also broaden the communications modes from one-to-one to one-to-many and many-to-many. With almost 2 billion subscribers in the early 2010's, mobility will be the common denominator for all communications.

In this period of technological evolution, it is important not to lose sight of what made mobile communications successful in the first place: ease of use, user control and low cost. The user requirements and development trends call for an easy method to connect service and content providers to mobile networks and the end-user.

The next few years will see the further convergence of mobile communications and the Internet, resulting in various new technologies and new pervasive products. This Mobile Internet will not be simply the Internet of today accessed from a mobile device. As users, we will not be browsing Internet pages for content as we do today. Instead, we will be using tailored applications and services profiled according to our personal preferences, time and place. Users will be able to download applications and content that fit their personal profile and lifestyle. The true challenge for the Mobile Internet's technical architecture is to provide this seamless user experience, making it as easy as placing an ordinary voice call.

The Mobile Internet vision assumes users will use different types of devices for connecting to many different services, via various access networks. In practice, this means that users must be able to access their personalized services through devices that best fit their needs, intentions and whereabouts. Users will expect that these underlying technical and architectural differences will not affect the actual usage experience. While there will be numerous applications, a small set of abstractions is sufficient to define the framework, necessary to build the foundation for seamless a user experience and architectural interoperability.

For example, to understand the application requirements for underlying network environments, the concept of different interaction modes has been created. Three interaction modes, rich call, browsing and messaging, categorize the applications in terms of variables such as end-to-end real-time response. Such architectural definitions will help to break the challenge into manageable tasks that can be solved and further specified by joint industry efforts.

To stimulate innovation and enable rich contribution by many different players, a shared architecture based on open technologies and standards is necessary. This will result in a rich offering of innovative and interoperable services and solutions for users and business opportunities for innovative companies.

Pertti Korhonen
Executive Vice President
Mobile Software, Nokia
February 2003

Foreword 1 to the First Edition

Our future is pervasive.

We were not meant to be chained to our computers in our Dilbert cubicles. Having millions of white-collar workers commute to work every day to sit for eight hours chained to their computer is something right out of a Kafka novel.

Before the cell phone, I can remember sitting at my desk waiting for calls to be returned. If I was traveling I made sure that I was all setup to place calls from public phone booths.

Today, I can't imagine having to look for a telephone booth! I can be reached on my tri-band digital cell phone anywhere in the world instantly. I also have a PDA that I carry everywhere. My PDA instantly synchronizes with my desktop computer, which in turn connects to a larger network of computers and web portals. The world is my neighborhood.

Soon digital devices will be working together hand in hand thanks to technologies such as Bluetooth. This includes devices such as Cell phones, Digital Cameras, PDAs, and desktop computers. The Wireless Internet is the catalyst for the convergence of digital communications, computing, and rich media. This is a mega trend that we call "Pervasive Computing".

What "Pervasive Computing" means to each and every one of us is the ability to take our own worlds of personal and business information with us wherever we go. Now we can always be in touch with our families, affinity groups, and co-workers.

The vision for "Pervasive Computing" encompasses universal and integrated instant person to person or person to business visual communications. This is a complete new media that will revolutionize the way we communicate. One good example is an instant wireless digital camera that can share photographs instantly with the rest of the world through the wireless Internet: point, shoot, and share… instantly. A picture is worth a 1,000 words!

As technologists we are delivering this vision through careful adherence to open industry standards such as JPEG 2000 and SyncML. Together with the worldwide build-up of a global infrastructure for the Wireless Internet these are the key ingredients to make our vision a reality.

The technology components necessary to deliver on this global vision are advanced and complex. However, they deliver an elegant and functional end-to-end solution. And as we know well, a world connected and synchronized is a world perfected.

Because of its universal appeal, the industry growth that will be generated by Pervasive Computing will be astounding.

"Pervasive Computing" is global and universal and focused on simplifying our daily lives. A very good thing!

Philippe Kahn
Chairman of the Board, LightSurf Technologies, Inc.
August 2000

Foreword 2 to the First Edition

Those of us old enough to have grey or missing hairs have lived through the aftermath of a number of major technological developments. These developments have resulted in the widespread deployment of, for example, black and white televisions, color televisions, VCRs, camcorders, quartz watches, mobile telephones, personal computers, and the Internet. In some of these cases (black and white television, VCRs, mobile telephones, personal computers, and the Internet), these technological developments have had a profound sociological impact. They have changed the way that people live their lives. In other cases (color television, camcorders, and quartz watches) they have been replacement technologies that have significantly improved the quality of an experience, but have had little or a less profound sociological impact.

We now stand on the verge of yet another major technological development – the embedding of inexpensive digital processing, storage, and communication capabilities into a vast range of devices both existing and new. These devices will span the gamut from home appliances (refrigerators, washing machines, stoves, etc.) to buildings (windows, doors, heating and air conditioning, etc.), vehicles (cars, busses, trucks, trains, airplanes, etc.), and new forms of personal electronic accessory (electronic organizers, electronic shopping lists, electronic books, web capable cell phones, etc.). What is not yet clear is how these capabilities will be exploited, if and how these devices will communicate with each other in practice, and what, if any, sociological impact their availability will have.

There are a number of preconditions that are necessary for the widespread deployment of devices that exploit new technological capabilities.

1. They have to deliver an easily perceived value, at a low enough price, to convince a significant number of individuals to part with their hard-earned cash.

2. They have to fit seamlessly into a widely deployed infrastructure.

3. They have to be sufficiently easy to use.

Without an easily perceived value, nothing will happen, but price is also a key element of any deployment equation. From the moment of their introduction, a huge number of people could see the value of owning, carrying, and using a cell phone. They didn't rush out and buy one because the initial prices of purchase and use were high, and could only be justified for high value calls. If one views cell phone deployment rates, especially in Europe, then there have been three pricing developments that have had a huge impact on deployment levels. a) The real cost of handsets came down, but perhaps even more important, most of the real cost of a handset was amortized by service providers over future subscription and call charges. This allowed handsets to be sold for an artificially low price, and often be given away for free. b) During evenings and on weekends, cellular call tariffs were priced significantly below residential wireline telephone tariffs. This made a cell phone a significantly cheaper alternative to a residential wireline telephone, especially for households which were away during the day. c) Handsets that employ prepaid smart cards (SIMs) were introduced. These allowed cellular users to both budget and cap their call charges on a call by call basis, and for parents to totally eliminate bill shock when providing their children with cell phones. Each of these pricing developments has had a profound effect on the willingness of very large populations to own, carry, and use cellular telephones. In the UK alone, the introduction of prepaid smart card cell phones turned an additional 1/4 of the entire population into cell phone users in little more than a year.

Once purchased, devices must have widespread access to a widely deployed infrastructure. This in practice requires at least national, if not the universal, adoption and deployment of de-jure or de-facto standards. If different television sets had been required to view different broadcast channels, deployment would almost certainly have been stalled. Attempts to employ a different parallel infrastructure, even if better, are also usually doomed to failure. Sony's Betamax VCR format is a well known example of this. Without near total coverage in the areas a user wishes to traverse, a cellular telephone becomes largely useless. It is just such a widely deployed infrastructure that has given GSM cellular telephones their extremely wide appeal. I personally carry a tri-band GSM2 cell phone that has worked in every continent I have traversed in the last year including large parts of the North America, Europe, Australia, and Asia. Perhaps the greatest testimony to the importance of a widely deployed and common infrastructure is the World Wide Web. Hypertext linking and packet switching are both thirty year old technologies. What gave the Web its growth power was the ability to use almost any browser on almost any computer to access an entire and mushrooming world of content and services.

Finally, devices based on new technologies have to be sufficiently easy to use. Note, I didn't say easy to use. The number of deployed VCRs that display a flashing time of 00:00 are testimony to this, but if the main things that a user wishes to do are too difficult they will just give up, and tell their friends and neighbors not to waste their money trying.

So where do devices that embed digital processing, storage, and communication capabilities stand with respect to these three criteria? The simple answer is that it is too soon to know. There have been some extremely successful pilots and even full scale deployments in relatively narrow spaces such as electronic shopping assistants and web capable cell phones, but these are hard to project from.

We know that in Japan, NTT DoCoMo are deploying more than one million new I-mode (web capable) cell phones a month, but projecting from Japan, a country in which there are only 3 million PC–based web users, almost all of them office based, to other geographies that already have large numbers of home–based PC web users, is extremely dangerous. Similarly, projecting from Japan's success in employing a pure Web (HTML 3.2 and HTTP 1.1 over a cellular bearer rather than TCP/IP) infrastructure to future European success in deploying an alternative WAP (non-HTML and non-HTTP) infrastructure is, I think, premature.

So what is the role of this book? It is twofold. It is to peer into the future by presenting a number of potential road maps, and it is to describe the technologies and infrastructures that will be required to pave the way. Some of these technologies and infrastructures are available at a suitable price now, some will be available at a suitable price soon, while others are still missing in action.

Like the early days of automatic teller machines (ATMs), early pervasive computing pilots and deployments are already having a significant bottom line impact for early adopters (see NTT DoCoMo above). With the passage of time the deployment of pervasive computing devices and infrastructures, as with ATMs, may well simply become a cost of doing business that delivers minimal bottom line advantage. For this reason, the time is now right for businesses to determine whether they wish to take the risks and garner the potential rewards of being early adopters. This book is a useful tool in coming to an informed decision.

Nick Shelness
Chief Technical Officer, Lotus Development Corporation
August 2000

Table of Contents

XIV   ■   Table of Contents
    ■
    ■

Preface

"...a billion people interacting with a million e-businesses
with a trillion intelligent devices interconnected ..."
(Lou Gerstner, IBM Chairman and CEO)

Information is the nucleus of today's interconnected economy. We need to be able to exchange and retrieve our personal information quickly, efficiently, and securely, at any time and regardless of our current physical location. Electronic storage, transmission, and access of information are common tasks we rely and trust on.

Information everywhere

Convenient applications for helping users manage information in daily life and business environments will be an essential growth factor in tomorrow's IT industry. Such applications will integrate software, hardware, infrastructure, and especially services and will focus on the needs and convenience of their potential users. Those who provide these applications will be able to offer added value, allowing them to differentiate themselves from their competitors. Traditional off-the-shelf software and even hardware will serve only as enabling components for these integrated solutions. These components will be standardized and exchangeable.

"Everywhere at anytime" ...

Pervasive Computing

This common slogan expresses in a nutshell the goal of *Pervasive* or *Ubiquitous Computing*. Both terms describe the visible and mobile front-end for the next generation of integrated IT applications. Pervasive Computing includes flexible and mobile devices like personal digital assistants, mobile phones, pagers, hand-held organizers, and home entertainment systems, which will access or provide a rich diversity of applications. As part of our daily life, these intelligent devices will offer an emerging number of private and professional transactions. They connect to worldwide networks without boundaries and provide quick and secure access to a wealth of information and services.

The early pilot applications of Pervasive Computing seem to originate from the pages of a science-fiction novel:

A new era of applications

A supermarket chain equips its customers with free mobile devices accompanied by an application for placing online orders. A salesman for agriculture seeds accesses, from a remote farm, his company's order systems via a wireless network. Theatergoers pay for their ticket using a mobile WAP phone. In a networked home equipped with intelligent appliances, the microwave and the air-condition are controlled through the Internet using a common browser.

Integration behind the scenes   Nevertheless, Pervasive Computing is far more than just plugging in new incredible devices for multiple purposes. Behind the scenes, an increasing number of these new devices must be integrated into the IT organization. Device management as well as application management are two issues back-end systems have to deal with. Gateways, application servers, and communication networks are just some of the components used for building Pervasive Computing solutions. Small footprint clients for common systems, like databases or web servers, have to be supplied for numerous platforms.

Services   Besides making the systems ready for Pervasive Computing, the service infrastructure has to be extended. Internet Service Providers must be prepared to react to a rapidly growing demand for network access by everyone. GSM has turned out to be the dominant wireless network for communicating anywhere, making it the first choice for accessing Pervasive Computing applications from any point. Upcoming Internet portals will act as an "anytime available" intermediator between users and providers offering payment facilities, shopping opportunities, or information access. Nevertheless, even more complex services, like a public key infrastructure adding enhanced security and privacy to transactions, are beginning to be established.

Common standards   To make all these things work together seamlessly, the industry is currently agreeing upon the required standards. SyncML is one promising approach for ensuring the possibility to synchronize all kind of data on different devices. XML is definitely the basis for standardized data exchange.

About This Book

Technology   This book explains the fascinating variety of front-end devices for ubiquitous information access and their operating systems. It also covers the powerful back-end systems, which integrate the Pervasive Computing components into a seamless IT world. Fundamental topics include such commonly used terms as XML, WAP, transcoding, cryptography, and Java, just to mention a few. Another focus is set on the evolving industry standards, like SyncML, WAP, on which these new technologies are based.

Development   The book presents an overview of the different development strategies and tools for different Pervasive Computing platforms. In order to ease the planning and development of new solutions, concepts and considerations are explained specific to a variety of different target environments, such as handheld computers, home networks, smart phones, and others.

Business   Finally, this book has a strong emphasis on the business aspects, such as the new generation of services, delivering added value to a growing number of customers. E-Business, private home, finance, or travel will not be the only industries which will change their face entirely due to a new class of mobile computing devices, making the customer omnipresent.

We try to give an impression of what Pervasive Computing is about and how it differs from conventional computing. We avoid losing ourselves in details, but provide a way through the jungle of terms, concepts, standards, and solutions instead. The major goal of this book is to put the various facets of Pervasive Computing together to a consistent and comprehensive view.

A red line through the jungle

The Audience of This Book

Giving a comprehensive and profound overview of Pervasive Computing makes this book very valuable for a wide audience of readers. Following the main thread of this book, they will find an easy and quick entry to the related topics.

Business managers will learn what impact Pervasive Computing has on economy and society. The knowledge about Pervasive Computing paradigms, new business models, and a new generation of applications will affect their work as well as their decisions. They will see where Pervasive Computing can help businesses to offer new services and new products or how to improve existing businesses to reach a new range of customers. This book gives an overview about the broad range of possible as well as already existing solutions in the field of Pervasive Computing.

Business Managers

Those software architects and project managers who extend their e-business activities to a new front-end will read about which components pervasive solutions are made of and how these building blocks are related with each other. This book gives an overview of state-of-the-art pervasive technology and shows which components are available as well as how they fit together to build a complete solution.

Software Architects and Project Managers

Application developers getting involved with a particular segment of Pervasive Computing will find a high-level but profound introduction leveraging the start before digging into mazes of programmer guides. They will learn how to rapidly enable applications for the use of pervasive devices. They will also learn some typical patterns of Pervasive Computing and how to trade off various desirable properties in making design decisions. Typical development processes and tools are described.

Application developers

No Need to Read the Whole Book

Most of us no longer have the time to read a book cover to cover. Therefore, we have broken this book into chunks that may be read in almost any sequence:

- "Introduction",
- "Part I, Devices",
- "Part II, Software",
- "Part III, Connecting the World",

- "Part IV, Back-end Server Infrastructure",
- "Part V, New Services", and
- "Part VI, Appendices"

At the end of each chapter, you will find a list of interesting links and suggestions for further reading.

Introduction
We set the stage by providing an overview of Pervasive Computing in general.

Chapter 1 **What Pervasive Computing is all about**

We describe the evolution from conventional computing via e-business to Pervasive Computing. We explain the paradigms and principles of Pervasive Computing, like decentralization, diversification, connectivity, and simplicity.

Part I ***Devices***

In this part, we cover the most commonly used pervasive devices, like handheld computers, smart phones, telematics, smart cards, and many more. We describe their characteristics from a user's perspective, explain the areas of usage, their basic applications, their look-and-feel and everything else you need to know to get a good understanding of these devices. This part groups the vast mass of devices into four categories, which are described in more detail within the different chapters.

Chapter 2 **Information Access Devices**

The first category covers handheld computers and smart phones, which are primarily used to access and communicate information. They are currently the most important group of Internet connected devices beside the traditional PC.

Chapter 3 **Smart Identification**

The second category comprises tiny labels and smart cards, which offer a huge possibility for identifying objects electronically in pervasive applications.

Chapter 4 **Embedded Controls**

This category shows how everyday devices, such as controls, refrigerators, or vending machines can provide additional services and applications while connected to a network. Another section describes the automobile as a versatile pervasive device.

Chapter 5 **Entertainment Systems**

Finally, the fourth category covers devices, which provide manifold sort of entertainment, such as game consoles and interactive television. For these devices analysts predict a promising future.

Software
Part II

In this part, we cover the most commonly used software components needed to build pervasive applications, like, Java, operating systems, middleware, and security building blocks. We also show which components are necessary to develop applications for these operating systems and how to start doing this.

Java
Chapter 6

Java is used in almost every area of Pervasive Computing, in some of the devices as well as in the server back-end infrastructure. The different flavors of Java, especially the ones for Pervasive Computing, are covered and explained here.

Operating Systems
Chapter 7

There is a wide variety of different operating systems used in the field of Pervasive Computing. In this chapter, we cover the most-widely used ones in hand-held devices: Palm OS and Windows CE, as well as EPOC, which is used within phones and smart card operating systems like Java Card. We describe the architecture and list the necessary tools to start developing for the respective platforms. This chapter also contains small sample programs for most of these operating systems.

Middleware Components
Chapter 8

Basic plumbing or middleware is necessary to leverage application programming for the devices themselves. Developing on top of common components helps to achieve independence from a particular Pervasive Computing device used by the customer. Middleware components are used to integrate the clients to their backend servers. We describe an exemplary selection of widely used components, like DB2 Everyplace, WebSphere MQ Everywhere, the JavaTV and JavaPhone APIs, as well as OpenCard Framework and PC/SC.

Security
Chapter 9

Security is an extremely important part of every mobile e-commerce solution. We explain the background of security and the cryptographic techniques used to secure Pervasive Computing and give an overview of the different standards, algorithms, and protocols used.

Connecting the World
Part III

Pervasive Computing and the different devices get really powerful if they are integrated with each other and the back-end infrastructure to form powerful solutions. This Part covers the important industry standards and technologies that provide connectivity and enable communication.

Chapter 10 Internet Protocols and Formats

As Pervasive Computing is somehow an extension of the Internet, most of the protocols and formats you might already know from the Internet are also used here. This chapter introduces HTTP, HTML, XML, XSL, XHTML, and XForms for those of you who are not yet familiar with these standards.

Chapter 11 Mobile Internet

The Wireless Application Protocol (WAP) and i-mode are today's standards for enabling wireless information devices and especially mobile phones to send and receive information and HTML-like pages in an efficient and performant way. We introduce you to the secrets of WAP as well as i-mode and describe to you the most important facts that you need to know.

Chapter 12 Voice

Ease of use is a very important factor in making pervasive computing application a success and voice is the natural way to communicate. In this chapter we introduce what is currently possible and what is underway.

Chapter 13 Web Services

An enormous amount of information is available on the Internet and web services provide a uniform and easy way to access this information for an application. In this chapter we introduce web services to you, show the benefits as well as the limit, and look into one specific example: a web service for remote portlets.

Chapter 14 Connectivity

We cover the different protocols used to connect traditional and Pervasive Computing devices. We explain the background of wired networks, infrared communications, Bluetooth, Wireless LAN as well as cellular and short-range radio wireless connections.

Chapter 15 Service Discovery

With a set of distributed Pervasive Computing devices offering different services to each other, a mechanism for service lookup and discovery is needed for an easy and automatic configuration of this complex environment. We explain the three mechanisms Jini, uPnP, and Salutation.

Part IV Back-end Server Infrastructure

Behind the scenes, the server systems need to be prepared for Pervasive Computing and its requirements, which differ significantly from the needs of today's PC-focused networks. We discuss the applied technologies and concepts.

Gateways *Chapter 16*

Gateways are intelligent interceptors between servers and specific classes of devices. We show how they can be used to prepare data and to establish own subnets, for example the wireless network of a particular service provider.

Application Servers *Chapter 17*

In most cases, Pervasive Computing devices are used to interact with data residing on back-end servers. This chapter describes the different additional jobs a back-end server has to do for Pervasive Computing front-ends. Load balancing, servlets, and Enterprise Java Beans are among the captions of this chapter.

Portals *Chapter 18*

Portals are the central points of access to services in a connected world. A portal must be very attractive, interactive, complete, and easy to use to make the customer come back. In this chapter we describe functionality that is common to most portals and explain some exciting examples.

Device Management *Chapter 19*

This chapter deals with managing the incredible amount of devices and applications that are deployed in the field. Examined aspects include customer profiles, accounting, and billing. Another topic is the management of device capabilities and life cycles. And of course, the distribution and maintenance of applications are an important issue covered within this chapter.

Synchronization *Chapter 20*

Keeping data consistent on the server and on various mobile devices requires intelligent synchronization mechanisms. Aspects of synchronization are detailled within this chapter.

New Services *Part V*

Pervasive Computing offers a vast amount of new possibilities to create services, streamline processes and to start new businesses. The following chapters provide an overview on these areas, which are the most likely ones to boom first. They describe some of the most known Pervasive Computing solutions, which are in use by the customers today already.

Home Services *Chapter 21*

Today a Personal Computer is the only computer that a user directly interacts with at home. We explain how Pervasive Computing offers a huge potential for new and additional services at home, like intelligent appliances, home automation, remote home health care, energy services, new communication services, and many more.

Chapter 22	**Travel & Business Services**

A person who is traveling is mobile per se. Offering to a traveler the services he normally uses from his office and additional ones that ease his travel, like checking-in from a mobile phone at an airport, are some examples. Pervasive Computing can also make the daily life in the office easier. This chapter describes the new service opportunities that service provides in these areas face.

Chapter 23	**Consumer Services**

This chapter describes the new services that businesses will offer to consumers, like offering the possibility to shop using mobile devices, or ordering stocks and checking the balance of a checking account from the road. In this chapter, we will explain what is available today and what will come in the near future.

About the Authors

Uwe Hansmann is currently release manager for WebSphere Portal – Express. He was the Secretary of the Open Services Gateway Initiative and a Board member of the OpenCard Consortium. Uwe received a Master of Science from the University of Applied Studies of Stuttgart in 1993 and an MBA from the University of Hagen in 1998. He joined IBM in 1993 as software developer and led the technical marketing support team for IBM Digital Library before joining IBM's Pervasive Computing Division in 1998. Since then he has managed various pervasive computing development projects for IBM.

Uwe is also the co-author of Smart Card Application Development Using Java, as well as SyncML – Synchronizing and Managing Your Mobile Data.

Scott Nicklous currently manages the WebSphere Portal and Pervasive Computing Solutions department at the IBM development laboratory in Böblingen, Germany. He joined IBM in 1984 as software development engineer at the IBM Böblingen lab. While at IBM, Scott has been involved as developer and team leader in numerous projects, mainly in the financial sector, including banking machine and image processing system development. He joined IBM Smart Card Solutions in 1997 to lead the OpenCard Framework development team. Later, he worked as development manager for the IBM WebSphere Portal product. He has been responsible for bridging the gap between development and customer organizations for Pervasive Computing technology products since 2002.

Lothar Merk is working for the IBM Data Management Division as team leader of a Data Management Infrastructure team. He joined IBM in 1995. While at IBM, Lothar has been involved in numerous service and development projects, mainly in the system integration area. Before joining IBM, Lothar worked in a Medical Imaging and System Integration Project for the European Union.

Lothar holds a Master's degree in Computing Science and Theoretical Medicine from the University of Heidelberg/Germany, a Master's degree of Artificial Intelligence from the University of Chambéry/France, and a PhD in Computing Science from the University of Tuebingen/Germany.

Thomas Stober received a masters degree from the University of Karlsruhe and a Ph.D. from the University of Stuttgart, Germany. After 5 years of research at the Fraunhofer-Institute IPA, where he focused on mobile computing, and information logistics, he joined IBM's Pervasive Computing Division in 1998. As a technical leader and architect, Thomas developed smart card technology, data synchronization solutions, and was a member of several related standardization activities. In the past 2 years he worked for the Lotus development team in Westford, Massachusetts. As software architect he pursued the extension of Lotus Notes and Domino to mobile devices. He was a key player in the development of the Websphere Everyplace Access Server. Thomas filed several patents and wrote numerous publications.

The authors can be reached at pvcbook@web.de .

Acknowledgements

We had a unique opportunity to work on several projects in the area of Pervasive Computing that helped us gain and broaden our experience. Without that we could not have written this book. We would like to thank the International Business Machines Corporation and, in particular, it's Pervasive Computing Division for having provided us with that opportunity.

Numerous people furnished us with in-depth reviews of the book, supported us, or provided us with their invaluable expertise. We are indebted to Alexander Busek, Frank Dawson, Kate Dueck, Hermann Engesser, Gabriele Fischer, Dorothea Glaunsinger, Klaus Gungl, Stefan Hepper, Philippe Kahn, Helmut Kehrer, Hans-Jörg Klein, Carl Kriger, Michael Moser, Noel Poore, Gregor Reichle, Stephan Rieger, Frank Seliger, Nick Shelness, Peter Thompson, Michael Wasmund, and Dirk Wittkopp.

We would especially like to thank Amber, Anna, Jakob, Maria, Melanie, Michael, Sandra, Sandra, Taddeo, and Ute for the borrowed time.

1 What Pervasive Computing Is All About

"Convenient access, through a new class of appliances, to relevant information with the ability to easily take action on it when and where you need to." (IBM's definition of Pervasive Computing)

1.1 Times Are Changing

The Industrial Revolution of the 19th century was a result of the steam engine, which was developed by James Watt. *Energy* was used to extend the power and strength of the workers and soon became a key factor of the economy. Since generated energy could not be transmitted over distances at those times, the engines and machines were concentrated at those locations where energy was produced. Large mills and plants were the typical symbol of that era.

Energy triggered the industrial revolution

A number of important inventions initiated a change in this centralized view of production and usage of energy: Werner von Siemens invented the dynamo in 1866. In 1882, Thomas A. Edison took the first power station into operation in New York. Energy could now be transmitted over power cables to various remote consumers, such as electric motors. The electric locomotive was only one of many applications using electric power, which had been made generally available. In 1892, Rudolf Diesel published his patent of the internal combustion engine, allowing energy to be generated at any place in an easy way. The motor car is probably the most visible application using that technology.

Decentralizing energy

Both combustion engine and electric motor changed our world significantly and introduced the second phase of industrialization:

Engines and machines were now *decentralized*. It became possible to provide energy in larger extend at almost every place in the world. Suddenly an increasing number of manufacturers now were able to take advantage of energy. This second phase of industrialization is characterized by an exploding wave of mechanization. Using more and more machines it became possible to multiply the productivity of manpower at every place. Automation, assembly-lines, and manufacturing machines dominated innovation in industry.

Providing energy everywhere at anytime

But ubiquitous availability of energy had also an impact on the daily life in industrialized countries: First, the new manufacturing processes allowed mass-production, reducing costs and even raising the wages of workers. Wealth and standard of living grew. For example, Henry Ford's famous automobile "Tin

A new life-style

Lizzy" was affordable for a large number of customers. Second, the availability of energy created a tremendous number of new applications, which began to spread into private homes. Besides the motor-car were the electric bulb, sewing machine, heating, refrigerator, lawn-mower, and television set – just to mention a few.

All these innovations expose a specific application to the user. They are distinguished by convenience and usability for a very specific purpose and they hide the underlying technology. The applications changed the world and made achieved technical progress visible – not the technology by itself! When using a CD player today you want to hear music. Nobody will care about the tiny electric motor which ejects the inserted disk and consumes electrical energy, which was produced somewhere.

Computing power triggered another economic revolution

During the second half of the 20th century the era of industrialization was followed by another era: *computers* started their impact on the economy. While the industrialization enhanced the *manpower* with the help of machines, computers multiplied the *mental capabilities* of man. With the help of the steam-engine it suddenly became possible to achieve great things, such as lifting tons of steel or driving a locomotive. The usage of computers allowed to process and use incredible masses of information, creating phone bills, payrolls, invoices, balance sheets, manufacturing data or weather forecasts.

The mainframe era

Like the steam-engines, computers began their existence as huge and mysterious black boxes. Centralized data processing centers improved information management of insurance companies, administrations or air traffic controls. Classical closed shop information technology was concentrating on the creation of a static environment for employees. Central management of equipment, applications, and user interfaces was key to increase overall productivity. Customers or business partners were hardly involved in the daily information processing operations. Tightly controlled data exchange interfaces or satellite systems were the only interfaces to the outside world.

Decentralizing computing power: The personal computing era

But, very much like the second phase of the industrialization a hundred years ago, when energy became ubiquitous, we can today observe a distinct move towards decentralizing information technology. Data processing is no longer a privilege of large enterprises operating their own computer systems. The microprocessor has burst the shrine of the mainframes, since the good old personal computer conquered offices and homes. Computing became an intrinsic part of today's telecommunication, entertainment, commerce, finance, and industry. Our high-tech culture increasingly relies on the electronic creation, storage, and transmission of personal and business information. Information technology penetrates and changes job descriptions, life styles, and business relations. The social and economic impact is unimaginable.

E-Business is the vision that today's organizations strive toward. E-Business means to perform business processes of an organization based on computer systems. E-Business improves productivity, efficiency, and vitality of an organization. Innovation cycles and time-to-market are reduced, allowing to react dynamically on changing market requirements. E-Business includes especially the challenge of interfacing directly with business partners and customers.

The *Internet* has turned out to become the standard communication platform for this purpose. The Internet connects a global community of billions of private and professional users, exchanging information, sharing applications, and providing content and services to each other. The rising number of Internet accounts prove that within the next years we will have the largest community computing ever had. Internet and e-Business convert a living room into a shopping mall and trading floor. Customers can compare product prices and features at home. Employees can access business information from remote. In commerce national borders vanish, competitors from all over the world face each other. This comprises retail as well as Business-to-Business sales.

Internet – a worldwide network for everyone

Since information is no longer available only through monopolized systems it became a prerequisite for growth and success in a revolutionized economic system. Information is intrinsically combined with applications and services to provide them. Small firms taking advantage of new technologies explode almost overnight into worldwide enterprises, changing the rules of economy and making the Internet to a global market place.

Information is the new currency of the global economy

1.2
Decentralization Continues

The decentralization of computing power continues: The computer is irresistible on its way to push all limits and is becoming omnipresent (Figure 1.1). Computers will be part of everyday life and an inevitable component when performing a variety of private and business related tasks. Beyond the era of personal computing, the era of *Pervasive Computing* begins: A new class of devices make information access and processing easily available for everyone from everywhere at any time. Users get enabled to exchange and retrieve information they need quickly, efficiently, and effortlessly, regardless of their physical location.

Everywhere at any time: Pervasive Computing

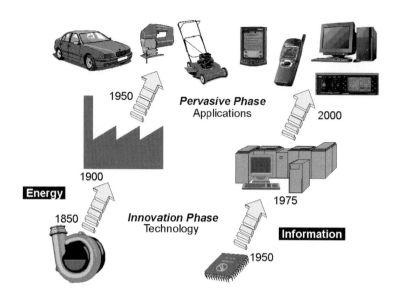

Manifold Pervasive Computing devices extend e-Business from the offices to ubiquity:

- Lightweight PC companions such as handheld computers provide a complementary mobile user interface to intelligent networks.
- Smart cards are ultra-thin security tokens for information access, fitting into every purse.
- Cellular phones achieve wireless mobile access to various computer systems and information-based services from almost everywhere in the world.
- Set-top-boxes, interactive television, and game consoles are the interface between home entertainment systems and entertainment providers.
- Intelligent appliances in a networked home allow the access to a variety of controls from a remote Internet browser. For example, tiny web servers in the size of a matchbox could be applied to manage the state of an air conditioning system. Telephones, light switches, refrigerators, and washing machines will be part of a residential network.
- Industrial controllers or network switches will decentralize intelligence of complex manufacturing systems.
- Embedded systems for cars such as cruise control, on-board phones, and directional assistance systems provide a range of in-vehicle diagnostic, communication, navigational, and security features.

This listing comprises only a few examples of these upcoming pervasive gadgets.

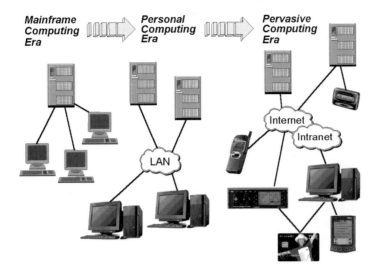

1.3
Applied Pervasive Computing

Today's emerging Pervasive Computing technology faces serious technical is- *Pervasive*
sues: Most devices have strong limitations on memory usage and processor per- *Computing today*
formance as well as tight constraints on power consumption. The footprint of
operating systems and software need to be reduced as much as possible. Mobile
devices must handle power shortages and their applications must be able to
resume again after a shutdown. Pervasive applications need to take care of various hardware and software platforms, as well as of very different form factors
and user interfaces. This obstacle strongly impacts portability.

However, the technology evolves at a tremendous pace: New software architectures are prepared to deal with that diversity and limitations. Form factors
shrink, while computing power increases and devices get less expensive.
Matchbox sized hard disks, tiny embedded processors, miniature energy cells,
small screens, speech and handwriting recognition extend the mobility and capabilities of those computer-like devices. The embedded processor of a smart
card already has the same power as the first PC.

Network technologies and especially wireless communication infrastructures *Pervasive*
like GSM vastly increase the accessibility of information. The bandwidth of *Computing*
wireless networks is rapidly catching up with today's wired connections. The *tomorrow*
Internet has evolved to an universal interface for accessing data and services.

All those manifold devices will soon outnumber personal computers as network connected information processing entities.

Pervasive Computing will have a strong impact on our society. There will be
a flexible and productive new work style. Lifestyle will be influenced by in-

credible communication possibilities, staying in touch with everyone from anywhere. There is endless demand for exchanging and sharing information. Information is accessed and used wherever it is needed in a convenient manner.

Pervasive applications and services

With new technical possibilities, entirely new kinds of applications and *services* arise, bringing benefits for individuals and businesses. Technology and services together will make up the environment in which Pervasive Computing evolves (Figure 1.3):

- Traditional telecommunication companies and phone manufacturers take advantage of this tremendous opportunity coming along: In the past they missed the Internet movement, allowing Internet Service Providers like AOL and others to grow. Now they are eager to be part of the next computing generation from the very beginning. They promote high bandwidth networks, wireless communication, email, voice, paging, SMS, and Internet access: the telecommunication and the IT industry merge silently, both looking towards a promising future. Value-added services far beyond simple voice communication are means of expanding market reach and retaining customer loyalty. New appliances and applications will rapidly increase the customer's demand for networking technology and capacity.

Figure 1.3: Pervasive Computing

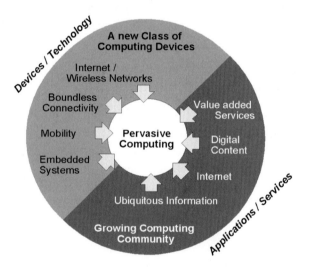

Environment

- For commerce and retail, Pervasive Computing opens up entirely new chances and possibilities of being successful in existing markets and reaching new markets. The owners of these spreading devices will be a large group of potential customers, which cannot be ignored. Mobile Commerce (M-Commerce) is the successor of today's PC based e-Commerce. New value-added services and differentiation are the magic words in the struggle for market share. A mobile shopping application running on an Internet connected handheld improves the consumer's satisfaction and convenience when retrieving product information or submitting orders. On the other hand, enhanced shopping opportunities create new demand, since the customer is enabled to buy more conveniently and at any time.

■ The entertainment industry is targeting a broad field of pervasive applications and services. Video-on-demand or interactive television can be exciting means for delivering content to a broad audience. Triggered by new technical possibilities, traditional broadcasting enterprises face new competition, growth, and profits.

■ The industry will quickly adopt Pervasive Computing to increase productivity. According to Palm, every second Palm OS based handheld computer is bought with corporate money. Pervasive devices will appear in manufacturing, logistics, management, and sales force, enabling ubiquitous access to enterprise data. Flexible responses to changing market requirements and decentralized organization structures need well-informed employees and workers. Information technology is deployed in the field, close to the customers, reducing cycle times and costs.

Despite computers becoming more and more ubiquitous in our life, they tend to get more and more invisible. Like the tiny motor in the tray of the CD player, which is just taken for granted, a computer system built into a mobile MP3 player, which downloads music titles from the Internet will be the most natural thing of the world. No one will care if there is Java inside and how the TCP/IP stack is implemented. Not the enabling technology, but the applications and the delivered services will have a strongly visible influence on our high-tech culture. New applications and services will integrate manifold devices and help users to manage and access information in daily life and business environments. Providing those applications and services will be a key growth factor in tomorrow's IT industry.

Hiding technology behind the application

1.4
Pervasive Computing Principles

Pervasive Computing postulates four fundamental paradigms, which will be detailed in the following sections:

Pervasive Computing paradigms

■ Decentralization
■ Diversification
■ Connectivity
■ Simplicity

1.4.1
Decentralization

The shift from a centralized view to a strongly decentralized computing landscape is the first paradigm of Pervasive Computing and has already been outlined in the previous section. During the mainframe era, powerful supercomputers provided their processing capacity to dumb terminals. With the upcoming personal computer the client-server architecture was introduced, which shifted computing power from server systems to the client workstations.

Distributed systems
Pervasive Computing goes one step further and distributes the responsibilities between a variety of small devices, each of which take over specific tasks and functionality. Each of these autonomous entities contributes to a heterogeneous overall computing landscape. They cooperate in an open mutual community establishing a dynamic network of relationships.

Synchronizing information
The ability to use applications and information on mobile devices and synchronize any updates with network based systems or other devices is a new task arising from that decentralization. Information sources and destinations are widely distributed in a pervasive world. Popular mobile devices, like handheld computers, cellular phones, pagers or laptops have to synchronize their data on the fly between each other as well as with desktop applications, such as calendars or address books. Databases on devices with different capabilities and storage capacities have to be kept consistent.

Managing applications
Pervasive devices and applications are often embedded into a service infrastructure, like a cellular phone network. Decentralization makes it necessary for service providers to administer their deployed software and deliver updates to the customer's devices from remote. They have to keep track of individual user profiles and different device capabilities. To deploy applications and manage devices in such an environment, the server software must be highly scalable and flexible. Back-end systems have to face millions of manifold pervasive devices travelling around the world instead of just thousands of traditional PCs resting peacefully in their offices.

1.4.2
Diversification

The second paradigm of Pervasive Computing affects the functionality of computer systems.

In today's IT world, the typical customer buys an universal computer unit from an arbitrary manufacturer and gathers up the software he needs: Internet browser, word processor, accounting system or whatever. The user usually performs all his tasks with one all-purpose workstation, applications are implemented by software.

Pervasive Computing introduces an entirely new view of functionality: There is a clear move from universal computers challenging performance, price, and functionality to diversified devices which aim at best meeting the requirements of a specific group of users for a specific purpose. A journalist working on an essay needs different tools than a stock trader ordering options. End users will most likely have a whole bunch of specialized computers (just as they have a bunch of electric motors). Those new gadgets appearing in these days in association with Pervasive Computing, such as WAP phones, screenphones, or handheld computers, offer only a highly customized functionality for a particular application context. Applications are a seamless integration of software and hardware. They are intended to be used in a specific situation and optimized for exactly that environment.

Targeting specific needs

A consumer will own and use several devices in parallel, which might have some overlap in functionality, but he will have preferred tools for each specific purpose. For example, he might want to use an Internet screenphone at the time he wants to surf the Internet from home and enjoy multimedia effects of web sites in the best possible quality. For mobile Internet access, a wireless connected handheld computer might be the best choice. Of course that device offers only reduced graphical capabilities and a small display, but he can still retrieve fairly informative web sites, such as travel directions, book catalogues or online newspaper articles. A third alternative for Internet access could be the usage of a WAP phone. It is ultra-light and very handy, at the time he needs specific information on travel, such as flight check-in data or schedule updates. Nevertheless he will not be able to enjoy colorful web sites. All four Internet access devices are complementary and the capabilities of each of them are optimized for a very specific application context (Figure 1.4).

Four alternatives to surf the Web

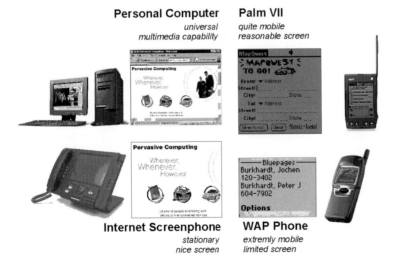

Personal Computer
universal
multimedia capability

Palm VII
quite mobile
reasonable screen

Internet Screenphone
stationary
nice screen

WAP Phone
extremly mobile
limited screen

Figure 1.4: Increasing Diversity

How pale is in contrary the PC business, using more or less only one operating system and one processor platform! The PC had the stickers "Intel Inside" and a Windows logo attached. Pervasive devices will apply the technology that matches best the accompanying usage purpose. It is the application that counts – not what's inside. Each device will take the chipset and operating system which is most appropriate for its individual constraints. An antilock breaking system facing tremendous price-cutting has 16 to 32 kB of memory to work with. It is obvious that this target needs a different operating system than a Windows CE device having a generous 32 MB.

Managing the diversity

One major challenge arising from the increasing diversity is how to manage the different *capabilities* of those manifold devices. Each delivery platform has its own characteristics making it difficult to provide common applications.

The user interface is probably the most obvious difference between devices: some have high-resolution color screens allowing windows-like GUIs while some screens are limited to small two line displays. A user interface designed for a 17″ color monitor is not applicable for a small Palm III display. Applications designed for different devices will have to do more than just zoom a GUI. To ensure readability (and usability) the displayed content has to be filtered and modified. For example a web-based shopping application will have to present product offerings differently on different classes of devices. While a screenphone is able to surf the webpage without limitation and interpret multimedia product presentations, a handheld computer should only visualize a brief overview of the product information including few simple graphics. A WAP phone might even want to reduce the received content to a short price list and disclaim all graphics.

But not only the display capabilities differ from device to device: The set of available data input mechanisms such as stylus, function key, speech or handwriting recognition make user interfaces device specific. Wireless devices having always access to a network require different usage scenarios than only intermittently connected PC companions.

1.4.3
Connectivity

A vision of boundless connectivity

The third paradigm of Pervasive Computing is the strong demand towards connectivity: Manifold devices are seamlessly integrated in an IT world without boundaries. They beam information to each other via infrared, are intermittently connected with plugs, use powerlines as an information carrier or communicate wireless with various back-end systems. Email can be exchanged between heterogeneous devices. Nomadic documents travel through networks, being accessed from anywhere. Cellular GSM phones benefit from international roaming agreements, allowing to connect to alien communication networks. A handheld computer collaborates with a cellular phone via infrared in order to synchronize data over a wireless network. Alternatively, the same handheld can connect via serial port to a LAN. Lou Gerstner, CEO of IBM, described his vision of connectivity as "Everybody's software, running on everybody's hardware, over everybody's network".

At a first glance, this seems to be contrary to the diversification described in the previous section. Platform specific issues are a major obstacle for application and information exchange: The available persistent storage of different devices range from few kilo bytes on smart cards to many Gigabytes on entertainment systems, which can download entire movies. Different processors induce different restrictions on performance and memory usage. Operating systems are numerous and often proprietary to a particular device. Form and shape of devices require different plugs. How shall such different devices as handhelds and set-top-boxes fit together?

Real life obstacles

One approach for achieving connectivity and interoperability is to base the applications on common *standards*. This results in an important task for the IT industry: Open standards have to be established which are prepared to face the demands of the described manifold and differentiated devices. Communication standards, markup languages, and cross-platform software must be integrated on a global basis of interoperability.

Agreeing on common standards

New standards like WAP, UMTS, Bluetooth or IrDA have been created by large cross-industry initiatives, defining the necessary communication protocols as well as the underlying physical connections. The Internet has evolved to be the backbone of worldwide private and public networks.

But it is not only connecting two devices with each other. Applications and data have to be exchangeable too. Java is one approach for achieving platform independence of applications: Java is a programming language concept which produces platform neutral code and can be applied to almost every device. This is an important characteristic in regard to the diversity of pervasive devices: Java code runs on smart cards, handhelds, and intelligent appliances. For the purpose of data exchange, XML is the upcoming de facto standard for platform independent representation of information and content.

Concepts like Jini or UPnP help devices to discover suitable services in a network to which they can delegate specific tasks. Components plugged into the same network communicate with each other and share their resources. The required and available capabilities are negotiated between them automatically. Devices make themselves available, they remove themselves and make an explicit device administration and configuration obsolete. Such plug-and-play capability implies self-explained and easy usage of network-connected utilities. With the help of Jini, the network turns into a dynamic and distributed system.

1.4.4
Simplicity

The flexibility of an all-purpose personal computer is certainly a technical achievement, but it has its price: Those computers we are used to are becoming increasingly complicated. Many of the features a state-of-the-art word processor offers confuse the majority of users and reduce ease-of-use. In spite of plug-and-play capability, the installation of new software is often a challenge for those who are not trained computer experts.

As already mentioned, pervasive devices are very specialized tools that are not optimized for general use. They perform the tasks for which they have been designed very well from a usability point of view. This lines up with the fourth paradigm of Pervasive Computing: Aiming at simplicity of usage.

Convenient, intuitive, self-evident

The magic words are availability, convenience, and ease of use. Information access and management must be applicable without spending significant time learning how to use technology. The user acceptance of user interfaces will have a major impact on the acceptance of services and products offered in the next generation IT landscape. A computer in any form factor should be a easily-accessible tool of everyday life, just as the telephone is today. While proper selection and education of user groups was required to manage the complexity of traditional computer systems, pervasive computers are intuitive to use and might not even require the reading of a manual. Processes that today require installation procedures and take a number of commands on a PC will only need the push of a single button or even accept spoken requests or interpret handwriting.

Mature human computer interfaces

Simple must not be confused with primitive. Pervasive Computing postulates a holistic approach: Hardware and software should be seamlessly integrated and target the very specific needs of an end-user. Complex technology is hidden behind a friendly user-interface. Achieving the intended easy usage requires substantial efforts for application design and development. In order to give quick access to functions, complicated hierarchies of menus and dialogs are not acceptable. A Palm user will appreciate every tap he can save when retrieving information. Nobody will use a handheld for his time management, when it takes more than an instant to boot the device and more than a few clicks to find today's calendar entries. While a Windows PC is still booting, the handheld already has displayed all requested information. Speech recognition, intuitive usage, one-handed operation, instant on/off or touch screens are just a few features of mature human computer interfaces. Providing all this in a small and cheap device is definitely a challenging task for developers.

1.5
Pervasive Information Technology

A quote from Sun Microsystems summarizes the complexity of a pervasive solution: Pervasive Computing aims at the "Convergence of Computers, Communication, Consumer Electronics, Content and Services". Figure 1.5 shows a generic schema that is applied somehow to most solutions and can be simplified as a three tier vertical structure:

- Device:
  The front-end of information technology is the wide range of pervasive devices, designed for creating and accessing information on the fly. These devices are the most visible interfaces to the user and penetrate our business and all day life.
- Workstation:
  Workstations form an optional middle tier. The traditional Personal Computer offers capabilities for working with complex information and managing local personal devices. Often, this layer is even omitted, since most pervasive appliances are able to access their provider's networks directly. Devices like set-top-boxes can replace or complement the personal workstation as a gateway between personal devices and public networks.
- Server:
  Web servers, enterprise servers and mainframes mainly focus on storing and processing large amounts of information using their strong computing power. Pervasive Computing introduces significant changes to software products.

Three vertical tiers ...

Behind this hierarchy of computing systems two underlying layers can be identified, which are of increasing importance:

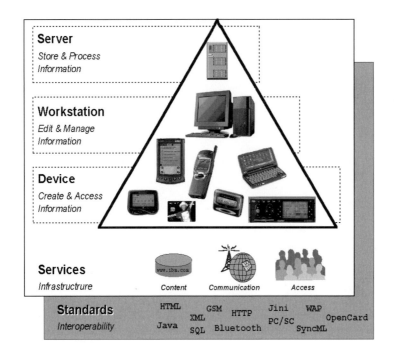

Figure 1.5: A generic view of pervasive information technology

- Standards:
  There is a broad basis of common standards on which the information technology is based. Standards ensure interoperability and connectivity of systems as well as information and application exchange. Since standards are an important issue for Pervasive Computing, they will be ubiquitous throughout this book.

- Services:
  Numerous kinds of services complete the Pervasive Computing landscape. They establish the infrastructure for the widespread usage of computing, since information is intrinsically combined with the accompanying services to provide them.

Part I
Devices

Pervasive devices combine the four paradigms we introduced in Chapter 1: They are strongly decentralized, diversified, connected, and simple to use.

Probably the first device following this philosophy became common long before Pervasive Computing was propagated: it was the ordinary calculator. This application exists in manifold different forms and offers a variety of different features for different purposes, such as a sinus function for scientists, larger keys for merchants, solar energy power for ecologists, miniaturized form in order to fit into a wallet. Calculators are intuitive to use – nobody needs a manual to get started.

Today we have many more pervasive devices offering applications for an information-centric world. And every new promising task one can think of will be followed shortly by new devices helping to fulfill that task. Searching for new killer-applications is definitely one of the challenges for the competing device manufacturers, software vendors, and service providers. *Horizontal and vertical applications*

Many devices are mass products, intended for a broad scope of users. They provide widely applicable personal and professional solutions, such as regular email, Internet access, games and many more. Other devices focus on a very particular usage segment, like an industrial controller device applied in a manufacturing process, handheld companions for parcel delivery services, or data capturing appliances for warehouses. These have highly customized form-factors and run very specific application software.

In order to get a structure into the incredible variety of pervasive devices we distinguish between four major categories (Figure I.1):

- Information access devices
- Intelligent Appliances
- Smart Controls
- Entertainment systems

Dominant in the category of information access devices are pocketsize handheld organizers called Personal Digital Assistants (PDA). They carry around relevant information and are able to intermittently plug into intelligent networks. Their applications are comprised as Personal Information Management (PIM) and include calendar, address book, and mail functionality. While today's usage is mostly restricted to schedule appointments and carry phone numbers around, PDA are evolving to powerful mobile network clients. They allow *Information access devices* immediate access to corporate databases and can be used as an e-commerce platform. Manifold form factors make PDAs sometime hard to recognize, especially when they merge with cellular phones, which are another large group of devices within this category.

Intelligent appliances

Intelligent appliances cover a broad scope of familiar appliances, which are enhanced by embedding more intelligence and connectivity using pervasive technologies. Examples are everyday equipment like point-of-sale terminals GPS navigation, industry controllers, or vending machines, using integrated microprocessor systems to enhance their capabilities and intelligence. Washing machines are connected to the Internet for downloading new program updates or requesting maintenance services. Retail kiosks and self-service terminals let customers place their orders from a online catalogue or print their train tickets including seat reservation.

An important group within this category are network connected information systems in cars. They provide applications, such as remote diagnostics interfaces, wireless communication, Internet access, and navigation. Custom-built machines designed for specialized tasks and filled with computing power are another group within this category.

Smart controls

Smart controls are represented by tiny intelligent controls for stoves, gas pumps, and thermostats. They are connected to networks and managed from remote. Web-pages and Java applets are used to navigate via Internet to a particular lamp in a house, or to set the recording time of a VCR. Mutually all these appliances will communicate with each other: For instance, the alarm clock can pass-on its new settings to set up the heater in the bathroom.

This class also comprises devices like miniaturized web servers, network switches, smart labels, and smart cards. Smart cards are plastic cards with an integrated microprocessor. They are used to store personal cryptographic keys and offer sophisticated cryptographic and identification functionality.

Entertainment systems

New kinds of entertainment systems will change the world of traditional broadcasting. Improved video and audio quality are just one facet. Interactive digital television, electronic programming guides, and video-on-demand are emerging fields of activities, today's entertainment providers are just beginning to enter. Set-top boxes are an enabling technology in this application domain. MP3 players, screenphones, Furby and game consoles are just a few more examples how entertainment industry makes usage of Pervasive Computing technology.

Of course, many devices offer a combination of applications and so for can be assigned to multiple of these four categories. Same as fax machines and printers complement each other, cellular phone capabilities are a logical extension to a handheld device. A handheld computer with an integrated MP3 player is an entertainment system as well as an information access device. White goods in households, consumer electronics, entertainment and telecommunication systems will no longer be independent entities. They will merge together with the accompanying services to seamlessly interconnected clusters.

The hardware technology built into these devices is an important key for making devices even more pervasive. Here are just a few exciting examples:

Hardware technology

- A single chip computer runs the iPic Internet web server, the world's tiniest implementation of a TCP/IP stack. The chip has the size of a match-head, and costs less than a dollar. It can be easily applied to publish the state of a connected light bulb to the Internet.
- The IBM microdrive is a tiny hard disk, storing 1 GB within a PCMCIA card size form factor. Digital cameras and handheld computer are targets of that technology.
- Motorola's new miniature fuel cells will provide ten times the energy density of conventional rechargeable batteries. They will be lighter and cheaper than the batteries we use today. They will make laptops, phones, and PDAs more mobile and less dependent on a power supply.

2 Information Access Devices

2.1
Handheld Computers

Handheld computers comprise the largest group of Internet connectable pervasive devices. They are small, lightweight, and fit into pockets. A stylus is used to tap on a touch screen to activate applications or enter data. Currently two major operating systems are applied on handheld devices: Palm OS and Windows CE.

2.2
Palm OS-Based Devices

Palm OS was developed by Palm Computing, a subsidiary of 3COM. It is the best-known and most popular operating system for handheld devices, having a market share of about 70 percent. Palm OS based devices are manufactured by Palm or other companies, like Symbol, IBM, Handspring, Acer, AlphaSmart, HandEra, Kyocera, Samsung, and Sony, which licensed the operating system and sell devices on their own.

The renown Palm III models were launched in 1998. They included infrared communication, version 3.0 of Palm OS and 2 MB of memory. The Palm IIIx had 4 MB of memory and an improved display. The heart of the Palm IIIx is a Motorola Dragonball EZ processor, which is more power efficient and opens the way for the use of less expensive memory upgrades. The Palm IIIc was the first Palm OS based device with a color display. It was released in 2000.

Some Palm OS devices

The latest Palm devices are the Models m105, m125, m130, m500, and m515. All of the models in this family have 8 MB memory size (except the m515 with 16 MB) and can be distinguished by their design, screens and batteries. The i705 model provides additional e-mail and internet capabilities. The models m500 and m515 are equipped with an extension slot that allows to add memory and back up data. The Palm m500 handheld is based on the Palm V platform.

The Palm devices can be plugged into a cradle, which connects them to the serial port of a PC. An external modem is also available, allowing them to access a network directly through a phone line.

Photos courtesy Palm Computing, Inc.

A cellular phone and a Palm within one device – that is the concept of Samsung's color communicator SPH-I300 and the Kyocera QCP 6035 smart phone, which also includes an MP3 player. These are full-functional Palms with additional mailing and Internet access capabilities.

Symbol's SPT 1500 and 1700 devices are specialized versions of the Palm III. Symbol's SPT 1500 and 1550 have built in laser-beam readers, and the SPT 1800 can transmit the scanned data using a wireless transmission module (Figure 2.2).

Figure 2.2:
Symbol SPT 1800

Photo courtesy of Symbol, Inc.

The last group of Palm OS devices we want to mention here is the Handspring family. The Handspring Visor family was developed by the team around Jeff Hawkins, who left 3COM in 1998 to found his own company. The advantages of the Visor concept are the very good performance, the improved date and address book applications, the USB interface and – maybe most important – the possibility to plug additional peripherals into a hardware interface called *Springboard*. The Visor family comprises about five variants with differences in display, memory size and batteries.

The latest products from Handspring are is the Treo Family. The Treo 180 and Treo 270 combine phone, organizer and internet services in one device (Figure 2.3). All Treo models have a built-in keyboard.

Figure 2.3: Handspring Treo 279

The most important native applications of Palm OS based devices are the Address Book, the Date Book, the To Do List, the Memo Pad, and the Mail application (Figures 2.4 and 2.5).

Palm applications

- The Address Book displays a master list of addresses sorted by last name and displaying the principal phone numbers.
- The Date Book is used to display and edit appointments in a daily, weekly or monthly view.
- At the time the Memo Pad application is started, it first shows an index consisting of the first lines of the memos. If you tap on a list element, the complete memo is displayed.
- With the Mail application users can read downloaded emails or generate new emails offline.

Many other applications are available as freeware, shareware or commercial products from various third-party providers. Among them are fax software, calculators, alarm clocks, drawing applications, games, travel guides, and many, many more.

Figure 2.4:
Address and Date
Book

Figure 2.5:
Mail and Memo
Application

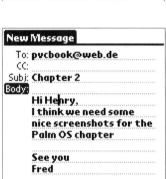

Text can be entered with taps on a small keyboard and displayed on the screen. Alternatively, glyphs can be drawn in a dedicated area below the screen. Glyphs are simplified characters, which can easily be interpreted by a simple handwriting recognition software called Graffiti (Figure 2.6). Graffiti input eliminates the complexity of different writing styles. Almost every letter can be formed by a single stroke.

HotSync The synchronization mechanism between a Palm OS based device and a desktop PC is HotSync. The changes made since the last HotSync are automatically updated on the Palm device and the desktop PC. If corresponding records are changed on the PC and the Palm OS device, HotSync will detect a version conflict.

Infrared port The infrared port offers a convenient possibility to exchange data between different devices. Business cards, calendar entries, and entire applications can be "beamed" from one Palm to another without the detour over HotSync and a PC. The infrared port can also be used to communicate with a cellular phone instead of an attached modem to connect wirelessly to a network.

The Palm allows to access Internet sites either through an Internet connected PC or directly by going online via modem. There are several Web browsers available. Similar to the capabilities of a PC based Web browser, the Palm Web browsers are able to cache Web sides. This feature is extremely important, since a Palm device usually doesn't have a permanent Internet connection. There are several providers for delivering Internet content, especially for handheld devices. One access point for synchronizing content is AvantGo: *Internet access*

- Internet content can be downloaded by the AvantGo software installed on a PC. AvantGo offers selected Web sites related to manifold topics, which are referred to as *channels*. Users can subscribe individually to particular channels of interest and receive updated content with every HotSync. The information is stored in HTML, thus also graphics, tables, and different fonts can be displayed. The Web pages are compressed before they are transmitted to the Palm OS device.
- Since version 2.0 of AvantGo was introduced, it is also possible to use AvantGo as an online Web browser. A proxy server at AvantGo.com modifies the Web pages in a way that they can be displayed on Palm devices.

Similar to the Web access, there are two ways of working with emails – via a desktop PIM application or directly via mail server.

- Choosing the PC way means that sending and receiving emails takes places on the PC. The email is synchronized between the Palm device and the PC whenever a HotSync with the email program takes place. An email written on a Palm has to be delivered to the PC mail application before it is sent to the mail server in a second step.
- For direct Internet access the Palm uses a modem connection and an email server running the standard POP3/SMTP protocols.

2.3
Windows CE-Based Handheld Computers

Windows CE is the second major operating system for handheld computers, taking almost the remaining rest of the market share. Windows CE has been developed by Microsoft and is applied by manufacturers like Casio, HP, and Compaq.

Windows CE versus Palm OS

Comparing Palm OS and Windows CE devices is somehow pretty philosophic: Microsoft wants to extend today's Personal Computer platform to mobile computers. Users should be able to do the same things in the same way on any computer alike device – as far as such is possible under the given constraints of form factors and reduced processing power. Windows CE provides the same consistent look and feel, both on the common PC and on the smaller handheld. For instance Word and Excel will welcome you on Windows CE and the mailing tool is an Outlook derivative, of course. From the Microsoft perspective, a handheld is nothing else than a miniaturized PC! This philosophy is reflected by terms as PocketPC and Handheld PC.

Palm propagates PC companions instead. They are different and complementary to PCs. Since the usage scenarios are different, the applications and their usage must be different too. Palm OS user interfaces are pretty straightforward to use and are optimized for quick mobile information access – even in a crowded subway.

In fact, the opinions about these different approaches are split: Some get started more easily with Windows CE because of the familiar user interface and because they have similar applications on their handheld, as they have on their PC. Others dislike the more circumstantial usage of Windows CE. There are more windows, more controls, and more taps required than a Palm needs for the same operation.

High-end Windows CE hardware

Compared to the competing Palm OS, Windows CE requires plenty of memory and processing power making devices much more expensive. The Casio E-200 (Figure 2.5) for example takes a minimum of 64 MB of RAM and has a 206 MHz processor embedded. The common Palm devices instead are shipped with up to 8 MB RAM and include a cheap 33 MHz processor, which keeps them easily affordable for a broad audience. Only the latest Palm device Tungsten has a 175 MHz processor and 16 MB of memory.

Personal Information Management

While Palm concentrates on state-of-the-art mobile information management, Windows CE devices are technological wonderboxes, generously equipped with fancy hardware features: They usually have a color display with a good resolution, enabling multimedia applications such as computer games. There is a built-in microphone for a voice recording facility and a MP3 player. A stereo output is provided to plug in a headset. USB and infrared support ensure connectivity with peripherals and other devices. A CompactFlash card slot can be used to attach memory cards for additional storage capacity or one of a wide variety of the other special-purpose cards such as a digital camera card, or a modem card. All these add-ons make Windows CE devices extremely versatile and justify the high price for those, who can take advantage out of them.

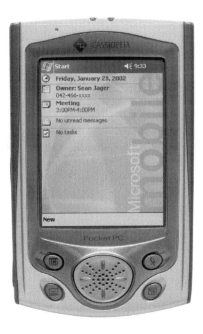

Figure 2.5:
E-200 PocketPC
from Casio

Delivered with Windows CE are the typical Personal Information Management (PIM) applications (Figure 2.6). Data can be entered with a displayed keyboard, which is operated with the stylus. Handwritten notes can be drawn on the touch screen and stored within graphic files or interpreted by a recognition software. Memos can also be recorded using the built-in microphone.

ActiveSync is the synchronization software, which Windows CE applications use for keeping the mobile data consistent and up-to-date with databases on the user's corresponding PC. This way, information like address book or calendar entries can be maintained both on the mobile device and on the workplace PC. ActiveSync supports communication through USB, infrared, and serial port, as well as via Ethernet LAN. ActiveSync cannot synchronize between two devices or between client and server directly.

Active Sync

ActiveSync comprises a Service Manager and multiple Service Providers. The Service Provider is the synchronization engine, establishing the connection, tracking changes, and resolving data conflicts. Service Providers are plugins contributed by application providers. They implement all application specific tasks, such as user interfaces or retrieving and storing changed data.

Most of today's Windows CE devices in the field run the PocketPC version of Windows CE, which is tailored for handheld computers. Products are the Casio Cassiopeia, the Compaq iPAQ, and the HP Jornado. For PocketPC Microsoft's developers have redesigned the user interface entirely, improving especially the clarity of the application layouts.

The performance of PocketPC has been doubled compared to the previous releases of Windows CE. The operating system itself is much more stable than in recent versions. Among the new features an improved handwriting recognition. Glyphs, script or print characters, written on the touch screen can be interpreted by that new software very reliable.

Figure 2.6: PocketPC Version of Calendar, Mail, Word, and Internet Explorer

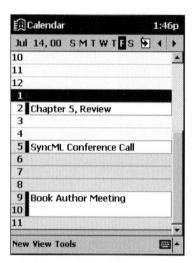
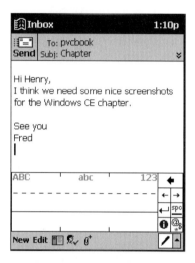
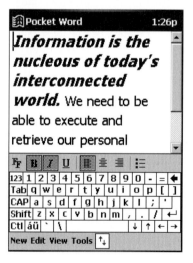
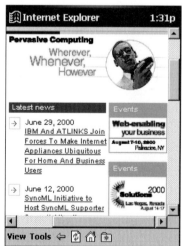

Part of PocketPC are pocket versions of Microsoft Word and Excel and the PIM application Pocket Outlook. Word and Excel will help viewing attachments (if there is any chance to download them onto the device). MS Money and a street mapping program are nice complimentary utilities.

PocketPC Office suite

The Pocket Internet Explorer is definitely one of the highlights. It supports online and offline browsing, ActiveX and JavaScript. While the predecessor Mobile Channel Viewer could only browse dedicated web sites, the Pocket Internet Explorer is free to surf anywhere in the net. A "shrink to fit" mechanism is an approach to deal with the smaller screen size: Web content is zoomed down, until it fits into the window.

Internet Explorer

A electronic book application and a media player reflect the aim of Microsoft to emphasize on content delivery complementary to providing just base technology like operating systems.

Multimedia access

The Microsoft Media Player can download an individual music library from major record companies onto a handheld device. The Windows Media format compresses two hours of music into 64 MB. That is about half as much as MP3 needs. Access management prevents copyright violations using cryptographic mechanisms.

Microsoft Reader is an electronic book software delivering books to PocketPC based devices. Well readable fonts, a built-in dictionary, and a large collection of book titles are the benefits for the user. The goody for the publisher is a corresponding copy-protection system preventing illegal reproduction.

While handheld computers are today mostly used by business people keeping track of their appointments, there will soon be a significant shift towards private and entertainment usage. Especially new multimedia computer games will target the PocketPC platform in close future.

2.4
EPOC Based- Handheld Computers

EPOC is a versatile operating system designed for usage in various mobile devices. Beside sub-notebooks, it is also applied for phones. EPOC has been developed by the Symbian consortium, founded by Ericsson, Motorola, Nokia, and Psion. EPOC based sub-notebooks are available from Ericsson, Oregon Scientific's OsariS, and Psion's 5mx (Figure 2.7), Series 7, and Revo. All of them are based on the current version of EPOC, EPOC Release 5, except for the OsariS, which still uses EPOC Release 4.

Photo courtesy of Psion

Remarkable is, that the EPOC operating system is very stable and users only experience crashes every few years. These devices contain a rich set of applications for mobile workers. Besides the PIM applications, they feature a word processor, a spreadsheet application, a web browser which supports Java applets, a communication suite with email, fax and SMS, as well as an application to backup and synchronize the device with a PC.

Data entered on an EPOC device can be exported to PC documents and imported in applications like Lotus SmartSuite, WordPerfect or Microsoft Word. There is plenty of additional third-party software available for EPOC devices. Maps, databases, dictionaries, tools for internet, email, fax and SMS, financial calculators, and accounting applications, just to name a few of them.

Table 2.1:
Comparing
Handheld
Computers

Device Class	Palm	Pocket PC	EPOC
A sample Product	Palm i705	Casio E-200	Psion Series 5mx
Operating System	Palm OS 4.1	PocketPC 2002	EPOC 32 bit
Size	8 x 12 x 2 cm	13 x 7.8 x 1.6 cm	17 x 9 x 2 cm
Weight	140 – 160 g	255 g	350 g
Data Entry	Stylus	Stylus	Keyboard
Display Size	7.8 cm	21 – 25 cm	12 – 14 cm
Display Resolution	160 x 160 pixels Monochrome or 256 colors touch screen	320 x 240 pixels 65000 colors touch screen	640 x 240 pixels Monochrome touch screen
RAM	8 MB	64 MB	16 MB
Processor	Motorola Dragonball VZ33, 33 MHz	Intel Strong ARM 1110, 206 MHz	32 bit RISC ARM710T, 36 MHz

Device Class	Palm	Pocket PC	EPOC
Battery charge	> 10 hours	12 hours	> 10 hours
Peripherals	Serial, Infrared, Extension Card slot	Serial, USB, Infrared, Compact Flash., Voice recording, Audio speaker,	Serial Port, Infrared, Compact Flash Voice recording
Applications	PIM, Email HotSync, Graffiti,	PIM, ActiveSync, Handwriting recognition, Internet Explorer, Office Suite, PocketOutlook Media player	PIM, Email Internet Browser, Database, Spreadsheet, WAP (MC218 only)

2.5
Sub-Notebooks

In terms of size, weight, processing power, and functionality, sub-notebooks or clamshells are somewhere in between a handheld device and a regular notebook. Most of a handheld computer's characteristics are valid for sub-notebooks too. But they have more memory, more CPU power and they provide a reasonable screen as well. Additionally, they have a small keyboard, making data entry much easier – a significant advantage, in case a lot of text has to be typed in. Although sub-notebooks won't fit into pockets anymore, they are still highly portable and a bit lighter than full-sized notebooks. The main differences to regular notebooks are that the sub-notebooks have much less computing power, cost less and that they are intended for mobile usage. Like handhelds, they have instant-power on, avoiding to boot an operating system first before usage. Their battery capacity exceeds the usage time of notebooks. Nevertheless, floppy and CD ROM drives are uncommon for sub-notebooks. And hard disks are rare. The operating system of these sub notebooks vary. EPOC, Windows CE as well as Windows98 can be found.

Something between handheld and laptop

But being less convenient to carry around than a handheld and having less functionality than notebooks, makes the sub-notebook class not very popular for a broad audience.

Probably the largest group of users are mobile sales forces. Sales representatives need to fill out contracts, orders, or lead sheets directly at the point of sale. They must have immediate access to all relevant data, such as customer profiles, order status or product catalogues. Sub-notebooks are very well suited for these kind of tasks: The keyboard allows quick data entry and the display is

Targeting mobile sales forces

large enough for product presentations. There is enough storage for local data repositories allowing to work offline while on travel and to synchronize only periodically with the company's back-end systems.

2.6
Phones

Modern phones create a high demand for new pervasive technologies. They have evolved from a simple person-to-person voice interface to powerful network clients. Fancy computing features like tiny hard disks and voice recognition are condensed into a tiny shell dissolving the demarcation to handheld organizers. Today, address books, calendar, memos, and games are state of the art applications, which are no longer restricted to high-end phones.

But it is more than just surfing the net with a screenphone or playing Tetris on a cellular phone while waiting for a train: address books and calendar entries on the device are synchronized with a Personal Computer. On the other side, data from server databases, like Enterprise Resource Planning (ERP) systems can be accessed, making phones valuable e-Business components.

2.7
Cellular Phones

An incredible diversity   The diversity of cellular phones is incredible; their functionality too: the endless list of features gets longer every day. There are manifold games, fancy form factors, built-in FM radio, and even PIM functionality. The operating systems used in cellular phones are quite manifold too. Mostly proprietary systems like RTOS from Siemens or GEOS from Geoworks are used. Others like EPOC are applied by several companies.

Entering text   One issue, cellular phones without a reasonable keyboard have to cope with, is how to enter text easily: Voice recognition is one sumptuous way. Instead of picking one of the miniaturized keys, commands and data can be spoken to the device. Stored phone numbers are dialed with one shout.

Predictive text input is another way to leverage data entry. T9, which stands for "Text on 9 keys", has been adopted by numerous phone manufacturers. The user types in arbitrary texts with the 9 keys of the numeric keypad of a phone. Only one keystroke per character is needed instead of the annoying multi-tap text entry. Since each numeric key stands for three or four letters, a software decodes the sequence of pressed keys. Within a linguistic database possible word variations are checked until the text is no longer ambiguous. There are databases in more than a dozen languages available.

The communication bearer between a mobile phone and the telecommunication network differs strongly from country to country:

Global System for Mobile Communication (GSM) is the accepted de facto standard for mobile wireless communication. A dense net of GSM base stations has been established by numerous telecommunication providers throughout more than 100 countries. Each station feeds the digital signals from cellular phones within its reach into the worldwide telephone network. The standard ensures interoperability from a technical point of view. Roaming agreements allow to use the phone in networks operated by other providers. This approach is strongly decentralized: all information required for user identification and authentication as well encryption of the communication is stored on a smart card, which is plugged into every phone. The so called Subscriber Identity Module (SIM) functions as a secure token.

GSM

There are several frequencies used for GSM: 900, 1800, and 1900 MHz. Triple Band phones support all frequencies, giving travelling users more flexibility.

In North America and parts of Asia wireless communication providers operate networks based on several other digital standards. Among these, Code Division Multiple Access (CDMA) is today's most common one. It also allows a secure communication using an encoding key, which is sent along with the voice data. Time Division Multiple Access (TDMA) is a competing standard, which works similar to CDMA, but has an improved bandwidth.

CDMA and TDMA

2.8
Data Transmission Capabilities

Besides the pure voice communication, there are several ways for transmitting and accessing data through a wireless network. Data communication is used to transfer information between devices or between a device and an IT network, like the Internet. Especially the connection to the Internet gives mobile users access to a standard resource for information.

Accessing the Internet

- Paging is one way of sending data to a recipient. Only a notification and a callback number is delivered.
- GSM data is a part of the GSM communication, which can be used for data or fax transmission.
- Short Message Service (SMS) allows up to 160 characters to be transmitted within one message from and to a GSM phone.
- Wireless Application Protocol (WAP) is a cross-industry standard used to deliver specially formatted content from a web server to phones and other wireless devices with a limited screen. A WAP browser within a device enables Internet interaction for retrieving flight schedules, weather forecasts, driving directions, brokerage services, or making hotel reservations (Figure 2.8). All these applications are definitely a promising opportunity especially for mobile users.

Multimedia
Messaging

Data transmission using a mobile phone is becoming more and more important – the emerging WAP applications make this evident. But WAP is only the very beginning of new data based communication technologies. Beyond SMS, Picture Messaging allows to send text and graphics. Multimedia Message Service adds digital imaging capabilities. Mobile Multimedia supports entire new content types, such as animations or video clips.

Figure 2.8 :
Some WAP
Applications

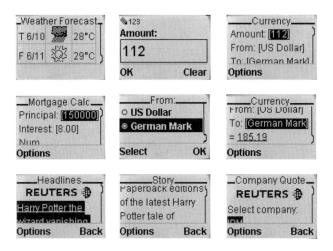

GPRS and UTMS

The most important obstacle for this evolution is the speed of the data transmission itself: GSM Data currently allows only 9,6 kilobits per second, which is the bottleneck for wireless information retrieval. From the year 2000 on, a new enhancement to GSM, called General Packet Radio Service (GPRS) is available. This packet-oriented standard will increase the bandwidth of mobile data communication up to 115 kbps. That will already be more than the speed of today's ISDN wired telecommunication nets. But GPRS will only be an intermediate solution. In 2003 the Universal Mobile Telecommunications System (UMTS) will be introduced as a worldwide common standard for wireless communication. UMTS will allow 100 kbps to 2 Mbps for data transmission! Phones based on these enhanced communication standards are often referred as third generation phones (3G).

But communication aspects of phones are not limited to the pure telecommunication:

Infrared

Mobile phones usually have an infrared port to exchange data locally with another device. For example in order to "beam" an address book entry to someone else's phone or to interface with a calendar on a PC. Or a simple handheld computer could use an infrared connection to a mobile phone to establish a wireless connection for accessing some server.

Bluetooth

Bluetooth is a convenient approach to let multiple devices collaborate wireless with each other within a short distance of a few meters. While infrared is a one-to-one communication, Bluetooth establishes a local wireless radio frequency network. Phones, handhelds, and personal computer can communicate with peripherals, like headsets or hard disks. A typical scenario could be a handheld, connecting via Bluetooth to a phone, which dials up to a network. Using a Bluetooth connection, the downloaded data can be sent to a standalone memory card.

2.9
Smart Phones

Smart phones combine a mobile phone with a handheld organizer into an all-in one communication system.

One example is the Nokia Communicator 9110. The device can be unfolded to access a small keyboard inside. The Communicator is based on the GEOS operating system and features an Intel embedded 386 class processor with 24 MHz. 4 MB of RAM are occupied by the operating system and the included applications. 2 MB is free for user data and another 2 MB is used for program execution. The monochrome display has a good resolution of 640 x 200 pixels. The communicator is based on dual band GSM wireless communication and provides fax, email, short messaging, Internet access, and Personal Information Management (PIM) applications. While PC connectivity via Infrared and serial port is a common feature, telnet and a VT100 emulation are an additional way of accessing data on servers of interest.

Nokia Communicator

Another smart phone is the Ericsson R380 (Figure 2.9). The PDA functionality is similar to the Nokia. The key pad can be unfolded to access the touch screen hidden behind. With 360 x 120 pixels that screen has a reasonable resolution. Dual band GSM, modem, and infrared port ensure connectivity, for example to synchronize with a PC. The R380 is WAP enabled and based on the versatile EPOC operating system.

Ericsson R380

Figure 2.9:
Ericsson Phone
R380

Photo courtesy of Ericsson

The near future smart phones of all manufacturers will take advantage of high-speed wireless communication networks like GPRS and UMTS. Besides the phone and organizer functionality, these devices will be able to exchange video and audio streams as self-evident as today's SMS (Figure 2.10).

Photo courtesy of Nokia

For instance Ericsson's future Communicator platform is a high-end device with a handheld computer look-and-feel (Figure 2.11). This EPOC device will support the three GSM frequencies 900, 1800, and 1900 as well as the new GPRS communication for high speed data transmission and Bluetooth. GPS is another feature Ericsson plans to include to their Communicator.

The MultiMobile is a smart phone prototype, developed by Siemens and Casio. It is based on a Cassiopeia PDA and runs the PocketPC version of Windows CE. Besides WAP and SMS, the MultiMobile can play MPEG-4 video files as well as MP3 audio files. This device can even be used for mobile video conferencing.

Figure 2.11:
Ericsson
Communicator
platform

Photo courtesy of Ericsson

2.10
Screenphones

Screenphones are the convergence of a phone and an Internet terminal. They have a reasonable screen, an optional keyboard and are dedicated to two single tasks: phoning and accessing Internet content and applications.

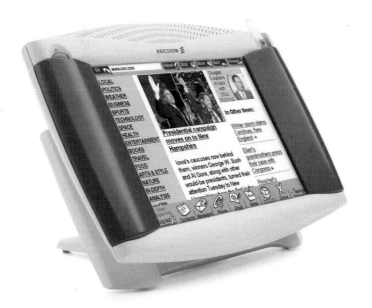

Figure 2.12:
A screenphone by
Ericsson

Photo courtesy of Ericsson

The main goal of screenphones is to offer simple and convenient usage. They provide a one-click connection to the net targeting especially non-technical users who would hesitate to install a PC on their own. The integrated browsers are capable of displaying multimedia web sites and usually can execute Java applications. Screenphones can be used for simple information retrieval, for executing e-commerce web applications, emailing, etc.

The device is delivered pre-installed and ready to be plugged into the telephone line. The connection to the Internet is established through an dial-up access provider. Mostly screenphones require an ISDN high-bandwidth connection, some use cable modems or even wireless connections. Usually a smart card reader is built-in, allowing secure financial transactions like home banking or e-commerce payments. Figure 2.12 shows a screenphone from Ericsson.

2.11
Further Readings

AvantGo

http://www.avantgo. com/frontdoor/index.html

Information about the AvantGo portfolio on different operating systems.

Ericsson

http://mobile.ericsson.com

The homepage of Ericsson's mobile phone activities.

Ericsson Mobile Phone R320

This white paper [ERI00] describes the features and functionality of the R320 smart phone from Ericsson. It includes the WAP configuration and a detailed glossary with mobile telephony related terms.

Handspring

http://www.handspring.com

Handspring shows the Visor Palm OS devices and some interesting add-on hard- and software.

Motorola

http://commerce.motorola.com/consumer

Motorola's consumer products.

Multimedia Messaging

http://www.nokia.com/press/background/index.html

A white paper from Nokia [NOK00] describing how SMS evolves to mobile multimedia.

Nokia

http://www.nokia.com

The homepage of Nokia.

Palm Central

http://www.palmcentral.com

Plenty of software to download for Palm OS devices.

Palm Computing

http://www.palm.com

Information and downloads for Palm OS computing devices.

PalmZone

http://www.palmzone.com

Online Magazine covering Palm OS related topics.

Psion

http://www.psion.com

Psion provides the largest variety of EPOC powered devices. Their web site provides information about Psion devices, support, and free downloads of various utilities.

Symbol

http://www.symbol.com/products/mobile_computers/mobile_computers.html

Symbol's mobile devices with laser beam scanning capabilities.

WAP Development

http://www.forum.nokia.com/wap_developer

Information about WAP development based on Nokia WAP phones.

Windows CE PocketPC

http://www.microsoft.com/pocketpc

The starting point when searching for Pocket PC information.

Windows CE Mobile Devices

http://www.microsoft.com/mobile

An overview of Windows CE devices. This site includes reviews, features, and comparisons of handheld and PocketPC devices. There are also developer and enterprise resources.

Qualcomm

http://www.qualcomm.com

The homepage of Qualcomm includes technology and products for wireless communications.

ZDNet

http://hotfiles.zdnet.com

Shareware and freeware for Palm OS and Windows CE.

3   Smart Identification

In the Pervasive Computing world, people need to be able to identify or authenticate themselves to computer systems. For payment, tracking, and security applications, computer systems need to be able to identify products, packages, and objects in general.

What would be a more natural way to accomplish this than allowing the computers to communicate with computer chips carried by individuals or affixed to packages? As it turns out, a special breed of computer chips has been developed to fulfill identification and authentication requirements such as these. These chips can be embedded within plastic cards, paper labels, key ring fobs, and many other forms. The following sections describe the technology used for identification applications.

3.1
Smart Cards

This section describes the different types of smart cards and explains their features and benefits. It introduces smart card hardware and software.

3.1.1
What Is a Smart Card?

Many people already use one or more smart cards in daily life. A smart card can function as a phone card, as a health card carrying insurance information or medical records, or as an electronic purse. The latter allows the user to store digital "money" for purchasing tickets or buying drinks from vending machines, for example.

Smart cards evolved from plastic identification and magnetic stripe cards through adding a secure, tamper-resistant single-chip microcomputer. The microcomputer chip has a size of only about 25 mm² at most, which keeps the cost down and makes it less vulnerable to breakage when the card is flexed.

What are smart cards?

In the tasks of very reliable authentication, electronic signature generation, and cryptography, smart cards are superior to traditional magnetic stripe technologies. The integrated processor allows smart cards to actively perform computing operations such as cryptographic operations. Sensitive data, such as secret keys or identifying codes, never have to leave the card. Smart cards can use

What can smart cards do?

these capabilities to verify the identity of a user based on a password entered on a terminal or based on a cryptographic message before releasing stored data. Electronic cash transactions can be performed this way without an expensive online connection to a host system.

Plastic card history

Plastic cards have a long tradition. The first credit cards just had the name of the owner printed on the front. Later, cards with embossed printing were introduced. The embossing made it possible to take an imprint of the cardholder information instead of copying it down manually. A few years later, the magnetic stripe, which carries the account information and the name of the cardholder, was introduced. This made the card machine-readable: Now the information could be electronically processed. Still, one problem remained: Everybody with the necessary equipment can read and write the data on the magnetic stripe. This led to a fraud problem.

The first smart card

In 1968, a patent for an identification card with an integrated circuit was filed, and the smart card was born [RAN00]. An important characteristic of a smart card is that the information on it cannot be copied. A credit card's magnetic stripe can easily be copied and then be misused. This could never happen with a smart card-based credit card. Therefore, smart cards are recognized as the next generation financial transaction cards [VIS02].

Usage examples

In the case of an electronic purse, the card can store the current balance and can tightly control increasing or decreasing it. Smart card based electronic purses can reduce the cost of cash handling. Since the eighties, the smart card industry has grown. German banks issued over 52 million of their electronic purse card GeldKarte smart cards to their customers [EPSO02].

Today every mobile phone that complies with the GSM standard contains a smart card that identifies the phone and authenticates the owner to the telephone system.

In a building access system, the card can be used to store the data required to open a door. The same data can later authenticate the employee to his computer or can be used for payment in the company's cafeteria.

In home banking applications, the card can be used as a secure token to authenticate the user over a public network between the user's computer and the bank's system. In this application, the issuing bank places cryptographic algorithms and secret keys on the smart card. When the user inserts the card into the card reader at his computer to identify himself, cryptographic messages are exchanged between the bank's host computer and the smart card. Since only correct card for the particular user can generate the required responses, security is improved compared with use of passwords alone.

In a multi-company loyalty scheme, the card can store the loyalty points that the customer already earned.

In a mass-transit system, the card can replace paper tickets. The fare can be calculated based on the distance. This can be done at the time the customer leaves the public transport system, and the fare can be deducted from the card on the spot. Using a contactless card, the traveler could even leave the card in his pocket.

With all these benefits of the smart card as a secure and portable access token, we expect to see smart card usage grow year by year.

3.1.2
Smart Card Hardware

3.1.2.1
Contact and Contactless Cards

The host computer, in our case the computer or terminal that runs the off-card smart card application, has to communicate with the processor on the smart card to exchange information and commands. Communication can take place either through the contacts on the card or via wireless ("contactless") transmission. A hybrid smart card combines both technologies and is able to communicate with the host system using either method. *Communication with the smart card*

Contactless smart cards are often used in situations requiring fast transactions or where only a limited amount of data has to be exchanged. Examples are public transport systems or access control for buildings.

3.1.2.2
The Computer on the Smart Card

As we explained earlier, the most important characteristic of a smart card is that it contains a computer with CPU and memory. Today's smart cards have approximately the same computing power as the first IBM PC.

The chip of a smart card (see Figure 3.1) consists of a microprocessor, ROM (Read Only Memory), EEPROM (Electrical Erasable Programmable Read Only Memory), and RAM (Random Access Memory). An EEPROM requires a larger surface than a ROM of the same size. This makes EEPROM more expensive and lets EEPROM size become an important factor for the price of a smart card. *Smart card chip*

In addition to smart cards, there are also memory cards that have only ROM and EEPROM. They do not contain programmable logic. The EEPROM can be secured by a hard-wired key, which is checked on every access. Such cards are cheaper but much less flexible than smart cards. *Memory cards*

Today, most smart cards have an inexpensive 8-bit microprocessor, but high-end cards can contain a 16-bit or 32-bit processor. *CPU*

An optional cryptographic coprocessor increases the performance of cryptographic operations. By performing signature operations on the card itself, the user's private key never needs to leave the card. *Cryptographic coprocessor*

Figure 3.1:
Example of a
Smart Card Chip
and its
Components

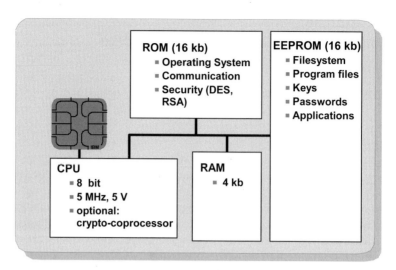

ROM
The information stored in the ROM is written during production. It is the same for all chips of a series. It contains the card operating system and maybe also some applications.

EEPROM
The EEPROM is used for permanent storage of data. Even if the smart card is unpowered, the EEPROM still keeps the data. Some smart cards also allow storing additional application code or application specific commands in the EEPROM.

RAM
The RAM is the transient memory of the card and keeps the data only as long as the card is powered.

Standards
The basic smart card standard is the ISO 7816 series [ISO02]. This standard details the physical, electrical, mechanical, and programming properties of smart cards. The ISO specification ensures a base level of standardization, but allow manufacturers to add proprietary capabilities. The EMV specification is based upon the ISO standard and is intended to be an industry-wide chip card specification that ensures interoperability between all chip cards and all chip-reading terminals, regardless of location, application, or manufacturer.

3.1.2.3
Hardware Security

The objective of smart card chip design is to provide high physical security for the data stored in the card. The processor and the memory are combined in the same chip. This makes it difficult to tap the signals exchanged between the processor and the memory.

Since smart cards typically contain sensitive information, smart card chip design must take many potential threats into consideration [RAN00]. These include slicing off layers of the chip to optically read out data, manipulating the voltage or clock to make the processor fail, attacks using high temperature or X-rays, and several others.

Sophisticated countermeasures are applied to guard the chip against the various attacks. For example, passivation layers are added to prevent analysis in combination with slicing off layers of the chip. Address lines and the memory cells of the chip are arranged in unusual patterns to make the physical examination harder. Furthermore, some chips have the capability to detect if the layer above the chip was removed, as it would occur if somebody were to examine the chip. Chips can detect unusual variations in the clock or in the voltage and react with shutdown of the operation.

Some newer techniques to spy on the information stored on the card try to do this by manipulating or observing the power supplied to the card. Newer versions of smart cards have been made resistant to these types of attacks.

3.1.2.4
Card Acceptance Devices

Card acceptance devices range from simple card readers to highly sophisticated, programmable payment terminals with several slots and user interface support (Figure 3.2). Readers can be attached to a PC via serial port, they can be integrated in a keyboard, or embedded into appliances such as banking terminals. Many pervasive devices like set-top boxes, cellular phones, or handhelds are equipped with smart card readers.

Figure 3.2: Smart Card Readers (TOWITOKO & Intellect)

3.1.3
Smart Card Software

Usually a smart card application consists of the following two parts:

- Off-card application
- On-card application

The off-card part of the application is the part that resides on the host computer or terminal connected to the smart card through a smart card reader device. For instance, the OpenCard Framework (OCF) is a framework that supports off-card application development using Java.

Off-card

On-card   The on-card application is a program stored in the memory of the smart card chip. This part can consist of data and in general executable code. If the on-card application has executable code, this code is executed by the smart card operating system and can use operating system services, such as encrypting or decrypting data. These functions can be used to make the smart card and the communication with the smart card very secure.

File-system cards   The majority of current smart cards have a file system integrated into the operating system. A file system on a smart card supports the storage and retrieval of all kinds of data and is useful for many types of applications. According to ISO 7816, a file system consists of directories (*DF*) and files (*EF*). The root directory is referred as *MF* (Figure 3.3). When multiple applications share one card, usually each application uses a different directory in order to separate their data from each other. In order to find the directory assigned to a specific application, the EMV-specification introduced a directory file, listing all applications present on a card.

Figure 3.3:
ISO 7816 File
System

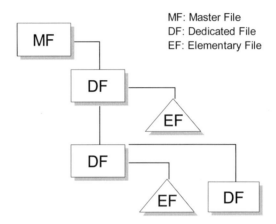

For most types of current smart cards, development of on-card executable code is not in the hands of general application developers. Development of such code can only be done by the card operating system developers. The code must be integrated into the mask for the ROM of the smart card before the card is manufactured.

Recently developed card operating systems enable application developers to create and download on-card application code on their own. The most important of these operating systems are Java Card and Multos.

An end-to-end smart card application involves several different players: Each player contributes software that corresponds to its portion of the complete solution.

■   The application provider creates the off-card and sometimes the on-card applications seen by the end user of the smart card solution.

- The card issuer is responsible for card initialization, personalization, and issuing.
- The card operating system provider creates the basic operating system on which the on-card application runs.
- The card reader provider contributes the devices that interface directly with the smart cards.

3.1.4
Communication Between the On-Card and Off-Card Parts

The protocol stack of the communication between the smart card and the host has several layers. On the application layer, the communication takes place between the off-card part of an application and its corresponding on-card part. The commands and data exchanged are specific to a particular application and cover tasks like read, write, decrease, and others.

The next lower layer is the layer of the Application Protocol Data Units (APDUs). The format of the APDUs is independent of the application.

One layer below we encounter low-level protocols with names such as T=0 and T=1. We briefly introduce these protocol layers in the following sections.

3.1.4.1
Application Protocol Data Unit (APDU)

Application Protocol Data Units are used to exchange data between the host and the smart card. ISO 7816-4 [ISO02] defines two types of APDUs: Command APDUs, which are sent from the off-card application to the smart card, and Response APDUs, which are sent back from the smart card to reply to commands.

There are several variants of Command APDUs. Each Command APDU contains:

- A class byte (CLA). It identifies the class of the instruction, for example if the instruction is ISO conformant or proprietary, or if it is using secure messaging.

Command APDU

- An instruction byte (INS). It determines the specific command.
- Two parameter bytes P1 and P2. These are used to pass command specific parameters to the command.
- A length byte Lc ("length command"). It specifies the length of the optional data sent to the card with this APDU.
- Optional data. It can be used to send the actual data to the card for processing.
- A length byte Le ("length expected"). It specifies the expected length of the data returned in the subsequent response APDU. If Le is 0x00, the host side expects the card to send all data available in response to this command.

Command APDU

Response APDU

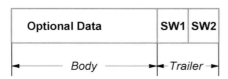

CLA, INS, P1, and P2 constitute the header of the command APDU. This header has a fixed form. Lc, optional data, and Le constitute the body of the command APDU. This body can have several variations, upon which we will not elaborate here. More information is available in ISO 7816-4 or in [RAN00].

Response APDU   A Response APDU contains:

- Optional data.
- Two status word bytes SW1 and SW2. They contain the status information as defined in ISO 7816-4.

The length of the optional data in the response APDU is as specified by the preceding command APDU. Should an error occur, no optional data might be returned in spite of the specified length.

The content that will hopefully most often be in the status word bytes SW1 and SW2 is 0x9000, indicating successful execution.

If you are using the OpenCard Framework, the framework will do most of the APDU handling for you. Nevertheless, it is good to understand what actually happens when OCF communicates with the card. If you ever need to run a trace, this knowledge will help significantly.

3.1.4.2
Protocol Layer

The protocols T=0 and T=1 are the two most-used variants of half-duplex asynchronous protocols defined in ISO 7816-3.

With T=0, each character is transmitted separately, while with T=1, blocks of characters are transmitted.

Most modern smart card readers are capable of transmitting with either one of these protocols. From the card's Answer to Reset (ATR), the reader can find out which protocol the card requests. The ATR is the first data block returned to the reader after a card became powered up. In addition to the protocol information, the ATR can contain data identifying the type of card. The ATR is specified in ISO 7816-3.

ATR

The T=0 protocol has been in use since the first days of smart cards. The GSM card is probably the best known application of this protocol. Its advantage is that it has simple and space efficient implementations. The price to pay for that simplicity is the incomplete separation of the transport layer from the layer above:

T=0

To retrieve data from a smart card, two command exchanges are necessary. In the first, the host issues the command and the smart card returns the length of the response that it will return. In the second (GETRESPONSE), the host asks for the expected number of response bytes and the card returns these.

The T=1 protocol can send a command and receive the response in the same exchange. It cleanly separates the application layer from the transport layer and is suitable for secure messaging between the host and the card.

T=1

One of the details that increase the complexity of asynchronous transfers is the error handling, especially the prevention of endless waiting. For this reason, a block waiting time (BWT) is specified, which indicates how long it is reasonable to wait for a response block. The BWT appropriate for a card is among the protocol information that the card returns in the ATR.

Block waiting time

For some commands, the smart card needs more time than typical, for example for complex cryptographic computations. To prevent the host from giving up waiting for the response too early, the card sends a preliminary response asking for a wait time extension. The card and the host communicate on a wait time extension using so called "S-blocks", while the standard command and response exchange is made with "I-blocks".

Waiting time extension

For a complete and excellent coverage of these protocols please see [RAN00].

3.2
Smart Labels

One of the major problems to solve in the Pervasive Computing world is the identification of packages or products in general. The current solution is through use of bar codes. Everyone has seen bar codes on product packaging. They have many advantages – they can be printed on labels, they are very inexpensive, and they can be reliably scanned. However, there are also a number of disadvantages:

Object identification

■ Since bar codes are scanned optically, they must be visible on the outside of the object.

Bar code disadvantages

■ Scanning takes place at a short range – a few centimeters.

- The objects must be separated in order to be identified.
- The information conveyed by a bar code is fixed when the bar code is printed and cannot be changed.
- The bar code itself is completely passive and any bar code reader can access its information, making it very difficult to fulfill security requirements demanded by some applications.
- If a bar code is hidden from the scanner, it cannot be read, making it useless for Electronic Article Surveillance applications.
- The bar code scanners are complicated – typically involving a laser, moving mirrors, and detection hardware – making them expensive.

Taken together, these disadvantages limit the usefulness of an object identification system consisting of bar codes and bar code readers.

Figure 3.5:
Smart Label
Block Diagram

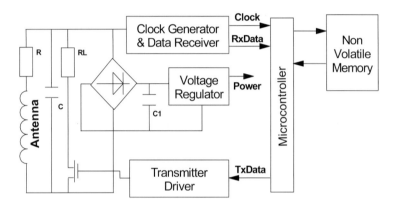

A more flexible and capable system has evolved from contactless smart card technology. A contactless smart card chip, sliced very thin and attached to a radio frequency antenna coil, can securely hold more information than a bar code. Just as with regular smart card chips, data can be written to them as well as read from them. Attached to a plastic carrier, these are known as smart labels or smart tags. Figure 3.5 shows a smart label block diagram.

Power supply and communication   The labels do not require a power supply – they obtain the necessary energy from the radio frequency field emitted by the reader device. An RF antenna captures the radio waves and special circuitry on the chip converts the radio frequency energy to appropriate voltage levels for chip operation. Data is superimposed on the carrier wave so communication to and from the chip uses the same antenna as the power supply.

Communication between reader and smart label takes place using open communication protocols at a frequency of 13.56 MHz in accordance with the FCC 15 part 3, ETSI 300 330, and ETSI 300 683 standards. Data is typically transferred at a rate of 26 kbps, although faster speeds are possible. This allows reading of up to 30 smart labels per second.

Smart label readers contain no moving parts. All that is needed is an antenna and an electronics module. This reduces the cost of the reader and opens the door for integration into many new applications.

Data can be read and written at distances of over one meter. Through use of collision avoidance algorithms, several smart labels can be accessed simultaneously. This means that separation of individual items before reading them is not required.

Companies such as Gemplus, TAGSYS, Philips, and Texas Instruments are pioneering this technology. A broad line of labels and tags with various packaging options are available for purchase today.

These devices can be laminated into an adhesive paper label for packaging, sewn into clothing, or embedded within a product. Typical dimensions vary from 14 x 14 mm with a thickness of 0.9 mm to 50 x 50 mm with a thickness of 0.5 mm.

Figure 3.6 shows three Texas Instruments Tag-it smart label inlays. These extremely thin inlays – measuring only 0.375 mm at the chip and 0.085 mm elsewhere – are mounted on a PET plastic carrier and are intended for incorporation into paper or plastic labels.

Through appropriate packaging, smart labels can also be made extremely rugged. Such smart labels can be molded into plastic objects or used in an outdoor environment.

Since data is read and written via radio waves, there is no need for the smart label to be visible on the outside of the object and there is also no requirement that the object be oriented in a certain manner relative to the reader.

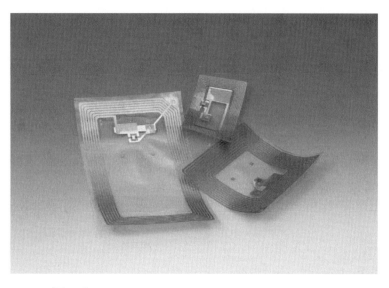

Photo courtesy of Texas Instruments

Data capacity   Smart labels contain control logic with non-volatile read/write memory for data storage. Data capacity ranges from 64 bits to about 2K bits. This may seem somewhat limited compared to a smart card chip or standard micro-controller, but in this application, size, cost and power consumption must be minimized. The smallest-capacity and simplest smart labels have a single, factory-programmed data block containing a unique serial number. The data cannot be changed. When signaled by the reader device, the entire data block is transmitted.

Security   More elaborate smart label chips have a larger, read/write memory organized into multiple blocks. Access conditions can be set on a per-block basis, allowing the data to be read or written only when presented with valid security credentials. It is also possible to write data to a block in such a manner that it can no longer be changed.

There are also units available that implement a challenge/response security feature. The smart label has a pre-programmed unique serial number plus storage for a user-programmable secret key. To test for validity of the smart label, the reader sends a command containing a random number to the smart label. The smart label encrypts the random number and returns it along with its unique serial number. The reader decrypts the message using the key associated with the serial number and verifies that it is the same as the random number sent. In this manner, the secret key is never transmitted between reader and smart label, and the response is different for every transaction, making it much more difficult to copy or imitate the smart label.

3.2.1
Example Applications

The security available through smart labels is useful in many scenarios. For example, when smart labels are used to identify items in a store, an EAS (Electronic Article Surveillance) field could be programmed into the smart label using a password. When the article is purchased, the smart label reader/writer at the checkout counter would use the proper password to mark the item as purchased. The security protection prevents non-authorized persons from doing likewise.

Since the smart label is accessed via radio waves, it can be hidden within the packaging or inside the product itself, which makes it much more difficult for would-be shoplifters to ply their trade. For high-value merchandise it would even be feasible to mount smart label readers on the display shelves or racks. Personnel could be notified as soon as a protected article is removed.

Smart labels can also be used in the shipping industry. In this application, the smart label contains identification and destination information. Each sorting station would read the destination information from the smart label and route the package accordingly. It would not be necessary to resort to complicated and relatively unreliable optical recognition schemes to sort the packages.

Since the smart labels are very thin, they can be integrated into a paper stick-on package. The destination information can be stored in the smart label when the shipping documentation is printed.

Inventory control is another area where smart label technology can be put to use. Some equipment, such as personal computers, is highly subject to theft. Smart labels can be placed within the equipment and smart label readers mounted at building exits. Whenever a piece of equipment tracked in this manner leaves the premises, the smart label reader would identify it and alert security personnel.

Inventory Control

If smart label or contactless smart card technology were used in employee badges, it would even be possible register the employee and determine whether the employee is allowed to remove the equipment before raising an alarm.

3.3
Smart Tokens

Smart cards have a convenient form for carrying in a purse or wallet and smart labels are very convenient for laminating within paper labels. However, many applications call for packaging that is more robust or in a different form factor than can be provided by smart card or smart label technology.

The need for robustness can be fulfilled by encapsulating the chips in plastic or metal. These packages can be made in any convenient form for the application in question. For example, a contactless access control token can take the form of a key chain fob so that it can be conveniently carried by the user. When the user approaches an access-controlled area, the token can be read at a distance to allow access. Smart tokens can also be molded into appropriate shapes with screw holes for mounting on factory trolleys. By reading and writing data to the tokens at factory workstations, a material tracking system can be set up.

Smart tokens can be based on either contactless or contact technology. Contactless technology is useful when the tokens need to be read at a distance. Contact technology could be called for in electrically noisy environments that would preclude the use of contactless tokens, or when the chip used in the token consumes more energy than can be transferred conveniently with contactless technology.

In some applications, it is necessary to read the token using contactless technology at a distance of several meters. For example, a tollbooth or gas station application must be able to identify cars with electronic passes at a distance, and in some situations, while the car is moving. Such applications call for a higher contactless transmitting power than can be supported by smart card or smart label technology. A long-range transponder can be made by packaging a contactless identification chip with antenna and a battery within the transponder module.

Smart tokens can also complement existing security systems. A contactless chip could be molded into the plastic head of an automotive ignition key. When the key is inserted into the ignition lock, electronics in the ignition system verify the token embedded in the key head using cryptographic mechanisms before allowing the vehicle to be started. This makes it much more difficult to pick the lock.

3.3.1
Smart Token Examples

Key Chain FOB As an example, Figure 3.8 shows a key fob from Gemplus containing a contactless identification chip and associated antenna. The chip contains 1024 bytes of EEPROM memory with security features that can be used to protect the data from unauthorized access. Data is read from or written to the chip in the same manner as a contactless smart card or smart label is read or written. The fob is ideal for carrying on a key chain in a pocket or purse and can be used in access control, consumer loyalty, or other applications.

Figure 3.8:
Gemplus
KeyFob8000

Photo courtesy of Gemplus

Another example of an extremely robust smart token is the iButton ™ from Dallas Semiconductor. The chip is packaged in a stainless steel can. The top and bottom of the can form the two electrical contacts needed to communicate with the chip. Communication is carried out using the serial 1-Wire ™ communication protocol at either 16 kbps or 142 kbps. The chip can be powered either through the communication line voltage or through a battery enclosed in the iButton case.

The iButtons are read by inserting them into special reader sockets. The reader sockets can be mounted at turnstiles or doors for access control applications, or can be attached to a wand used with portable equipment for inventory applications.

IButtons can contain microprocessor chips or memory (NVRAM, EPROM, or EEPROM) chips. A cryptographic chip implementing the JavaCard 2.0 standard is also available.

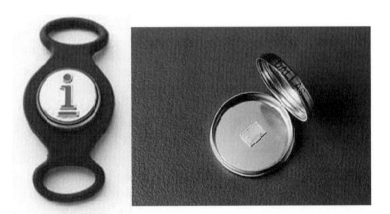

Figure 3.9: Dallas Semiconductor iButton

Photos courtesy of Dallas Semiconductor

The top of the can is hermetically sealed to the bottom through a plastic grommet in a manner similar to a button cell battery. The rugged button can be immersed in water, dropped on the floor, and even walked on without damaging the chip. The iButtons can be mounted on personal objects such as key fobs or rings for authentication applications, or on shipping containers or tanks for material tracking applications.

3.4
Further Readings

Dallas Semiconductor's IButton

http://www.ibutton.com

The IButton is a security module in form of a ring. Its programming interface is very close to the Java Card specification.

On this web site, you can order your IButton including accessories on-line. The 1-Wire for Java™, the latest development tool for the IButton supporting OCF, is also available for download.

Gemplus

http://www.gemplus.com

Gemplus is one of the leading providers of Smart Cards and Smart Card readers.

This web site contains information about smart cards in general and about products from Gemplus of course. You can also order smart cards like Gemplus' GemXpresso online.

Gemplus Developers

http://www.gemplus.com/developers

This web site contains the latest information for developers, like a forum, latest trends, and current drivers. It also provides access to OpenCard CardTerminals and CardServices for hardware from Gemplus.

TAGSYS Radio Frequency Identification

http://www.tagsys.net

This site provides information on TAGSYS smart labels, tags, and associated reader devices. TAGSYS is an independent company formed from a former division of Gemplus and is one of the world's leading manufacturers of RFID equipment.

IBM

http://www.ibm.com/pvc

IBM Pervasive Computing provides a wide range of technology and solutions for Pervasive Computing and smart cards. IBM, a member of the OpenCard Consortium, developed the OpenCard Framework reference implementation.

On this web site, you can find an overview of the solutions IBM offers in the area of Pervasive Computing and smart card technology.

Java Card Forum

http://www.javacardforum.org

The Java Card Forum is the group working on the Java Card Specification. On this web site, you can find the list of its members, its charter, information on the membership, minutes of recent meetings, and technical documents.

Mondex

http://www.mondex.com

This is the web site of Mondex, the organization behind the Mondex electronic purse. Here, you can find information how Mondex works, which devices are available to be used with Mondex and how it can be used for payments over the Internet.

Multos

http://www.multos.com

On this web site, the Mondex International provides information about its Multos smart card operating system and how you can obtain implementation and application licenses. Developers find technical manuals, technical bulletins, as well as training modules.

OpenCard

http://www.opencard.org

OpenCard Framework provides a Java application-programming interface to smart cards. It is supported by the OpenCard Consortium which has founded by members from the IT industry, smart card industry, and payment system industry. The OpenCard Framework reference implementation is publicly available free of charge.

On this web site, you can find the current version of OpenCard Framework with source code, binaries, documentation, and samples. The OpenCard Consortium also maintains a discussion list for interested parties.

PC/SC

http://www.pcscworkgroup.com

The PC/SC workgroup specified an interface to smart cards in a Windows environment. Microsoft offers an implementation of this interface for the current versions of the Windows operating system.

On this web site, you can find the PC/SC specifications as well as a list of compatible products.

Philips Semiconductors

http://www.semiconductors.com/index.html

Enter the Philips Semiconductors main site and search for "I-CODE" to find information on Philips Semiconductors I-CODE smart label devices.

Texas Instruments TIRIS

http://www.ti.com/mc/docs/tiris/docs/index.htm

This site provides information on the Texas Instruments TIRIS vehicle tags.

Texas Instruments Main Site

http://www.ti.com

To find information on the Texas Instruments Tag-it smart labels, begin at the TI main site and search for the term "Tag-it".

4 Embedded Controls

This chapter describes how pervasive information technology will be used in embedded home and automotive settings. Although most of the examples and scenarios described are oriented towards consumer use, many can be extrapolated for use in a commercial environment.

The network includes more and more objects of everyday life. Standard household appliances will soon be able to communicate with one another and with external infrastructure. Heating systems will adapt themselves to the habits of the residents. Power outlets can be deactivated when no valid electrical device is attached to increase safety for small children. A computer-controlled power grid in the home will also allow effective power management.

Also, networking technology is being built into cars and other vehicles. The various control units and information devices in automobiles can communicate with one another and with external networks.

This section describes some of the developments that will make this come about.

4.1
Smart Sensors and Actuators

Think of a video camera that can be accessed through the web. A PC that is connected to the Internet controls the camera. By clicking on the appropriate web page, any Internet user can access the camera and see what is being displayed.

By building on the idea of the web camera one can imagine smart sensors. *Smart sensors*
They are built to collect information about their surroundings and are small, self-contained, and connected to a network.

Smart actuators, on the other hand, accept signals from the network in order *Smart actuators*
to act on their environment accordingly.

In both cases the network connection can be through a wireless connection, although they are more usually connected via wire. Twisted-pair and power line signaling are commonly used for wire-line networking.

Figure 4.1 shows the general internal structure of a smart control (a control *Smart control*
being either a sensor or actuator). *internal structure*

A **transceiver** component connects the device with the network. It produces and interprets the physical signals on the network medium and handles the low-level protocols for data transfer.

The **control logic** is usually a microprocessor that provides the actual intelligence in the device. It interprets the commands from the network, processes them, and sends out the appropriate response. During command processing, the control logic will generally read the sensor level or carry out the requested action. If the network protocol supports asynchronous messaging, it might also be possible for the control logic to generate unsolicited alert messages when critical situations occur.

Figure 4.1:
Smart Control
Structure

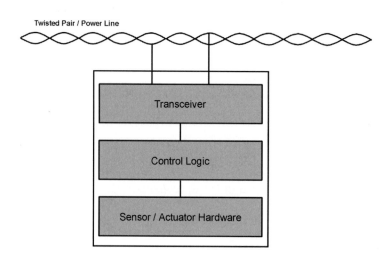

The **sensor / actuator hardware** either senses its environment or activates motors, solenoids, and the like to carry out an action.

There is a wide range of smart controls available for home and industrial use.

Sensors
Sensors are available for air temperature, liquid temperature, pressure, and to detect various gasses. Outdoor and indoor motion detectors enhance security systems and light level sensors control lighting installations in the evening.

Actuators
Actuators are sometimes just controlled electrical switches that are used for switching lights and appliances on and off. However, there are also special-purpose actuators such as valve or window shade control units that contain special feedback mechanisms to allow exact positioning.

Example network
Figure 4.2 shows a simple network for external lighting control. This network contains an ambient light sensor, a presence sensor, a security controller, and a lighting actuator.

The controller is responsible for the overall system function. It checks the ambient light level to determine when it is dark enough to activate the lighting. It also checks the external proximity sensor to determine if a person approaches the building. If someone approaches the building when it is dark outside, the lighting controller turns on the outside lights.

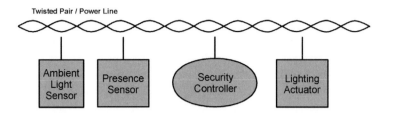

*Figure 4.2:
External Lighting
Control*

This is admittedly a very simple example, but it is easy to imagine significant extensions. Sensors could be added on the doors and windows to detect a break-in. Actuators for an alarm or for police notification could be added so that the security controller can take action if a break-in does occur.

Other types of controls can share the same physical medium with the lighting network. For example, temperature sensors and actuators for heating, air conditioning, and ventilation can coexist with the lighting components. This is illustrated in Figure 4.3.

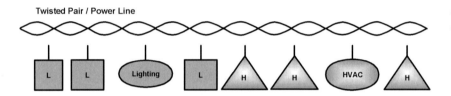

*Figure 4.3:
Multiple Control
Types*

When one imagines a network of controls communicating with each other over the power line, for example, a number of questions arise.

- How are the controls addressed in the network?
- How will network configuration be performed? In other words, how will the controllers "find" the sensors and actuators?
- And what about security – if power line networking is used, will the neighbor be able to turn on **my** lights?

These and related topics belong to the area of **Control Networks** or **Home Networks**.

Control networking concerns itself with communication between smart sensors, actuators, and controllers in general. Control networks may be used in industrial as well as residential settings. In an industrial setting, dedicated wiring is generally used for reliable, high-speed communication between the controls.

Control Networks

As an example, one can imagine a control network being used in a chemical plant where typical actuators would be valve and pump controls. The controller uses information from temperature, pressure, and chemical concentration sensors to make decisions about actuator settings.

The topic of smart controls in a residential setting is a subset of the larger home network theme. Home networks in general will be covered in more depth in Section 12.3 "Home Networks".

For the purposes of this section, a home network can be thought of as a control network in a residential setting. The main emphasis is on economical implementation. At present, not even many new homes are wired with provisions for networking lighting, heating, and entertainment devices. The cost associated with laying new wires would be enough to discourage homeowners from installing networked controls if dedicated wiring were required. Fortunately, other solutions are available.

Power line networking For connecting sensors and actuators in a residential setting, power line networking is often used. This works out very well, since most devices to be controlled consume power and are therefore connected to the power grid. Data packets are sent from device to device through the power grid using a high-frequency data signal superimposed on the power waveform.

Proximity or presence sensors and actuators for lighting and appliances can simply be plugged into existing wall outlets or can easily be wired into the power grid. The controller is often a general-purpose device that can be programmed through use of a personal computer.

The actuators, controls, and controllers must be able to address one another in both home and control networks. The way the configuration is performed depends on the underlying technology used.

Simpler systems require physical switches (typically rotary switches) to be set on each control in order to program the network address. Since the danger of signals propagating into adjacent apartments or houses is present, the address is composed of two portions: the house address and the unit address. If the user notices interference, the house address portion must be changed on all controls.

The devices used in more sophisticated schemes can have factory-set unique addresses or serial numbers that can be used to select a particular control. The controller uses a discovery protocol to find all of the controls of a particular class in the system. During system setup, the operator uses software menus to associate specific controls with one another.

Assuming smart light switches and smart lighting actuators in a residential setting, a home could be wired without calling for a direct electrical connection between the light switch and the ceiling lamp it controls. The switch and the lamp would both be connected to the power grid in the most convenient manner. When the light switch is touched, it would send a message over the power line to the lamp, which would then turn on, turn off, or dim accordingly. This is illustrated in Figure 4.4.

To configure such a system, the controller broadcasts messages into the system requesting all devices of each device class to report. The operator uses the resulting lists to associate lamps with switches or intrusion sensors with alarms. The controller stores this information in some type of non-volatile storage so that the control network in the home could reconfigure itself after a power outage.

Traditional
Wiring

Wiring with a
Home Network

It would be easy to add new lamps controlled by the same switch or new switches for the same lamp simply by wiring the new unit to the most convenient location in the power grid and configuring the system accordingly.

The controls could be dynamically reprogrammed to implement specific behavior. For example an "on vacation" function could note typical lighting patterns while the family is at home and then simulate these patterns when the family is on vacation. It could go so far as to turn on the radio in the kitchen in the morning and the television in the living room in the evening!

Security is another important element of control networking. It is especially critical to provide for adequate security in a home network setting where controls are interconnected through the power line. Since adjacent homes and apartments are generally connected to the same power distribution grid, they could conceivably send signals over the power line into your home to disarm your security system when you are away, for example.

On the other hand, sometimes communication with devices in the home is desired. If you leave on a trip, drive 300 miles, and then have the sneaking feeling that you left the oven on, it would be nice to be able to dial into your home network with your cell phone and make sure that the oven is turned off.

A residential gateway provides the necessary connectivity for authorized users but prohibit access by others.

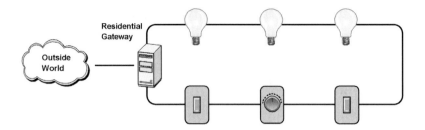

Figure 4.5 shows how a residential gateway protects the home network from unauthorized access. The residential gateway itself is a computer that is programmed to allow access from authenticated sources but deny access to all others.

In addition to providing access control for the residents, the residential gateway may also allow authorized service providers access to the home network. Power companies may want to read the electric meter through use of power line networking, for example.

4.2
Smart Appliances

In this book, the term "smart appliance" refers to devices used to increase comfort or convenience that are endowed with a certain computer intelligence and networking capability. These devices are useful not only in the home, but also in many service settings such as restaurants and hotels. Examples of smart appliances include refrigerators, freezers, stoves, microwave ovens, washing machines, and heating units.

Currently, most appliances are not smart – though they may employ a microprocessor control unit, they are not networked with other devices in the home, and they tend to have, let's say, a very clear, but limited, user interface.

A fridge, for example, basically has just a handle for opening and closing, and a knob for controlling the temperature. This will, for better or worse, likely change very shortly. Actually, the direct interface presented to the user will probably remain much as today, but embedded computing, networking, and sensing capabilities will add a new level of secondary functionality not available with current appliances.

4.2.1
The Smart Clock

Smart clock   One very useful smart appliance will be the smart clock. A clock will become a provider of time services in the home network. Other networked devices can base their operation on that time signal. For example, a smart VCR could listen in the network for a time service in order to automatically set its time, eliminating a chore that challenges many of our contemporaries.

Your alarm clock also knows some very interesting information about you – when you plan to get out of bed in the morning. The alarm clock could make this information available to other household appliances to enable them to maximize energy efficiency while at the same time maintaining your comfort. The heating unit might listen to the household alarm clocks to determine the earliest riser and then activate itself so that the bathroom is nice and warm just in time to accommodate her.

On the weekend or during vacation when the alarm clock is set later, the heater will automatically wait longer before warming the house, thus conserving energy.

4.2.2
Heating, Ventilation, and Air Conditioning

Current heating, ventilation, and air conditioning (HVAC) installations use dedicated wiring between the thermostat and the heating and air conditioning unit. This means that wires have to be run to each room in which a thermostat is installed. Due to the cost and complexity of such an installation, this usually limits the number of thermostats per home to one.

HVAC

By allowing communication through home wiring, for example, thermostats could easily be installed in any room of the house. The heater could find its thermostats through appropriate broadcast messages, and the thermostats would report their settings and temperatures.

This would allow a much finer-grained control of the heating system and would contribute to energy savings. There are usage patterns for each room in a home. The heating system could adapt itself to these usage patterns using information from the room thermostats and from the other devices in the room. The living room would be heated in the evening until the television is turned off and the lights are all turned out.

A smart thermostat could also exchange information with other household appliances. The temperature information could be made available to the television set for display. The thermostat could also use information from the security system to automatically lower its target temperature when the house is unoccupied.

Smart thermostat

Additional savings can be achieved by controlling ventilation and heat distribution to each room separately. In the morning, the bathrooms and kitchen could be heated while the living room remains cool. In the evening, the living room could be heated while the target temperature in the kitchen is reduced.

An external weather service provider could also push projected daily temperature information to the home network. An intelligent heating system could use this information to optimize its operation. On a cool morning of a projected hot day, the heating could remain off, for example.

Water heater   The water heater is another example of an appliance that could adapt its operation to the habits of the inhabitants. Standard water heaters keep a large tank of water at a constant high temperature 24 hours a day. This is not really necessary since most people sleep during the night and are often away from home during the day. A smart water heater could measure flow rates to determine typical use patterns. It would adapt its operation to minimize energy consumption while insuring that hot water is available when needed.

Power management   Power management is another application of smart devices. The city electric network comes under extreme stress at time of peak load. On a hot day in a large city, all residences and businesses have their air conditioners going. Peak demand has even caused citywide or area-wide electric system blackouts.

If the power companies were able to regulate power consumption by controlling appliances in the individual businesses and residences, it would, in many cases, be able to avoid a brown or blackout in a manner that minimizes the impact on any one residence.

In a time of need, the power company might send signals over the power line to turn the air conditioner off for ten minutes out of each hour in all homes in a round-robin fashion, for instance.

Homeowners would naturally have to agree to allow this type of control to be carried out. On the other hand, companies might also be willing to pay for the right to control appliances in this manner.

The heating system, air conditioning system, and water heater are all examples of appliances that might benefit from this type of power management.

4.2.3
White Goods Appliances

The capabilities of washing machines, ovens, dryers, refrigerators, and ranges can be enhanced by using pervasive computing technology.

4.2.3.1
e-Maintenance

Remote diagnosis   A large cost factor when servicing appliances is associated with diagnosing the problem and obtaining the proper spare part. Savings can be achieved through remote service or remote diagnosis. The service team could access the defective appliance through the network to diagnose the problem. Service personnel could then take the correct replacement part along on the first maintenance visit.

Modern appliances are often controlled by a microprocessor. In a washing machine, the microprocessor controls the various motors, pumps, heating units, and sensors within the machine to execute the desired wash program. A great deal of the washing machine behavior is actually determined by the computer programs, or microcode, executing on the microprocessor.

Everyone knows that bugs are constantly being found and fixed in regular computer programs for personal computers. The same situation exists for microcode running in washing machines as well. Until now, however, consumers had to make due with the microcode installed in the appliances at the time of purchase. With network-capable appliances it will be possible to download new version of the microcode into the machines, adding new functions and updating old ones as software progress is made. *Microcode upgrades*

Consumers may also be willing to pay for significant new functionality. A new spin cycle program that causes much less vibration might be such a chargeable upgrade.

4.2.3.2
Article Awareness

Section 3.2 "Smart Labels" explained how smart labels can be used to identify objects. If we imagine a time when smart labels are in prevalent use in product packaging, additional possibilities in for the automated home arise.

Smart labels can be made robust enough to withstand multiple washing cycles and flexible enough to be incorporated into clothing tags. A washing machine could contain a smart label reader in order to be aware of the type and required washing program for each article of clothing put into the drum. The washing machine could sound an alarm if incompatible articles are put into the same washing load.

By reading the smart label on the detergent packaging, the washing machine or dishwasher could also be made aware of the type of detergent being used. The washing machine could optimize washing parameters based on the detergent type.

The refrigerator and kitchen cabinets will also become aware of their contents through smart label readers. They could keep track of product inventory and expiration dates. It could inform the consumer via phone or email that a product expiration date is about to be reached or when a particular product is empty. A smart appliance could also detect consumer usage patterns – for example, "the average bottle of milk is removed from the refrigerator six times before it is empty" in order to predict replenishment need.

A new type of grocery subscription service would become possible. The consumer could subscribe to milk or other product delivery based on actual need. The intelligent devices in the home would recognize when new supplies need to be ordered and would signal this information to the grocery service. The consumer could define whether the goods would be delivered automatically or only after confirmation of the order.

These ideas would also be useful in a commercial setting. If a large, walk-in restaurant freezer or refrigerator is aware of its contents, it can help insure optimum stock rotation by making sure that the oldest product is always used first.

4.2.3.3
Remote Access and Operation

Web-capable appliances will allow access through the Internet or through a telephone. The consumer will be able to call into the home network to check if the oven is off or to start a washing machine program.

If the consumer wants to prepare something special for dinner, she will be able to select a recipe through the Internet and query the shelves and refrigerator if all necessary ingredients are present.

Remote access is also an interesting function for commercial use. Suppliers could access smart shelves, refrigerators, and freezers in order to implement just-in-time restocking.

4.2.3.4
Real-World Smart Appliances

The Merloni Elettrodomestici Corporation from Italy offers a line of advanced, web-enabled household appliances. These appliances can communicate among themselves as well as the consumer and customer service organizations through the Internet. These appliances do not implement all of the features described in the preceding sections, but do provide significant network functionality.

Figure 4.6 shows one of the world's first digitally enabled washing machines – the Merloni Margherita2000. This machine is a member of the Ariston Digital line of network capable appliances.

The Margherita2000 can accept commands or display operating status through a cellular phone or through the Internet. It can interact with external telecommunications networks through use of a special device known as a Telelink that manages communication between every Ariston Digital appliance and the outside world.

The Telelink uses either a standard analog modem or a GSM cell phone modem for communication external to the home and uses power line networking techniques for communication with other Ariston Digital appliances. The appliances communicate among themselves and with the Telelink through use of WRAP – Web Ready Appliance Protocol. WRAP currently uses the LonTalk™ power line carrier technology from Echelon Corp. as a data transport medium.

Services offered through the Telelink vary by appliance, and include remote diagnostic, customer support, and power management services. For example, the appliances can communicate with a power-metering device and can be programmed to activate in off-peak hours when the electricity rate is cheaper.

Photo courtesy of Merloni Elettrodomestici

Figure 4.7 shows two more members of the Ariston Digital line. The Leon@rdo is an interactive kitchen monitor that can communicate with the Internet and with other Ariston Digital appliances. The Ariston Digital Oven can download cooking instructions from the Internet.

With the Leon@rdo monitor, the cook of the household can browse the Internet to visit the http://www.aristonchannel.com web site. Recipes and household tips can be retrieved and read. Selected recipes on the web site contain cooking instructions for the Ariston Digital oven. When it is time for "e-baking", the cooking instructions are automatically loaded into the oven.

When baking complicated dishes, the oven temperature must be optimized for each phase of the baking. To get a nice crust some dishes need to start out at a high temperature and be baked at a reduced temperature until done inside. These types of cooking instructions can be downloaded into the Ariston Digital oven.

Figure 4.7:
Ariston Digital
Appliances

Photo courtesy of Merloni Elettrodomestici

4.3
Appliances and Home Networking

There are a number of different possibilities for networking smart appliances, and at present, it is completely unclear which, if any, of these ideas will win out.

Associated with each choice are compatibility and usability issues. These needs are easy to understand. If a consumer buys a smart clock of brand X and a smart VCR of brand Y, then the VCR should still understand the transmitted time signal. From the usability standpoint, it is clear that consumers will not use the technology if it results in a great deal of added complexity for them. If the consumer would have to set the network addresses of the smart clock and smart VCR, he might as well just set the time on the VCR himself.

The issues of network addressing, configuration and security touched upon in Section 4.2 "Smart Appliances" are also valid, if not more acute, for smart appliances. From the consumer point of view, system configuration and network address setup must happen automatically. Setting up adequate security must also be very easy. The typical homeowner will not want to deal with creating a public-private key pair and exchanging certificates between the refrigerator and other devices just to get her new refrigerator working.

The following discussion will illustrate several different ideas for appliance networking and examine the characteristics each. A consumer wishing to ask his refrigerator over the Internet whether there is any Dijon mustard will serve as a use case.

4.3.1
Residential Gateway

In this scenario, the refrigerator is connected through a home network. This could be done using power line networking, since the refrigerator must be connected to power anyway.

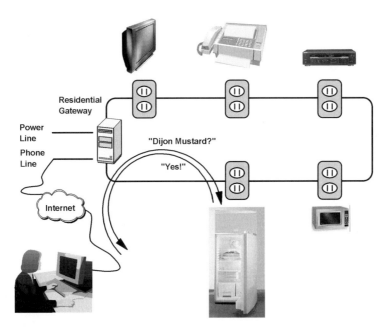

*Figure 4.8:
Networked
Appliances*

Figure 4.8 shows a possible home network configuration that would allow for such communication. The consumer would address the residential gateway, authenticate herself, and then communicate with the intended appliance. The residential gateway would block access by unauthorized callers.

Either the consumer could dial into the residential gateway, or the gateway could be connected to the Internet through an always-on connection. The appliance would make the information available in standard web format using an embedded miniature web server.

Use of a residential gateway as an interface between the outside world and the Internet has a security advantage, but requires the presence of a properly configured computer to perform the gateway function.

Somebody has to set up the residential gateway and make sure it knows how to route requests to the refrigerator. If the consumer must configure the gateway, this would likely lead to acceptance problems.

4.3.2
Cellular Communication

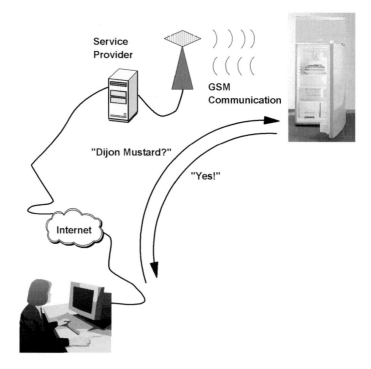

This scenario calls for each appliance to have its own cell phone number. The consumer would browse the web site of a service provider, who would dial up the refrigerator and obtain the requested information. This is shown in Figure 4.9.

There are a number of problems with this idea. First of all, home appliances are sometimes placed in locations where radio reception is poor. The consumer will not want to push his refrigerator around until the service provider obtains good reception!

Also, if the appliance could only communicate through the cellular net, it would be inconvenient to have appliances communicate among themselves.

Security could also be a problem in this scenario. If the appliances are accessed using a standard cell phone number, there is nothing to prevent anyone from dialing in. The appliances could be protected using some type of password or security code, but if the password is left at its default setting, it would not provide much protection.

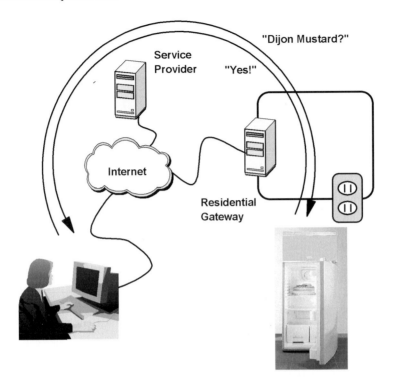

4.3.3
Service Provider and Residential Gateway

In this instance, the consumer browses a service provider site. The service provider accesses the appliance through the residential gateway. This is illustrated in Figure 4.10.

The "home network" service provider would actually be responsible for the setup and operation of the residential gateway. This would relieve the consumer of knowing all the details about the networking going on inside his house. The residential gateway would be used as a central access point. It would have a security as well as a routing function.

In essence, this type of service provider would be providing network administration for the home network. This would be an advantage for the consumer, since he would be relieved of a difficult chore. It would also be an advantage for the service provider, who would be in a position to offer content that would directly appeal to the consumer's interests.

4.4
Automotive Computing

4.4.1
Intelligent Controls, Sensors, and Actuators

*Pervasive
Computing in
vehicles*

For a couple of years, there has been an emerging demand for new electronic systems in automobiles. Controllers, computers, and in-vehicle networks collaborate to make driving easier, safer, and more comfortable.

Sensors are detecting all possible parameters, like rainfall, pressure in the tires, or intensity of sunshine. All this data is transmitted to different processing units in a car, which trigger an action if required. For instance, wipers are activated automatically as soon as rain starts. A warning can be displayed on the dashboard, when the tire pressure is below a minimum threshold. And the air conditioning can increase its cooling, if the car is exposed to extreme sunshine.

Beside safety equipment, which helps to avoid accidents or protect the passengers, there are manifold convenience features. These have in common that they replace manually operated parts by electronic modules, such as motors and switches. Instead of letting the driver crank the window, automatic window openers take over that job. Mirrors, seats, and door locks can be easily operated using a panel.

The next paragraphs give some examples of electronic modules in today's cars:

*Powertrain
controllers*

Powertrain and transmission controllers adjust the engine and the gear for best performance. These sophisticated systems apply powerful 32-bit processors running modern real-time operating systems. They process sophisticated algorithms, which help to make the engine more efficient, save fuel, and minimize emissions. Only with the help of theses electronics, it became possible to achieve the energy savings, required by environmental protection laws. Parameters and measurements are accessible through common diagnostic interfaces such as defined by ISO 9141. A service engineer can retrieve and change these values to detect and fix problems with the engine.

Among the real-time operating systems often used in such automotive controller systems is OSEK. OSEK has been specified by European automobile manufacturers as a standard operating system platform for embedded controllers. OSEK includes communication protocols and network management for automotive environments.

Antilock braking

The antilock braking system comprises a smart controller, which surveys the rotation of all four wheels and the motor, as well as the speed and gear of the vehicle. If the controller detects a mismatch of these parameters, the associated brakes are inactivated for a short duration.

*Tire pressure
sensors*

There are two different ways to determine the tire pressure. The first method uses a controller that calculates a change in the tire pressure from the vehicle speed and the tire rotation speed. All of these parameters can be retrieved from the antilock braking system. The second method applies a sensor within the tire, which is connected wirelessly to the car's body electronics.

The airbag is another good example of how multiple electronic subsystems collaborate: The pre-crash sensor triggers the airbag activation process. A dedicated sensor inactivates the airbag, if a children's seat occupies the front seat. Additional sensors detect, if the passenger sits in an improper position, in which the airbag could cause more harm than benefits. For example, if a person puts his feet on the dashboard or leans his head against the side windows, the system will detect an "Out-of-Position" situation and will not inflate the airbag.

Airbag

The adaptive cruise control pays attention to the traffic in front of the vehicle. An infrared sensor detects other vehicles in front, and reduces or increases speed automatically in order to keep a minimum distance and avoid collisions.

Adaptive cruise control

There are several approaches for keyless access. Contactless smart cards and other devices replace the traditional key and offer additional functionality. For instance, smart cards can hold personal preferences, such as the seat position, preferred radio stations or climate control settings.

Keyless entry

When electronics and computers get pervasive within cars, one impact is the increased electric power consumption. 800 watt is already needed for today's cars. This is equivalent to about 1 liter of fuel per 100 km. In the next decade the average power consumption will reach up to 4 kilowatt. Moving to higher voltage is one way to handle this issue.

4.4.2
On-Board Computing Systems

Intelligent controls and appliances in automobiles comprise far more than just controllers, sensors, and actuators. Vehicles extend their traditional interface to driver and passengers and offer navigation, telematics, personal communication, and infotainment systems. Similar to other pervasive computing technologies, cars will see a convergence of Internet, multimedia, wireless connectivity, consumer devices, and automotive electronics.

4.4.2.1
Navigation Systems

Navigation systems use digital maps to guide the driver to his destination via the optimum route. By comparing the car's current geographic position with the map, the system informs the driver about turns and crossings (Figure 4.11).

The standard way to determine location is through use of the Global Positioning System (GPS). A GPS receiver listens to signals emitted from multiple GPS-satellites. Each signal carries a precise timestamp indicating when the satellite sent it. The receiver compares the timestamp with its own internal time and uses this information to calculate the distance to the satellite. At least three satellites are required to determine the current location. For positioning pur-

Global Positioning System (GPS)

poses, there is a network of 24 satellites, which are arranged to give worldwide GPS coverage. Originally deployed by the US military, GPS is now applied for civilian purposes as well. Today, GPS receivers are built into phones, PDAs, and of course into automotive navigation systems.

Car navigation systems can achieve a precision of about 100 meters. When coupling the GPS receivers with other additional information, such as digital maps or tire rotation speed, the precision can be improved to 5 meters of tolerance.

Figure 4.11:
Blaupunkt
Navigation
Systems

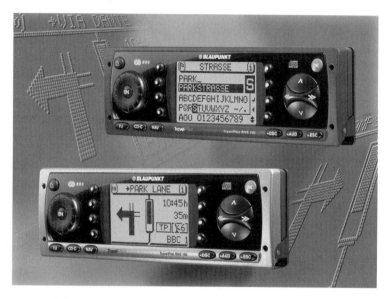

Photo courtesy of Blaupunkt

4.4.2.2
Telematic

Wireless links between the car systems and the outside world open up a wide range of telematic applications. Automotive systems are no longer limited to information located on-board, but can benefit from a remote networks and service infrastructure.

Dynamic
navigation

Dynamic navigation is a telematic application complementing simple navigation by retrieving up-to-date travel information from a content provider. The route proposed by the navigation system takes into account traffic jams, road construction, and even bad weather conditions that may make changes of the itinerary advisable.

The navigation system can retrieve relevant traffic information in different ways:

- Traffic Message Channel (TMC) is part of the Radio Data System (RDS). RDS is an additional information packet broadcast above the audible bandwidth of the regular FM signal. Each TMC message consists of location, event, and quantifier codes, describing the obstacle. One limitation is that the location code does not directly include a geographic coordinate, but identifies particular reference points instead. These predefined points need to be mapped by the navigation system. TMC always sends all available information to everyone. Unfortunately, in many countries there is only a limited coverage of radio channels offering RDS and TMC.

  Traffic Message Channel (TMC)

- Short Message Service (SMS) is used by travel service providers to send traffic information to subscribed users. The navigation system notifies the provider's systems about an intended itinerary. Every time new traffic information becomes available, a SMS message is created automatically and distributed to all registered navigation systems that are impacted by that event. After receiving the SMS, the route can be adjusted accordingly.

  Short Message Service (SMS)

- Wireless Application Protocol (WAP) is the third way to feed the navigation system with latest news for route planning. With a WAP browser, a wide range of Internet based route planning services can be accessed. Beside traffic jams, other information like directions to the next gas station, recommended restaurants, and the current cinema program can be displayed on the dashboard (Figure 4.12). Using WAP, the navigation system initiates requests to a traffic information server, which returns the available data.

  Wireless Application Protocol (WAP)

Figure 4.12: TravelPilot DX-N WAP Enabled Navigation System from Blaupunkt

Photo courtesy of Blaupunkt

TMC requires a regular FM tuner. For SMS and WAP, the navigation system must integrate cellular phone circuitry. TMC is delivered to everyone as a free service of the radio station broadcasting the FM carrier. Connection fees apply to SMS and WAP service, and each contacted travel service provider may charge individual subscription fees.

Of course, radio stations and other service providers offering individual traffic and route information need to collect the data about the traffic flow first. For this purpose, two methods can be applied.

Traffic jam sensors
One common way is to install traffic speed sensors at highway bridges. If the speed falls under a certain limit, the sensors use a build-in mobile phone to call the computing center and send data about the sensor's position and current traffic situation. This approach works fast and reliably, but is quite expensive. Often it is only financially feasible to cover the major freeway system.

Floating Car Data (FCD)
Another way to collect the required data is a system called Floating Car Data (FCD). In this case the cars contain, besides the navigation system and a GPS system, a unit which informs the travel information center in regular intervals about the position and the speed of the car. The big advantage of the Floating Car Data system is that it is not limited to the highways surveyed by traffic jam sensors. This approach is reliable for a particular region as soon as enough cars are equipped with FCD systems. In Germany, there were about 1000 systems in use as of early 2000.

More telematics
For dynamic navigation, content is delivered from a back-end system to the car. However, telematic applications can also take advantage of a wireless connection the other way around, as the following two examples show:

■ Why shouldn't the car send data to the car manufacturer or closest repair shop as soon as a control system detects a problem or reports a warning of a likely failure. Allowing maintenance systems to access the internal body electronics of a vehicle enables remote diagnostics. This gives the repair station the possibility to arrange the necessary parts before the car arrives. Even over-the-air engine controller reprogramming is a technically possible scenario, although it is not very likely to become reality in the near future.

■ After triggering the airbag inflation, the pre-crash sensor can notify the board computer system. Using a built-in cellular phone, the board computer can automatically originate an emergency call to a service center. The current location of the vehicle can be retrieved from the GPS and included into the emergency call.

4.4.2.3
Infotainment

Another area of Pervasive Computing in automobiles is infotainment, the combination of entertainment and information applications. Digital audio broadcasting (DAB), TV, and Internet are beginning to extend their mobile usage to cars.

Soon data, video, and audio streams will be delivered via high-bandwidth wireless links to the driver to provide travel-related content as well as e-commerce capabilities, games, and movies.

The Internet turns out to be the dominant platform for such infotainment offerings. Some of today's in-car Internet browsers use WAP for transferring data over a wireless connection. Such systems allow the users to surf the Internet as from a mobile phone, but feature larger screens and better input capabilities. WAP-based traffic services as described in the previous section are only the very beginning. Internet multimedia will be the content as soon as the required wireless bandwidth becomes available. Already today, satellite links help to provide a better download performance.

Internet access

Figure 4.13:
Multimedia
Enhancements
Implemented by
BMW

Photo courtesy of BMW

Concept cars from various manufacturers show how extensive usage of computing devices within a car can look like: flat screens are built into in the center of the dashboard or into the rear of the front head restraints for use by the driver and passengers. Various mobile devices like handheld computers and wireless keyboards have access to the car's systems through infrared transceivers. The center armrest includes a cradle for a PDA (Figure 4.13).

Multimedia usage in vehicles implies new requirements to human computer interfaces in order to allow hands-free and eyes-free operation of the devices while driving.

A speech interface to the vehicle's functionality is one way to achieve this: Voice recognition accepts user-spoken commands, for example, when requesting travel information from a web server, or while programming the navigation system. Text-To-Speech (TTS) technology is used for presenting the user with synthesized verbal output such as driving directions.

4.4.2.4
Automotive Computer Systems

Applications as described in the previous sections run on powerful computing devices, which require an extensible software and hardware architecture. The Microsoft Windows CE-based Auto PC and the Motorola Mobile GT Architecture are two examples for high-end automotive application platforms.

Windows CE   Microsoft follows a very PC-centric approach when postulating its Windows CE operating system as the universal automotive software platform. An Auto PC based on Windows CE is an information, productivity, and entertainment system for vehicles. Bi-directional speech interface, Personal Information Manger (PIM) , cellular phone, and digital audio systems are part of it.

The prerequisites for the automotive version of Windows CE are considerable: 60 MIPS processor, 256 x 64 pixel color screen, 8 MB RAM, 8 MB ROM, a compact flash slot, CD ROM drive, preamplifier, AM/FM tuner, microphone, USB port, and infrared port.

Mobile GT   Motorola's Mobile GT Architecture is a universal and modular computing platform for driver information systems. It can be applied to a range of infotainment systems embedded into vehicles.

Mobile GT is based on a PowerPC processor running the QNX real time operating system. Applications are written in Java and execute on an IBM Java Virtual machine (Figure 4.14).

Figure 4.14:
Motorola's
Mobile GT
Architecture

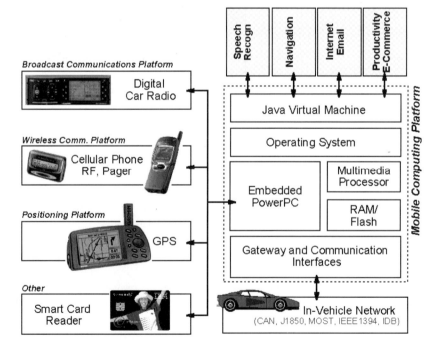

4.4.3
In-Vehicle networks

Only a decade ago, cars had little electronic equipment aboard. Compared to the mechanic parts, electronic components had only minor impact on the production costs. Design cycles were long enough to develop proprietary electronics for a specific car model.

As the usage of interconnected electronic sensors, controllers, and devices increased, the automotive industry faced three problems:

A bunch of electronic components

- The traditional spider web-like wiring of the electronic components increased the weight of the car, increased the complexity of assembly, and became too expensive.
- The electronic devices themselves made up a more and more significant part of the production costs. Complexity and dependencies made development more difficult. Customized electronic components for a specific a car model became unaffordable.
- While development of a new car model takes about three years and the model is sold for about six years, the cycle for electronic devices is much shorter, with replacement products available typically every year. When the first car of a new model is assembled, its electronic systems are already one or more generations behind state-of-the-art. Its electronics cost more and have fewer features than comparable products on the market.

These issues motivated automotive industry to work on generic bus systems that allow interconnection of state-of-the-art electronic components. The benefits of a vehicle bus system, which serves as a backbone for the collaborating electronics, are obvious:

The solution: in-vehicle bus systems

- Instead of complex, peer-to-peer wiring, each component is plugged into the in-vehicle network.
- Components have well-defined interfaces to the bus. This leverages reusability, cuts down development costs and allows integration of standardized third-party products.
- Due to the common interface, components can be developed independently from the car itself. They can even be exchanged or extended by improved products after the car is assembled.

The migration from the first proprietary in-vehicle networks to open standards for universal automotive bus systems is still ongoing. Several industry consortiums define and agree on specifications, which can be applied for mission-critical body electronics, as well as new kinds of telematic and infotainment systems.

Open standards

4.4.3.1
J1850 and ODB-II

Among the first industry specifications for vehicle bus systems was the J1850 standard defined by the Society of Automotive Engineers (SAE). With J1850, only limited interoperability was achieved, since the specification allowed the manufacturer to implement it in multiple proprietary ways. Beginning in 1996, the US Environmental Protection Agency (EPA) required a standard diagnostic interface to the J1850 bus in order to retrieve emission data. The Onboard Diagnostics (OBD-II) interface can also be used to provide a standardized access to some of the internal electronic systems.

4.4.3.2
Controller Area Network

While US manufacturers favored the J1850, the European automotive industry adopted the Controller-Area-Network (CAN), which now gains worldwide acceptance as the common technology for the mission critical networks (Figure 4.15).

The CAN bus uses two wires, usually twisted pair, and is suitable for harsh environments. Error checking mechanisms make the communication reliable and fault tolerant.

Message-based   Communication is based on messages broadcast by a node onto the bus. There is no explicit address mechanism for direct communication between two subsystems. All connected nodes receive each message, and must decide if that message is relevant and must be processed or can be ignored. An identifier is used to label the type of a message uniquely. All nodes connected to the bus are equivalent. New nodes can be added without changes to the system as long as they require no new message types.

Priorities   This message-triggered approach allows priorities to be assigned to each message type. When multiple nodes simultaneously transmit data, the lower priority node aborts the transmission. After the transmission of the higher priority message is completed, the interrupted message is repeated.

Message format   The format of a CAN message is defined by the ISO standards 11898 and 11519-2:

- A 11 or 29 bit identifier is followed by a *remote transmission request* bit (RTR). The RTR indicates whether the message is used to request data from another node or if it sends data itself.
- Four bits hold the length of the data field. Up to eight bytes can be transmitted within one message.
- A message is terminated by a 15 bit *cyclic redundancy check* code (CRC), a two bit *acknowledge* field and a seven bit *end of frame* field (EOF).
- After three bits of *Intermission* (INT), the bus is idle and can be used to transmit the next message.

Mercedes-Benz S-Klasse: Daten-Kommunikation in drei elektronischen Netzwerken
Mercedes-Benz S-Class: Data communication in three electronic networks

Figure 4.15:
CAN Bus
Network in
Daimler
Chrysler's
S-Class

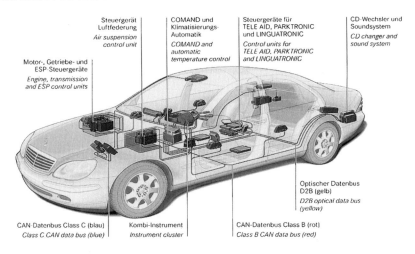

Steuergerät
Luftfederung
*Air suspension
control unit*

COMAND und
Klimatisierungs-
Automatik
*COMAND and
automatic
temperature control*

Steuergeräte für
TELE AID, PARKTRONIC
und LINGUATRONIC
*Control units for
TELE AID, PARKTRONIC
and LINGUATRONIC*

CD-Wechsler und
Soundsystem
*CD changer and
sound system*

Motor-, Getriebe- und
ESP-Steuergeräte
*Engine, transmission
and ESP control units*

Optischer Datenbus
D2B (gelb)
*D2B optical data bus
(yellow)*

CAN-Datenbus Class C (blau)
Class C CAN data bus (blue)

Kombi-Instrument
Instrument cluster

CAN-Datenbus Class B (rot)
Class B CAN data bus (red)

Photo courtesy of DaimlerChrysler

Based on the communication speed, three groups of CAN bus systems are classified: *Transfer rates*

Class A networks transfer messages with less than 10 kbps and can be used for various convenience features, like climate control.

Class B networks have a transfer rate of up to 125 kbps and can be applied for body electronics and diagnostics systems, like light control or door openers.

Class C refers to high-speed networks of up to 1 Mbps, which operate mission-critical vehicle dynamic systems, such as antilock braking or transmission control.

Although the CAN bus supports up to 1 Mbps, it can reach its technical limits when used for time-critical applications like airbag systems. Since an airbag sends only short notifications, the transfer rate is less important but the message must be delivered within a guaranteed time. The CAN has widely varying latencies due to its concept of messages, which are handled in the order of their priorities. This makes it difficult to design extremely reliable systems requiring a defined response time.

Time Triggered Protocols (TTP) can be a solution in such time-critical cases. *TTP*
TTP assigns each node an explicit time-slot in which it can transfer data. The available bandwidth is distributed between all participants during system design. Latency is constant and predictable, although not necessarily optimized since the time slots cannot be changed dynamically according to the communication traffic in a particular situation. Another disadvantage is the significant higher price compared to CAN.

A further limitation of CAN is that some systems need self-powered bus systems. Firing an airbag needs a higher voltage (around 25 V) that cannot be delivered on a CAN bus.

4.4.3.3
Local Interconnect Network

Local Interconnect Network (LIN) is a single-wired class A serial bus targeting low-costs applications where performance and complexity are not required. Interoperable LIN nodes can replace proprietary analog code switches and multiplex solutions. Costs per node can be expected to be one-half to one-third that of a comparable CAN network.

Hierarchical network architecture

LIN also allows implementation of hierarchical vehicle networks. Smart sensors, motors, pancls, and actuators can be connected to a local LIN sub-network. Communication between the sub-network and an in-vehicle CAN bus is carried out through a gateway (Figure 4.16). This approach reduces traffic on the CAN bus since part of the communication can be handled inside the LIN subsystem. Typically LIN is applied for assembly units such as doors or seats.

*Figure 4.16:
Typical
In-vehicle
Network*

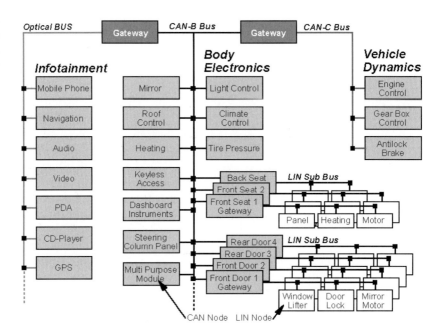

LIN is an open standard driven by Audi, BMW, DaimlerChrysler, Motorola, VCT, VW, and Volvo. The specification includes the transmission protocol, the physical layer of the transmission, and software programming interfaces. Additionally, interfaces to development tools are standardized to achieve interoperability during design. The specification was first released in 1999, and the first automotive applications are expected in 2001. Although LIN is an initiative of the automotive industry, the standard is applicable for a wide range of other applications, such as industrial electronics or even household appliances.

4.4.3.4
IDB-Bus

Three industry initiatives are working together to define the next generation of automotive bus systems. These systems will be ready for plug-and-play of interoperable devices inside the vehicle, and will allow connectivity to the outside world. The initiatives are:

- Intelligent Transportation Systems Data Bus (IDB) Forum
- Automotive Multimedia Interface Collaboration (AMIC)
- Telematic Suppliers Consortium (TSC)

The IDB Forum has created the Intelligent Transportation Systems Data Bus (ITS Data Bus or IDB). The IDB is intended as a standard interface for connecting consumer devices such as GPS receivers, phones, controllers, or diagnostics systems, within motor vehicles. The first version of the specification was the IDB for telematics (IDB-T), an RS-485 serial bus with 115.2 kbps. A reworked specification calling for 250 kbps CAN Bus technology was published as IDB-C.

IDB-T
IDB-C

Currently the specification of a plug-and-play mechanism for devices connected to the IDB-C bus is in progress. Electronic devices will be able to be added to the self-configuring bus at any time. Peripherals can be installed depending on customer needs and using technology that is state-of-art at the time of the upgrade – not at the time of the initial car design. The benefit of an open standard for in-vehicle networks is obvious: one universal version of an automotive peripheral, supplied by an arbitrary third party manufacturer, could fit all cars.

Plug-and-Play

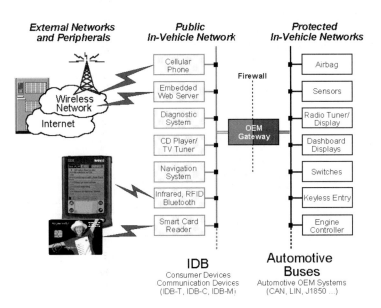

Figure 4.17:
IDB Technology

The IDB approach to achieve interoperability is to define standardized messages, which can be exchanged between the devices connected to the bus. For instance, there are messages to request the car's current location, read a sensor status, trigger an actuator, display text on the dashboard, get speedometer readings, retrieve gear position, and operate door locks or seat positions.

In order to allow external devices and systems to access the body electronics of a vehicle, a gateway is used as a bridge between the public IDB and internal networks such as CAN or J1850. Through this *OEM gateway*, user-installed third-party systems can access original equipment provided by the car manufacturer. For instance, an additional board computer can implement its user interface using the available switches, controls and the dashboard displays. The gateway filters transmitted messages and ensures that only authorized message traffic enters the private networks inside the vehicle. This security mechanism is similar to firewalls used for regular LAN. It protects the mission-critical systems and meets safety and reliability concerns of the automakers.

While the OEM gateway connects IDB with the interior, an embedded web server acts as a wireless gateway to networks outside of the car. Cars become nodes on the Internet and are accessible using a standard web browser. Telematic applications can connect to the IDB and perform various tasks, such as invoke functions, retrieve information, or download software. Demonstrations by IDB include switching on the light from remote, locking the door, and reading the current geographic position of a vehicle.

Short range radio-frequency networking, Bluetooth, and infrared technology allow attachment of intelligent peripherals to the IDB in an ad hoc manner. Beside personal devices such as PDAs, various other appliances like maintenance tools or toll stations can exchange information with the vehicle. A demonstration of the IDB Forum shows a robot filling the fuel tank. The robot accesses the IDB via a wireless radio connection to read the vehicle identification number, fuel type, and fuel level. The robot opens the fuel filler lid by sending a message, which triggers a corresponding actuator.

A 100 Mbps bus called IDB-Multimedia (IDB-M) is the next challenge for the IDB Forum. As audio and video streams need to be supported by the in-vehicle networks, there is a demand for high-bandwidth connections based on an optical carrier. IEEE 1394 and Media Oriented System Transport (MOST) from Oasis could be the starting point for the development of IDB-M. Due to higher costs IDB-M will be applied additionally to the existing IDB-C network and only deliver multimedia data between selected nodes.

Complementary to the work of the IDB Forum, the Automotive Multimedia Interface Collaboration (AMIC), specifies programming and hardware interfaces for the IDB OEM gateway. AMIC states its mission as "developing a set of common specifications for a multimedia interface to motor vehicle electronic systems in order to accommodate a wide variety of computer-based electronic devices in the vehicle". Members of AMIC are DaimlerChrysler, Ford, GM, Renault, and Toyota.

The AMIC specifications base on the IDB-C bus technology and target:

- simple devices (like sensors, actuators, switches, and displays)
- controllers running dedicated tasks (like radio, navigation systems, and embedded processors)
- hosts systems running user applications (like an Auto PC)

The third collaborating group is the Telematics Suppliers Consortium (TSC). TSC works on connecting the in-vehicle networks to the outside world. Remote application and host systems should be enabled to access an IDB-C bus in a standardized and interoperable way. Bluetooth and WAP are means how intelligent peripherals like handheld computers and phones could link to the car systems.

Members of the TSC are providers of telematics, consumer electronics, and wireless equipment.

Telematics Suppliers Consortium

4.5
Further Readings

Merloni Elettrodomestici Main Site

http://www.merloni.it/eng/default.htm

This site provides information on the Merloni Ariston Digital line of web-ready home appliances. Follow the links under "New Products", "Aristonchannel", and "Margherita2000.com".

The Web Ready Appliances Protocol Home Page

http://www.wraphome.com

This site describes the Web Ready Application Protocol used for communication between household appliances.

The Home Automation Association

http://www.homeautomation.org

This is an interesting web site for home automation topics.

Smarthome Systems

http://www.smarthomeusa.com

This site provides information about X10-based home automation products.

Automotive Multimedia

http://www.DaimlerChrysler.com

From the DaimlerChrysler main site follow the links "Products" and "Research and Technology" to find interesting articles about use of advanced electronics and other technologies in automobiles.

Blaupunkt

http://www.blaupunkt.de

Blaupunkt is a manufacturer of navigation, telematic and audio systems for cars. The homepage links to their product descriptions. A simulator allows operation of the devices online with an Internet browser!

IBD Forum

http://www.idbforum.org

The IBD Forum homepage. A section with public documents includes technical papers and several overview presentations. Most of the specifications itself are only accessible by members.

Microsoft

http://www.microsoft.com/windows/embedded/ce/default.asp

Further information about Windows CE can be found here.

Motorola

http://www.motorola.com/automotive

Motorola's starting point for automotive information retrieval. There are links for LIN, Mobile GT, and many other technical topics.

Society of Automotive Engineers

http://www.sae.org

The Society of Automotive Engineers (SAE) homepage is a portal for manifold information related to automotive technology. There are plenty of links to events, education material, and specifications. A mail order service delivers books and technical papers.

5 Entertainment Systems

This chapter covers pervasive devices, which are used mainly for entertainment purposes. Interactive television, set-top boxes, and enhanced broadcasting technologies will be one focus of interest. An overview of game console characteristics is the second topic of this chapter.

5.1
Television Systems

5.1.1
New Applications

In today's households usually two separate networks coexist: First, there are entertainment appliances, such as television and radio. These are traditionally connected via cable, satellite or air to mostly analog broadcasting information. Personal Computers and other computing equipment make up the second group: they interact with the Internet using digital communication over phone lines.

It is predictable that both networks will merge soon. Combining broadcasting with the flexibility of information processing will create new fascinating applications. Digital television, interactive entertainment, and Internet services are delivered over the same broadband networks into homes. The International Data Corporation (IDC) expects that television systems will represent about one third of the Internet connected devices in 2004. More pixels and better resolutions of digital High-Definition Television (HDTV) are only minor aspects of the anticipated success story. Instead, new usage scenarios promise innovation and thrill to the broadcasting industry as well as to the consumer electronic manufacturers.

Merging broadcasting and computer networks

- Electronic programming guides are interactive applications running on the television itself. They offer individual program information. The search for specific entertainment offers is leveraged by the usage of keywords and categories. Internet links can lead to detailed background information or related topics. An individual program selection can trigger reminders on a handheld computer or be forwarded directly to the VCR for automatical recording. According to the personal preferences or viewing habit, current tips can be highlighted.

New applications

- Enhanced video broadcasting allows multiple input streams of video, audio, and text content. Users can manipulate these incoming streams and arrange or overlay multiple windows (Picture in Picture). Translucent displays show individual information on top of the regular TV pictures. User requested information, like stock prices, can be displayed while watching a regular movie or a player resume can be laid over a football game broadcasting.
- Interactive video enables the viewer to influence TV content by himself. In case of an one-way interactivity the viewer is only allowed to filter the predefined content he has received from the broadcaster. For example, he can control the camera individually by himself choosing between multiple camera angles which are offered in parallel. More sophisticated interactivity is possible if a back channel is available. The back channel returns user selections and requests to the broadcasting provider. This allows viewers for instance to participate at game shows. Content related web sites can be offered through which the user can navigate with his remote control. Interactive applications can be delivered to the view, which look like television, but behave like computer games.
- Hyper video tries to map the classical hypertext concept to video experience. Digital video clips are extended with hotspots, which represent links to related clips or other kinds of information. For example, the viewer can trigger the plot of a story himself, by choosing different links.
- Video-on-Demand, Time-Independent TV, and Pay-TV are other applications, which are enabled by a two-way communication between user and provider.
- Personal Video Recording and Personalized TV uses hard disks to buffer broadcasting content. The viewer can pause, record, and replay the television program. Based on the viewers personal profile or viewing habits, the TV can record related programs and can create a personalized TV offering.
- Internet based video games.
- Internet TV provides web browsing capabilities on a television system. Other Internet based services such as home banking, email, chat, and e-commerce are just a few samples of value-added services, that broadcasters will offer to their clients. The television set evolves to a computing appliance.
- Interactive television exposes the viewers entertainment preferences configured in the TV itself and can track his selection of programs and viewing habits. This is a tempting data source for the content provider's advertisement strategy. Advertisement could be tailored to the individual and his situation and mood at a very specific moment. This will result in a conflict between grasping new business opportunities and protecting data privacy.

5.1.2
Analog and Digital Broadcasting

In general there are different possibilities to deliver interactive content:

- "Full signal" uses existing analog or digital broadcasting networks for transmitting audio, video, and data in one stream. Among the existing standards for signals are NTSC, PAL, and SECAM. The extensions for sending interactive information are often proprietary.
- "Dial-in + signal" combines and synchronizes the broadcasting signal for television content with a dial-in Internet connection to an online provider for accessing web content.
- "Full IP" uses a standardized high-speed Internet connection between a content server and the client TV. Data is cut into multiple packets, which are routed through a network to the destination address. The applied TCP/IP or HTTP protocols support two-way communication and work on almost any kind of physical network.

Means of transportation are cable, satellite, phone lines, and over the air. Today's broadcasting is dominated by satellite transmission, but cable TV will catch up in the long-term. One reason is that a satellite transmission allows one way communication only. For interactive usage a second transportation media is always needed to establish the back channel. Mostly a slow phone dial-in is used for that purpose. Besides being bi-directional, higher bandwidth, and speed are the other big advantages of fiber optic cable lines. With up to 27 megabits per second, cable beats satellite (about 4 Mbps) and ISDN telephone modem (up to 64 kbps) clearly. High speed Internet access from home is a welcome side-effect of powerful cable networks.

Communication bearers

While the methods to deliver interactive television differ, the offered services will be very similar.

In an analog television system the additional information needed for any new enhancements is transmitted within unused portions of the analog signal. The Vertical Blanking Interval (VBI) is a black stripe at the top and bottom of a TV picture. In case of the North American television standard NTSC the VBI comprises 21 of the 525 lines which are transmitted every second – each line having 427 pixels. The VBI can be applied to broadcast any kind of encoded data, which has to be interpreted by the receiving television system. Commonly, line 21 is reserved for sending closed captioning, teletext, and HTML data. For example, today TV schedules, stock tickers, and news headlines are transmitted this way embedded into the regular broadcasting signal. Using VBI, an impressive speed of four Mbps can be achieved! But nevertheless, large amounts of multimedia content cannot be delivered this way fast enough.

Analog television

With the advent of the digital broadcasting, such workarounds become obsolete: the Digital Signal Television (DTV) was already designed for communicating two-way entertainment and information. One incoming signal can be demultiplexed into separate audio, video, and data streams. Because of the required huge investments in new broadcasting facilities and new television receivers, the introduction of digital television will happen within an eight years period until 2006. In combination with High-Definition Television (HDTV), digital television features higher resolution and better audio quality than today's analog signals NTSC or PAL. HDTV doubles the visualized resolution to 720 or 1080 lines. Since these improvements imply more data, much higher bandwidth is needed than with existing standards. HDTV operates between 20 and 22 MHz, while the current PAL System uses 5.5 MHz. Data compression based on the standardized MPEG-2 format is applied to solve that issue.

5.1.3
Set-Top Boxes

Interactive capabilities can be either implemented in the television set itself, a personal computer or separately in a set-top box. Set-top boxes are the interface between broadcasters and the consumer's television set. They sit on top of a TV set and connect it to an entertainment network. They receive either analog or digital broadcasting from a satellite dish, a cable or a rooftop aerial and forward that information to an attached television device. Set-top boxes can even translate digital input into analog signals, allowing the support of older television devices (Figure 5.1).

Figure 5.1:
Receiving Digital
TV Signals

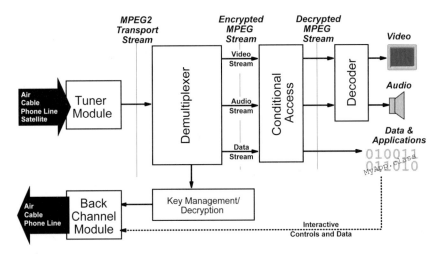

Photo courtesy of General Instrument

But beyond the basic tuner functionality they offer much more features. State-of-the-art digital set-top boxes are supporting multimedia games, HTTP browsing, application download, external printers, and SSL connections to a server. They include a smart card slot, an USB port, a CD ROM, and a wireless keyboard, as well as a telephone modem for the return path. Several gigabytes of hard disk, 8 or more MB of RAM, and a 233 MHz processor are characteristics of a high-end set-top box.

Features

Signal encryption and access validation modules ensure that only authorized viewers can use particular services. Smart cards are used as an access key for controlled usage and allow sophisticated billing schemes. Content and price of delivered entertainment and services can be part of an individual "ordering" process, upon which the broadcasting provider and the subscribed customer agree.

Very similar to the business model of telecommunication companies, broadcasting providers supply their viewers with set-top boxes for free or at little cost to attract and bind customers. The profit is made with broadcasting fees and other online services instead. This requires set-top boxes to be very cheap

5.1.4
New Players in the Entertainment Business

In Europe, interactive television pilots are dominated by proprietary networks, operated using "full-signal" transmission via satellite. The French Canal+ and the British Sky Broadcasting are among the pioneers. In the USA most broadcasting companies base their initiatives on standard Internet protocols either transmitted via satellite or cable.

While most of today's television initiatives base on existing analog broadcasting networks, on long-term the cable and satellite providers will have to migrate to digital transmission. They need to make investments in their infrastructure to enhance speed and bandwidth. Cable operators must upgrade their transmission networks to be prepared for the demand-driven advanced digital interactive services.

The IT industry is struggling about the underlying technology in order to ensure a good position for the starting race. WebTV, owned by Microsoft, and Liberate, related to Oracle, as well as OpenTV, founded by Sun Microsystems are among the leading technology providers.

While the Internet community achieved a very high degree of interoperability using open standards, the next generation of television is highly proprietary. Each provider uses a different hardware and software technology and exposes a different end user interface.

Heading towards open standards On long-term, agreed standards and portability across the manifold TV devices and set-top boxes will open the market for content providers. To leverage the development of extended television applications, a standardized JavaTV programming interface was specified by the Java community. Another standard was created by the Advanced Television Enhancement Forum, which specifies the content formats and delivery mechanisms of interactive television.

5.2
Game consoles

In the US, every fourth household has a game console today and the number is continuously increasing. Modern game consoles are good for playing games, but not only this: They more and more evolve to multifunctional entertainment systems, which conquer a place in the living room next to television sets. They might soon replace CD player, VCR, and in some areas even the PC, as today's generation of game consoles is able to play DVDs, audio CDs and, most important they are currently extended with the capability to connect to the Internet to play online games and partly surf the Internet. A large portion of the Sony Playstation 2 buyers in Japan for example say that they mainly are buying the Playstation 2 because of its build-in DVD drive.

Today's game consoles contain features like built-in modem, Internet connection, DVD player, USB port, IEEE 1394 port (Firewire), and hard disk. They provide real plug-and-play and are as easy to use as a TV set; their users do not have the same hassle as today's PC users. People who think that PCs and their game software are too complicated to install and maintain will probably buy one of these next generation entertainment systems.

Sega's Dreamcast was the first of these new generation entertainment systems on the market. In spring of 2000, Sony launched its Playstation 2 in Japan. Microsoft's Xbox followed on November,15th 2001 and Nintendo's GameCube on November, 17th 2001.

The manufacturers of game consoles currently make their money by getting license fees for each game sold. The game console itself is usually sold at a price that is sometimes not even recovering the cost. Compared to a PC with an average price of 800 to 1200 US $, a game console with a price of 149 to 199 US $ is a cheap alternative. The business model is based on making money on the games and services, as they only fit for one model. Customers are locked to the manufacturer as soon as the bought the game console. There are already thoughts of giving the console away for a even lower price and recover the costs by selling services to the buyer. The forecasts also say that a household will have several game consoles, like there is today often more than one CD player or more than one television set per household.

One of the most visible performance characteristics of a game console is the number of polygons the graphics processor can process per second. The surface of a 3D-object is build out of triangles, therefore the more polygons a graphics chip can process per second, the more powerful the console's graphics engine is. In a good game, some figures can consist out of several thousand polygons. To model a human body sometimes up to 500 000 polygons are used.

Graphics performance

A PC with a high-end 3D graphics adapter, like the nVidia GeForce 4, can process about 75 million polygons per second. Sega's Dreamcast can handle about 5 million polygons. According to Sony, the Playstation 2 can process about 75 million polygons, Microsoft claims the Xbox can render about 125 million polygons, and according to Nintendo the GameCube is can handle 12 million polygons per second. The number of polygons is not the only operation a game console has to perform; therefore also the whole system design is important for the overall performance of the system.

5.2.1
Sega Dreamcast

With Dreamcast, Sega had the first new generation entertainment system on the market (Figure 5.3). The Dreamcast console contains an analog modem with 56 kbps and enablement for Cable or DSL modem is underway.

Dream Key is the build-in Internet browser that Sega shipped in Europe. URLs can be typed in using a virtual keyboard on the display or by attaching the optional keyboard. Dream Key can display frames, tables, and flash animations, can execute JavaScript and handle cookies. More advanced web browsers for the Sega Dreamcast are Planetweb and DreamPassport. The browser from Planetweb even supports 128-bit SSL encryption.

The user can also send and receive emails. They are written with an HTML-based mail-client, which needs to be connected to the Internet. With the Dream Key browser, the Internet connection is provided by one of Sega's partners. The dial-in number is hard-coded inside the game console and unfortunately can't be changed to use any other ISP. Parents can enable a filter to ensure that their children do not access any unsuitable Internet sites. Planetweb and DreamPassport allow the user to choose his own Internet Service Provider.

Email

Photo Courtesy of Sega

Table 5.1 shows the technical components of the Dreamcast system.

Table 5.1:
Dreamcast
Hardware

CPU:	Hitachi 128 bit engine with on-board 64-bit RISC processor SH4 (200 MHz, 360 MIPS, 1.4 GFLOPS).
Graphics CPU:	NEC Power VR2DC, rendering up to 5 million polygons per second.
Main RAM:	16 MB.
Video RAM:	8 MB.
Sound:	45 MHz Yamaha chip with ARM7 CPU core supporting 64 channels.
Modem:	56 kbps.
CD-ROM:	12x speed, also able to read Sega's proprietary GigaByte Disc.
OS:	Sega's operating system and a tailored version of Windows CE with Direct X support.
Ports:	Four controller ports.

Windows CE   In addition to the Sega OS, the Dreamcast console can also run a tailored version of Windows CE 2.0. This should make it easier for developers of PC games to port these games over to the Dreamcast system.

The Power VR2DC graphics chip is the heart of the Dreamcast entertainment system. Saturn, predecessor of the Dreamcast console, already had Hitachi CPUs, the SH-1 and SH-2. The SH-4 CPU contains a 64-bit floating-point unit with a 128-bit vector graphics engine (Figure 5.4).

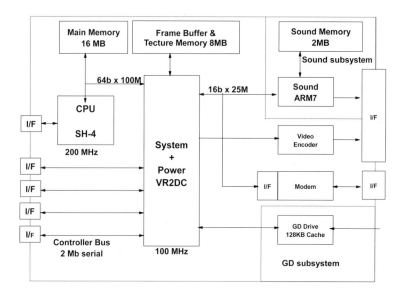

5.2.2
Sony Playstation 2

The Sony Playstation 2 marks the next step in the evolution from game consoles to entertainment systems (Figure 5.5). It contains a DVD player, two USB ports, one IEEE 1394 (Firewire) port and two memory card slots. Since August 2002, Sony provides an Ethernet network adapter which also allows dial-up with 56 kbps.

Figure 5.5:
Sony
Playstation 2

Photo Courtesy of Sony

The "Emotion Engine" is the heart of the system. It was developed by Sony in close cooperation with Toshiba and it can render up to 75 million polygons per second. This is about 15 times more than Sega's Dreamcast and about the same as a PC with a Nvidia GeForce 4 card (more than two years after the PS2 was released to market!). The Playstation 2 has only 4 MB of Video RAM (VRAM), compared to 16 MB in the Dreamcast and 8 MB in Sony's Playstation 1 (Table 5.2).

Table 5.2: *Playstation 2* *Hardware*		
CPU:	Toshiba 128 bit Emotion Engine running at 295 MHz.
Graphics CPU:	Graphics Synthesizer with 16 pixel processors working in parallel with 147 MHz. Renders 75 million polygons per second.
Main RAM:	32 MD direct RDRAM.
Video RAM:	4 MB.
DVD	4x DVD / 24x CD-ROM drive.
Modem:	optional.
OS:	proprietary Sony operating system.
Ports:	two USB ports, one IEEE-1394 port (Firewire), one Type 3 PC-Card slot.

The Emotion Engine is a 295 MHz RISC processor, which consists of about 13 million transistors. It has two integer units (IU), one 128-bit Single Instruction Multiple Data Unit (SIMD), one floating-point Unit (FPU), two floating-point Vector Units (VU1, VU2), and an image-processing unit (IPU), which is used for MPEG-2 decompression (Figure 5.6).

The USB ports as well as the Firewire port can be used to attach several external devices, like digital cameras, harddisks, and modems. Unfortunately, the required Playstation 2 device drivers are not yet available for most of these devices.

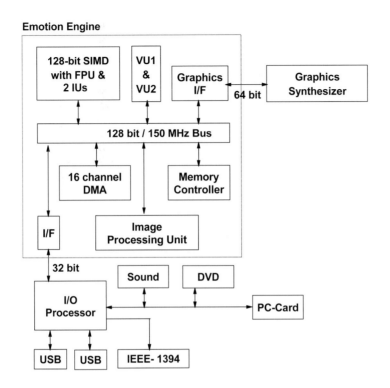

Emotion Engine

5.2.3
Nintendo GameCube

The GameCube is the successor of Nintendo's N64 system and has currently about 20% market share. During the development of the GameCube, Nintendo partnered with IBM. IBM designs and produces the processor for the Dolphin consoles called "Gekko". It is based on a 485 MHz PowerPC and is manufactured using IBM's leading copper technology.

The ATI graphics chip has 162 MHz and can render about 12 million polygons per second. It is manufactured by NEC.

CPU:	IBM PowerPC with 485 MHz, codename Gekko. 256 kb of L2 cache.
Graphics CPU:	162 MHz graphics chip from ATI which renders about 12 million polygons per second.
Main RAM:	40 MB.
Drive	3 inch Nintendo GameCube disk based on Matsushita's optical disk technology with approximately 1.5 GB capacity.

Network:	Optional modem and broadband adapter.
Ports:	Four controller ports, two memory card slots, analog AV output, digital AV output, two high-speed serial ports, and one high-speed parallel port.

Figure 5.7:
Nintendo
GameCube

Photo courtesy of Nintendo

The GameCube is with currently around 149 US$ the best priced game console, compared to for example 199 US$ for the Microsoft Xbox or the Playstation 2 from Sony.

5.2.4
Microsoft Xbox

On November, 15th 2001 Microsoft entered the game console market with a device called Xbox. The Xbox is part of Microsoft's strategy to gain a leading position in the video game console market. It is based on PC hardware and is using a Windows derivative as operating system. The Xbox is the first time that Microsoft developed, manufactured, and deployed consumer devices in large scale on its own.

Table 5.4:
Xbox Hardware

CPU:	Intel Pentium III with 733 MHz.
Graphics CPU:	nVidia NV 25 with up to 125 million polygons per second.
Main RAM:	64 MB DDR RAM (unified memory architecture).
Sound:	I3DL2 sound chip supporting 64 channels.

DVD:	5 x DVD drive.
Harddisk:	8 GB.
Network:	10/100 Mbps Ethernet adapter.
OS:	Tailored Windows version.
Ports:	4 controller ports, USB port, plug-in modem, and support for HDTV.

A game is started from CD-ROM, like it is the case with the Playstation 2. The hard drive is used for caching Internet sites, saving game data, or storing add-ons.

Figure 5.8:
Microsoft Xbox

Photo courtesy of Microsoft

Nvidia was chosen by Microsoft as the designer of the graphics engine, which was at the time the development started in 2000, three generations beyond nVidia's current chip in 2000, the GeForce 256. The codename for that chip is NV25 and it is able to process 125 million polygons per second. The GeForce 4 handles today about 75 million polygons per second.

The unified memory architecture allows to use the 64 MB of 200 MHz DDR RAM (Double Data Rate RAM) with no separation between video and main memory and therefore no AGP bridge. This flexible architecture allows memory either to be used as graphics or system memory, whatever is currently needed by the application.

The variety of ports makes the Xbox a good platform for further extensions. The Ethernet port enables the Xbox to connect two or more Xboxes directly. In addition to this, the Ethernet port can of course be used for an Internet connection via DSL or cable.

5.3
Further Readings

A boob tube with brains

http://www.usnews.com/usnews/issue/000313/webtv.htm

This article gives an overview of competing players in the interactive TV business [YAN00].

Digital TV: The Future of E-Commerce. E-Commerce Times

http://www.ecommercetimes.com/news/articles2000/000320-1.shtml

This article examines the future role of television in e-commerce.

Enhanced Television

http://www.itvt.com/etvwhitepaper.html

A comprehensive overview of broadcasting technology, economic potential, standards, and engaged companies [SWE99].

Futurefile.com

http://www.futurefile.com/broadcast.htm

A list of links to various articles related to broadcasting and interactive television.

Microsoft's TV activities

http://www.microsoft.com/tv

The Microsoft resource library for TV solutions.

Microsoft Xbox

http://www.xbox.com

This is Microsoft's official Xbox homepage providing information like the hardware specification, the games, access to the online game server as well as all kinds of other information around the Xbox.

Nintendo GameCube

http://www.gamecube.com

This is Nintendo's official GameCube site with all sorts of information interesting for users and developers, like details about games, screen savers, and hardware descriptions.

PlayStation

http://www.playstation.com

See what Sony has to say about the Playstation 2.

The Next Generation of Gaming: Forecast and Analysis, 2000-2005

http://www.idc.com

This IDC report includes a market analysis and forecast about the usage of game consoles till 2005 [IDC01].

Sega Dreamcast

http://www.sega.com

A variety of information about the Sega Dreamcast system can be found here, like latest games, hardware details, news, previews, contests, and release lists.

Part II
Software

As the previous chapters have emphasized, pervasive applications have to face different hardware characteristics of devices and changing usage environments. When developing software there are several particularities to be aware of:

- Mostly application developers have only a limited choice of programming languages. Java is widespread used because of the platform independence of the compiled code. Many Java language features focus explicitly on small devices. For instance, the Java Card is a smart card that is capable of executing Java code! C and C++ are other important languages, because very small native code can be produced, which is highly optimized for a particular platform. Often devices with tough constraints on memory do not allow any alternative to C/C++.
- While desktop computers load applications into RAM before executing, pervasive devices usually process executables and data directly from the location in RAM or ROM where they are stored persistently ("*Execute in place*").
- The amount of memory in such a device is only a fraction of which a normal PC is equipped with today, not to forget that these devices usually don't have hard disks.
- Since processing power and memory is limited, applications must be programmed efficiently. For instance, all user interfaces should operate asynchronously so that applications are notified when operations complete rather than having to wait or poll for completion. It is important to avoid loosing user data in situations, where battery power runs low.
- Often operating system and applications are not upgradable within the device's lifetime.
- An important characteristics of many devices is the lifetime of their applications. A PC is regularly rebooted and after each reboot, the applications are newly started in a clean environment. Applications on pervasive devices are usually not rebooted, and can therefore run nearly forever. It requires a very stable operating system and applications with very careful memory management to meet these requirements.

Several operating systems have been developed to meet the specific needs of Pervasive Computing: Windows CE, Windows for Smart Cards, Palm OS, EPOC, QNX, GEOS, and many more. They all aim at small memory footprint and are mostly designed to run on very different processor platforms.

Operating Systems

Middleware   On top of these operating systems, middleware components provide an abstraction of operating system specific dependencies (Figure II.1). Common interfaces between applications and applied software components leverage portability. Besides proprietary libraries and components, industry initiatives are
developing a growing number of standardized frameworks and APIs targeting
pervasive systems.

*Figure II.1:
Software Layers
on a Device*

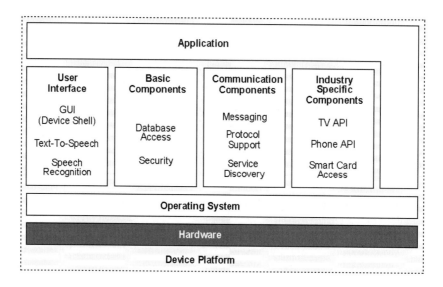

Security   Outside the firewalls of enterprises, security mechanisms are another important aspect, when examining software technology for Pervasive Computing.
Data and applications on ubiquitous consumer devices which are connected to
an open network are strongly exposed. Tasks like electronic payment, digital
signature, and data encryption need to be implemented in many solutions.

6 Java

C programmers know how annoying it gets when they have to compile their code for different target systems. Although the standard C language is well-defined by ANSI, there are always several platform specific issues to solve. No common library offers a complete set of functionality for all common operating systems. Therefore code versions for different platforms need to be maintained in parallel. Such problems are not limited to development only, but impact also the distribution of applications on multiple targets.

The Java programming language is intended for the development of platform independent software. Executing the same Java program on multiple targets requires no porting efforts.

Platform independent

How does this work? First, the Java source code is compiled into a standardized and platform neutral byte code. Each target operating system uses a Java runtime environment (JRE) to interpret and execute the compiled byte code on the fly. The byte code interpreter is also called virtual machine (JVM) and translates the generic instructions into native commands of the specific operating system or processor the application is currently running on (Figure 6.1). A Java program can run on any system providing a virtual machine without being recompiled. The range of computers which are able to interpret Java byte code spans from smart cards to enterprise servers. "Write once, run anywhere" – that's the slogan of Sun Microsystems, who initially created the Java technology for software development targeting consumer electronics. But Java had to wait until it was applied as an enabling Internet technology before becoming really successful. Today, Java sees its renaissance as a programming language for Pervasive Computing devices.

Byte code interpreter

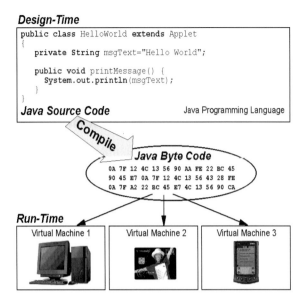

Design-Time

```
public class HelloWorld extends Applet
{
    private String msgText="Hello World";

    public void printMessage() {
       System.out.println(msgText);
    }
}
```

Java Source Code   Java Programming Language

Compile

Java Byte Code

0A 7F 12 4C 13 56 90 AA FE 22 BC 45
90 45 E7 0A 7F 12 4C 13 56 43 28 FE
0A 7F A2 22 BC 45 E7 4C 13 56 90 CA

Run-Time

| Virtual Machine 1 | Virtual Machine 2 | Virtual Machine 3 |

6.1
Language Characteristics

Simple   An important achievement of Java is that the language is kept very simple and the code is well readable. One way for simplification was to leave away features that have caused many pitfalls in other programming languages, such as pointers. Java takes care of referencing and dereferencing memory objects. An automatic garbage collection mechanism manages allocation and freeing of memory. Array indexes, and object types as well as casting between different objects are always validated at compile-time and run-time. Illegal overwriting of memory and corrupting of data can be prevented this way. Such friendly features leverage development, avoid potential bugs, and help the programmer to concentrate on the application instead.

Secure   Java has been designed as a secure language. Three mechanisms help to prevent intended or unintended damage to the system: First the byte-code is verified by the virtual machine ensuring that no illegal code is included. Additionally, the code could be signed, which means that cryptographic algorithms are applied to ensure that the code originates from someone the user trusts. The third security mechanism is the so called "sandbox"-model: The Java program runs in a memory area of the system which is controlled by the virtual machine. The program is allowed to perform any operation within the sandbox, but has only limited access to any system resources outside. For example unauthorized code cannot read or write any data on the system.

Last, but not least, Java is an object-oriented language. Classes are the bricks   *Object-oriented*
programs are built with. A class contains data and methods, which describe its
current state and its behavior. Interfaces describe the functionality of a class, independent from its implementation, leveraging reuse of software components.
Dynamic linking and class loading allow assembling the required components
during runtime (enabling downloading of components on demand).

Java classes are identified together with an unique package name. While package names starting with "java" and "sun" are reserved, developers should choose
package names based on the Internet domain of their organization. The full qual-
ified name of a class `UserId` could be for example `com.ibm.db2e.UserId`.

Java programs can be distinguished as follows:

- *Applications* are stand-alone programs. They are typically used when programs are distributed to run offline on a user's system, for example an
  office suite.
- *Applets* are mini-programs running within a browser. Typically, they are
  embedded into a web page. Whenever a client accesses such a web page the
  applet is downloaded together with the HTML document. The applet is
  executed directly on the client machine. To avoid security problems,
  applets have limited access to system resources, for example they cannot
  read or write files from the hard disk.
- *Servlets* are also related to web sites. While applets run on the client side,
  servlets are executed by the Web server. Typically, servlets are used to
  process a request of a web client. For example a servlet retrieves data from
  a database or handles an Internet order.
- *JavaBeans* are Java components which developers can integrate into their
  applications, applets or servlets. There are plenty of libraries containing
  JavaBeans for almost any purpose one can think of. This approach is
  intended to speed up application development.

6.2
Java Class Libraries

On top of the base language constructs Java provides a set of general-purpose
classes bundled into several packages, allowing the developer to take advantage
of given broad functionality. The following table 6.1 shows an extract of the
Java core packages.

package	Contents
java.applet	Classes for creating applets running within web browsers.
java.awt	Classes for creating graphical user interfaces. These classes are referred to as Java Foundation Classes and extend the original java.awt from the first Java releases. They include fonts, color, drag and drop, handling of user interface events, working with graphics, images, and 2D geometry, as well as printing.
java.beans	Classes for JavaBeans development.
java.io	Classes for system input and output, accessing a file system, and using persistent objects (serialization).
java.lang	Fundamental Java classes (such as Object and Class) and classes for obtaining reflective information about classes and objects.
java.math	Classes used within mathematical algorithms, such as BigInteger.
java.net	High level support for networking. e.g. URL class, extensive library for TCP/IP based protocols like HTTP and FTP.
java.rmi	Classes supporting Remote Method Invocation (RMI).
java.security	Classes of the Java security framework, including certificate management, RSA, and DSA interfaces.
java.text	Classes for handling text, dates, and numbers (e.g. String).
java.util	Manifold miscellaneous classes, such as collections framework, event model, internationalization, string tokenizing, data compression, and many more.

Beside the core packages listed above, there are four optional class libraries (Table 6.2).

package	Contents
javax.naming	Classes for accessing naming and directory services (including LDAP support).
javax.sound	Classes for the processing of audio data.
javax.swing	Classes for a more powerful and platform independent user interface than java.awt.
org.omg.CORBA	Provides the Corba API, including an Object Request Broker and IDL support

6.3
Java Editions

Its networking features and its platform independence made Java a common way of providing applications for the Internet. Using the extensive libraries for networking, Java software can access objects across networks as simple as reading from a local file system. Today applets and servlets are part of most sophisticated websites. Almost every browser can execute Java byte code and most webservers can execute servlets. With the increasing diversity of devices, the platform independent Java emerges to a key technology. Many handhelds, set-top boxes, smart cards, and other consumer electronics already provide a built-in Java Virtual Machine. While on one side, the number of platforms supporting Java byte code increases, on the other side, the complexity and the number of features of the Java programming language grows with each new version. In order to accommodate the very different requirements of different classes of computers an important split has been made:

Instead of one all-purpose Java, Sun provides now some different flavors of the language specification: The Standard, Enterprise, and Micro Edition as well as the Java Card (Figure 6.2). *Flavors of Java*

The *(J2SE) Standard Edition* (J2SE) is the default language specification for workstations and small servers. The Java 2 Software Development Kit (SDK) is a product from Sun, which implements that specification. The SDK includes all core packages of the language and the compiler required for creating Java byte code. Tools for debugging and documentation are also part of this edition. For non-programmers who just need to execute software it is sufficient to just install the Java 2 Runtime Environment (JRE). Developers can distribute it freely bundled with their product. Both, the Runtime Environment and the Development Kit are available from Sun's web site and run on Windows, various Unix systems, and Mac. All versions include a virtual machine for executing Java byte code. In future this VM will be replaced by Sun's new "Java Hot Spot Client VM", promising better performance while having a smaller memory footprint. *Standard Edition*

The *Java 2 Enterprise Edition* (J2EE) offers additional features on top of the Standard Edition, such as servlets, JavaServer Pages, JDBC, Java Message Service, Java Naming, and Directory Interface, Java Transaction, JavaMail, XML, CORBA, and Enterprise Java Beans. The Enterprise Edition targets large server systems. Currently a Software Development Kit for this edition is only available as a beta version for evaluation purposes and requires the Standard J2SE SDK as a basis. In addition to the virtual machines coming with the standard Java, the "Java Hot Spot Server VM" is currently developed by Sun especially for high-performance servers running Solaris or Windows. It will be available with the next release of the Enterprise SDK. *Enterprise Edition*

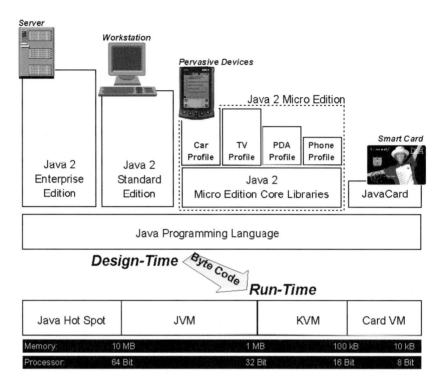

Java Hot Spot	JVM	KVM	Card VM

Memory:	10 MB	1 MB	100 kB	10 kB
Processor:	64 Bit	32 Bit	16 Bit	8 Bit

Micro Edition   The *Java 2 Micro Edition* (J2ME) has been introduced to meet the limitations and peculiarities of Pervasive Computing devices. It is a platform on which very small and flexible Java application environments can be defined. In order to minimize the required memory for the Java class libraries, J2ME introduces profiles, which specify language subsets for different groups of devices. There is a minimum core functionality, which is mandatory. Beyond the core libraries, only those Java classes, which are actually needed to support typical features of a device group are included into a corresponding profile. For example, devices without an user interface would exclude the entire AWT package from their particular profile.

Java Card   The *Java Card* is the smallest and most limited version of the Java family. Only a subset of the full language is available for smart cards. Restrictions are for example, that no strings are supported. And of course the smart card has no graphical user interface. The used Java language subset is specified by the Java Card Forum industry initiative. The corresponding Card VMs are provided by several card operating system manufacturers and shipped on the ROM of the smart card chip itself. Usually the Java applications loaded onto a smart card are controlled by a card issuer. Chapter 7 goes into more detail on this topic.

Third party Java   The core editions of Java are controlled and distributed by Sun Microsystems. Independent from Sun, several companies offer manifold Java APIs, for example JavaBeans libraries. There are virtual machines available from third-party providers, like the GhostVM for the Palm OS or an EPOC JVM from Symbian. Other compilers compete with the Sun SDK, for example the very fast Symantec Java compiler or the Cygnus GNU compiler. Cygnus provides also a Java-to-native compiler.

Beyond the scope of the core Java family, the Java Community Process defines a standardized way how the industry can continuously add new functionality and create new Java APIs. JavaMail, JavaTV, and Java Phone are such examples of consumer-oriented APIs, sitting on top of the regular Java class libraries and targeting specific industry segments. Likely extension APIs for Pervasive Computing devices will be integrated into profiles of the Java 2 Micro Edition. For example the JavaPhone package will be part of the wireless profile.

Java Community Process

Besides the activities within the Java community process, other standardization workgroups develop APIs and specifications based on Java, for instance OpenCard Framework, San Francisco Framework, or Open Services Gateway initiative.

6.4
Micro Edition

Java 2 Micro Edition (J2ME) includes virtual machines, several libraries of APIs, as well as tools for deployment and device configuration.

Choosing subsets of the Java language particular for a specific device class is the basic concept of J2ME. Each subset is called a profile. For example the *Mobile Information Device Profile* (MIDP) defines the Java functionality and APIs, which fit the screen size and memory constraints of PDAs. If an application requires more functionality than specified within the device profile, the missing libraries can be retrieved dynamically from a network: application code can be downloaded together with Java class libraries.

Profiles

Each profile is based on one of two *Configurations*:

Configurations

■ *Connected Device Configuration* (CDC) is intended for devices which have at least 512 kilobytes of memory available to run Java and have a high bandwidth network connection.
■ *Connected Limited Device Configuration* (CLDC) is intended for devices with only 128 kilobytes of memory for running Java and with limitations in the user interface, network bandwidth, and power supply.

The following figure shows the architecture of a J2ME software stack. Using configurations and profiles.

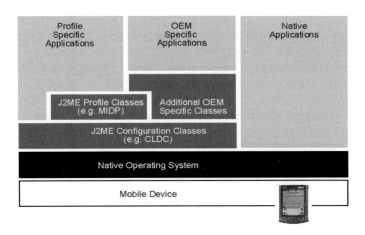

Profiles and Configurations are specified by the Java Community Process. Working groups currently develop profiles for

- wireless mobile phones and communicators,
- point-of-sale terminals,
- handheld computers,
- automotive systems,
- set-top boxes and interactive television, and
- generic network-connected consumer devices having a GUI.

KVM Targeted for the CLDC class of devices is the KVM, an extremely small footprint version of a virtual machine. About 128 kilobytes are enough for operating the KVM with the J2ME class libraries. The KVM comprises only 40 K of object code!

6.5
PersonalJava and EmbeddedJava

Personal and Embedded Java are predecessors of the Java 2 Micro Edition. All of them use a similar approach: they include only the actually needed subset of the Java class libraries, to reduce the required memory footprint. While the new Java 2 Micro Edition is part of the latest Java 2 family, Personal and Embedded Java base on the older Java version 1.1. In order to migrate PersonalJava to Java 2, it will be repackaged and will end up as a "PersonalJava" profile of the Micro Edition.

Personal Java The PersonalJava Application Environment focuses on smart phones, set-top boxes, game consoles, PDAs, and other resource constrained network-connectable consumer devices. Typical device characteristics are 32 bit processors running a real-time operating system with 2 MB ROM and 1 MB RAM available for the PersonalJava environment itself.

PersonalJava is scalable, modular, and configurable. This is achieved by distinguishing between core and optional class libraries. The developer can take advantage of the entire functionality of standard Java, but has the option to use only what is really necessary. While the optional APIs are configurable, a defined set of core functionality must always be part of every PersonalJava to ensure at least a minimum interoperability of applets on different PersonalJava devices. Based on the available classes, new applications can extend the functionality of a device during its life-time, for instance when downloading applets from a network.

The Java classes have been re-implemented keeping in mind a small footprint and the peculiarity of the targeted category of consumer devices. For example the underlying implementation of the AWT supports low-resolution displays and alternate input mechanisms for keyboardless devices. Since PersonalJava is a feature subset of the standard Java API version 1.1, applications are upward compatible, without being recompiled.

Development tools and lean virtual machines for several real-time operating systems are part of this product. Supported operating systems are from Acorn, Chorus Systems, Geoworks, Lucent Technologies, Microtec, Microware, QNX, and Wind River Systems.

It is also possible to compile the Java byte code of a PersonalJava application into executable code, which can be placed directly into the ROM of a device. A tool removes all unused parts and does some code optimizations. For example, redundancies are eliminated or symbolic references are resolved. Up to this point the compacted application environment is still platform-independent and could still be easily migrated to other real-time operating systems! After this point, another tool compiles and links the Java code into a native executable for a specific device platform. Beside being small and highly optimized, this executable doesn't require an additional VM!

The EmbeddedJava Application Environment targets very small devices *Embedded Java* with very strict memory limitations, such as industrial controllers, pagers, telecommunication switches, or low-end phones. While PersonalJava addresses sophisticated displays, Embedded Java is designed for devices having only character-based or even no display at all. The available virtual machines are designed for real-time operating systems running on a wide variety of microprocessors. EmbeddedJava preserves the benefits of real-time support offered by the final hardware.

Each application environment consists out of a fully configurable set of class libraries. There is no mandatory core functionality within EmbeddedJava, which reduces the memory footprint! Since not all classes are supported on each device, the interoperability of applications is limited. Usually EmbeddedJava is applied on devices that do not support applet downloading. They only need to provide those Java classes which are required for tasks known at device conception. But nevertheless, the used Java subset is upwards-compatible to Java version 1.1.

To leverage the creation of highly optimized software with high performance Sun provides a set of tools. Applications can be condensed and executable code for a particular target processor can be generated.

6.6
Development Tools for Java

Facing increasing complexity and decreasing life-time of software, it is an important issue to make application development as easy and fast as possible. Many vendors provide tools to leverage Java development. Wide spread and renown tools are Visual Cafe from Symantec, JBuilder from Borland, Microsoft's Visual Studio J++, and IBM's Visual Age.

Visual Programming

Visual Age for Java supports the visual programming method: Software components like JavaBeans can be assembled in a visual composition editor almost intuitively to complex applications. User Interface components can be logically connected with databases, servers, or distributed objects in a network. In the visual composition editor, visible objects like buttons or text fields as well as any other invisible objects like classes, variables or beans are represented by icons.

Figure 6.4:
The Visual
Composition
Editor of Visual
Age for Java

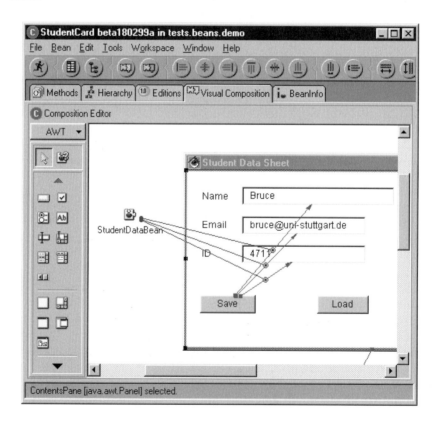

For example, the button of an user interface creates an event when it is pressed. This event triggers several text fields to be filled with associated data. The application logic can be plugged together visually by drawing an arrow from the button `Save`, which fires the event "button pressed" to the method `set-Text(value)` of the text field `Email`, as soon as the given event occurs. The value itself is retrieved from an object `StudentDataBean`, which encapsulates the corresponding data. To achieve this, another line between the method `getE-mail()` of `StudentDataBean` and the first visual connection must be drawn (Figure 6.4).

After composing the application out of a set of components, the visual development tool automatically builds valid Java source and executable Java byte code for the programmer.

6.7
Further Readings

Core J2ME

This book details the J2ME editions of Java [MUC02]

IBM Alphaworks

http://www.alphaworks.ibm.com

This website offers a vast of new beta software, mostly written in Java for download and evaluation.

IBM

http://www.ibm.com/java

IBM's entry point for Java. A library featuring news, education material, articles, and links to products and downloads.

Java in a Nutshell

David Flanagan's book is an excellent reference book for Java programmers [FLA99].

Microsoft

http://www.microsoft.com/java

Besides extensive online material, Microsofts publishes on this Java homepage software downloads, such as Java developer kits for their product portfolio.

Program Java Devices

http://www.javaworld.com

This article from the JavaWorld magazine gives an overview about how to apply Java technology for small devices [DAY99].

Sun's Java Resources

http://java.sun.com

Sun's Java site is the access point to specifications, development kits, articles, books, newsgroup, as well as related products, and other kinds of online information material. Sun's development kits and runtime environments can be downloaded from here.

Sun's Java Developers Connection

http://developer.java.sun.com/developer

The developer's zone includes additional detailed technical information.

7 Operating Systems

The following sections describe some exemplary operating systems, which are applied in Pervasive Computing.

7.1
Windows CE

The diversity of programming interfaces and the inhomogeneous tools for various processors and proprietary operating systems are an tremendous obstacle for the development and maintenance of Pervasive Computing applications. To leverage interoperability, the Java community postulates the standardized Java byte code as an integration layer between different operating systems and platform-independent applications. In contrast, Microsoft tries to achieve such standardization at an operating system level (Figure 7.1). As Windows is the de facto standard for personal computers, Windows CE was introduced 1996 to evolve as the common operating system for handheld computers, industrial controllers, data terminals, entertainment systems, embedded appliances, and wireless communication devices. The Win32 API defines a consistent interface for the application, while the Windows CE components encapsulate all hardware specific functionality in a platform independent manner.

Windows CE extends the idea of a Digital Nervous System to pervasive devices. The nerves and veins of Microsoft powered systems and devices should embrace the entire computing community from enterprise servers to smart cards. Featuring the widely-used Win32 API and well-known development tools, Windows CE is complementary to the large PC operating systems, Windows 2000 and Windows 98 and targets the large Windows community. The end-user will find the accustomed Windows look-and-feel on his PC companion and can use common Microsoft Office tools, even on a handheld computer. Programmers will find a subset of the familiar Win32 API, popular development tools, and languages, like Visual Basic and Visual C++. Additionally, Microsoft provides tools, which device manufacturers can leverage while creating their own device specific versions of Windows CE easily.

Windows look-and-feel on a handheld

Figure 7.1:
Comparing Java
with Windows CE

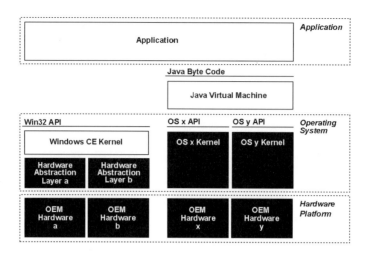

Unlike Windows 98 or Windows 2000, Windows CE is not an operating system to be sold from the shelf to the end-users. Microsoft just provides a given set of functionality. Each device manufacturer needs to configure the Windows CE operating system for the specific processor platform of his custom device. Usually the operating system is delivered on a ROM, which is built into the device.

7.1.1
Operating System Configurations

At the time Windows CE was developed from scratch, the development team at Microsoft had the ambitious goal to cover anything smaller than a PC. They achieved this universality by offering modular building blocks, which can be linked together by a particular device as needed. For example, a handheld computer would be able to include more functionality into its individual Windows CE version than a smart phone. This way the size of the operating system could be kept consistent with the size of the device. This flexible approach is similar to the configurable Java subsets of the Java 2 Micro Edition.

Depending on the included functionality, three groups of operating system configurations are distinguished:

- Handheld Professional PC (H/PC Pro) and Handheld PC (H/PC) are sub-notebook class devices equipped with keyboard, mouse, full-size VGA display, PC Card slot, and USB port. Some manufacturers include a voice recorder and text input. Available memory is larger than 16 MB RAM. Pocket versions of Word, Excel, PowerPoint, Access, Internet Explorer, and Outlook reside in the ROM.

- Pocket PC (P/PC) is a handheld or palm-sized computer, featuring built-in Voice recorder, MP3 player, and text input. Typical memory footprints are 8 to 32 MB RAM and 2 to 8 MB ROM. Simple PIM applications are preinstalled in ROM. Since Version 3.0, pocket versions of the office tools and the Internet Explorer are available too. This class was formerly referred to as Palmheld PC.
- Automotive PC (Auto/PC) provides a speech interface, CD-ROM for data and music, USB, infrared, AM/FM tuner with preamplifier, a small color screen, 8 MB ROM, and 8 MB RAM. Auto/PC applications are PIM, navigation, maintenance, diagnostics, and entertainment.

This universal approach differs fundamentally from the competing Palm OS operating system, which has been exclusively designed and highly optimized for exactly one target class of devices.

7.1.2
Memory Management

Windows CE makes no conceptual difference between the various kinds of storage. Mass storage devices like flash memory cards or hard disk drivers, ROM or RAM are all treated just the same way. All these memory resources are managed consistently by Windows CE and can be accessed through the Win32 API. This is important since each device balances the different memory types individually to optimize costs, performance, and flexibility according to the specific requirements: For example, ROM is quite cheap and small in size compared to RAM, while RAM allows more flexibility concerning application loading and updates.

The available memory is split into two separate blocks: program and storage memory.

- The storage memory includes the persistent data, which a common PC would usually store on hard disks. One part of the volatile storage memory is referred to as object store. The system registry, directory, all applications, and user data are kept within the object store. Data can be saved in file objects or databases. Files are automatically compressed. The size of the object store is limited to 16 MB. Outside of the object store, additional databases and file systems can be installed. Both can be placed in memory anywhere on the device: in ROM or RAM as well as on external storage devices. Each file system can be divided into multiple volumes, which are represented as folders of the files systems root directory. Each volume can be mounted separately.
- Program Memory is allocated for the stack and the heap of all running applications. Additionally, non-system applications must be loaded from storage into the program memory for execution. RAM is needed for the

stack, heap, and non-system applications. The operating system itself and the built-in system-applications, like Pocket Word, are stored and executed directly from ROM memory. Executing applications "in place" (XIP) spares loading them into RAM first. If all applications would be placed into ROM, theoretically only a minimum amount of RAM would be needed for heap and stack storage as well as the user data itself.

Internally, Windows CE manages the following memory access mechanisms:

- cacheable paged memory
- non-cacheable paged memory
- non-pageable non-cacheable memory
- non-pageable cacheable memory

Virtual memory   Program Memory is allocated and managed as virtual memory. A Memory Management Unit (MMU) within the kernel maps any virtual memory address to an actual physical address. This classical concept abstracts memory allocation from physical constraints. It simulates unfragmented, continuous memory, regardless how the data is actually spread within RAM or ROM. For instance it could be distributed on several flash memory cards or on a hard disk. Actually, Windows CE takes advantage of the MMU capabilities of some CPUs and performs paging (when hard drives are present). The page table is not maintained by the operating system, but by the hardware. Applications can allocate discrete pages of virtual memory at a time, each with a size of 1 or 4 kilobytes. The page size depends on the Windows CE implementation for a specific target system. When a process refers to pages which are not loaded into program memory, the kernel retrieves those pages from their physical location.

The total virtual address space of Windows CE is 4 GB. From that address space, 32 memory slots are defined, having 32 MB of size each. These slots are assigned to up to 32 simultaneous processes – one slot for each process. An additional slot is used by the currently active process.

Within the address space assigned to one process, the heap is created and each thread will get a stack allocated, with a minimum size of 1 kilobyte. All dynamic-link libraries, application code, and data are stored in that process slot too. The remaining virtual address space is reserved for the kernel or used for sharing memory between different processes (Figure 7.2).

System states   A Windows CE device knows three different states:

- "Dead": the device is not powered, all volatile RAM data and code is lost.
- "Suspend": the clock is running. All applications and persistent data in RAM are maintained. Peripherals and other unused parts of the system are disconnected from the power supply. This is the default state, when the device is unused.

- "On": the device is fully operational. If the processor is idle, its frequency is lowered in order to save battery power.

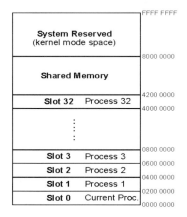

System Memory Map

Process Memory Map

Figure 7.2: Memory Maps of the Entire Virtual Address Space (left) and the Virtual Address Space of one Process (right).

7.1.3
Processes, Threads, and Interrupts

Windows CE is a 32 bit multitasking operating system. The kernel itself supports up to 32 processes. Within each process, an unlimited number of threads can run simultaneously (as long as there is free RAM available).

Threads are used to monitor asynchronous events. They can check for an incoming hardware interrupt or an user activity. Each thread owns a message queue, into which other threads, processes or system resources can place messages (for handling). For example, the kernel can send the message WM_HIBERNATE indicating that available memory runs low. An application should immediately free its allocated memory when receiving this message.

Threads

Since all threads of a process run independently, they need to be synchronized somehow. Typically, when common data is shared between threads, the manipulation of that data needs to be coordinated. This is done with basically the same mechanism as in other Windows operating systems, although some differences have to be obeyed. Synchronization objects, like critical sections, events, and mutex objects are used to block the execution of a thread until the "wait object" turns into a defined state for resuming. Since these synchronization objects work on kernel level instead of application level, they are well optimized. Continuous status polling of an application thread would waste battery power.

Synchronizing threads

Priorities   A thread has one of eight priority levels assigned, which range from "idle" to "time critical". A highest priority level thread never gets interrupted and is used for real-time processing and device drivers. Applications and the OS kernel will assign medium priorities to their threads. Lower priority levels allow background operations. Based on its priority each thread gets time-slices of the execution time. Threads with lower priorities have to wait until all higher priority threads are finished. Same priorities will share time-slices. Threads with the highest priority do not share processing-time, but continue until finished. The priority of a thread cannot be changed. Except, when a lower priority thread blocks a resource, which is needed by a higher prioritized thread. In this case, the low-priority thread in converted to higher priority until the resource is released.

Interrupts   Interrupts are used to notify external events to the operating system, like events originating from peripherals. Handling a hardware interrupt request (IRQ) in Windows CE takes two steps:

Each IRQ is assigned to a Interrupt Service Router (ISR). The ISR is a very fast kernel-mode routine which maps the request to a corresponding Interrupt Service Thread (IST) and passes the determined thread identifier to the kernel. The kernel sends a message to the waiting IST, which initiates the processing of the interrupt itself. The Interrupt Service Router can already process the next interrupt, immediately after the service thread has been started. Delays in interrupt handling are minimized this way.

Since an Interrupt Service Thread is just a normal thread, it can have a priority assigned. Usually high priorities are used, to speed up the event handling.

Separating the more time-consuming interrupt processing, from the interrupt routing makes the time for every system call predictable and independent from the current state of the system. This allows applications to respond to external interrupts within a specified latency. Further Windows CE guarantees a maximum time for passing the next processing time-slice to a thread with the highest priority.

This way Windows CE fulfills the requirements of a real-time operating system. This opens up the wide range of time-critical applications, such as telecommunication switching equipment, manufacturing process controls or navigation systems.

7.1.4
User Interface

The development and usage of the user interface is very similar to other Windows operating systems. There are the familiar windows, folders, menus, files, and shortcut icons. Some restrictions concerning the provided styles and options of user interface objects can be noticed. For example, a window cannot be resized.

In Windows CE version 2.11, application menus and tool bars are combined to a taskbar, placed at the top of the screen. The taskbar offers frequently used commands within an application. With the taskbar, windows can be switched and minimized. A taskbar can be hidden, to increase the usable size of the display. The familiar Windows Start menu and some status information appears at the display's bottom.

Version 2.11 user interface

For the latest version 3.0, the user interface has been completely redesigned (Figure 7.3). The taskbar with the application menus were moved to the bottom, while the status bar including the Windows start menu was placed as title bar at the top of the screen. Among the new features of 3.0 is a handwriting recognizer accepting glyphs, script, or print letters.

Version 3.0 user interface

Although minimum operating system configurations do not have any user interface at all, every application must have at least one window – even if it can't be displayed. This mandatory window is required, because all messages an application sends or receives are passed through the message queue of the corresponding window thread. The entry point of each application is a WinMain function that implements a message loop, which handles messages, like user events, and dispatches them to the appropriate application routine.

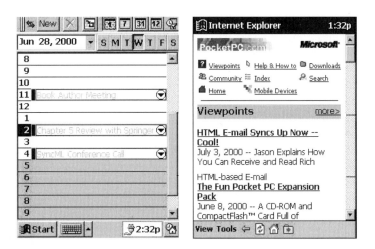

Figure 7.3: User Interface of Version 2.11 (left) vs. Version 3.0 (right)

Windows CE user interface programming is done with the Graphics, Windowing, and Event Subsystem (GWES). Beside windows, controls, bitmaps, text, fonts, shapes, lines, colors, curves, and menus, this API includes messaging and power management functions.

GWES

Unicode is the native text format. Each character is stored using two bytes (a "word") instead of the ASCII one byte. A string is consequently terminated by a zero-word (i.e. two zeros instead of just one). Unicode allows the representation of international character sets, which is important for world-wide sold consumer devices.

Figure 7.4:
The AutoPC User
Interface as it is
Presented in the
Emulator

Auto/PC has some standard ActiveX controls designed for the special auto-motive environment (like a list with track numbers of the compact disk, an vol-ume indicator, and cross hairs to set two-dimensional values (Figure 7.4).

7.1.5
Communication and Networking

Since Pervasive Computing devices have strong needs for connecting with each other, Windows CE supports a variety of communication technologies:

- Serial communication is a one-to-one connection. For establishing a serial communication, the COM port is opened, and data is "streamed" from one device to the other. Data transfer is very similar to reading or writing a file. Typically, two devices are connected by serial cables, but also infrared transceivers or modems can be used for this purpose. Serial communication protocols like Serial Line Interface (SLIP) and Point-to-Point (PPP) pro-vide more reliability than the raw serial connection.
- Network connections use one or many devices and provide more sophisti-cated and robust protocols, preventing the loss of data. Additional hardware like LAN adapters or radio transceivers can be applied. The most common networking connection types supported by Windows CE are Internet proto-cols such as HTTP or FTP and file exchange protocols like Remote Access Service (RAS), Windows networking or Remote File Access.

HTTP and FTP communication is provided by the WinInet API. Secure Sockets Layers (SSL) and Private Communication Technology (PCT) is sup-ported for implementing secure connections between a server and a client. These Internet protocols are based internally on the Windows Socket interfaces defined in the WinSock API. WinSock uses an underlying TCP/IP protocol. While WinSock can also be accessed directly, TCP/IP is not accessible by an application. Nevertheless some TCP/IP parameters can be modified for optimi-zation purposes. For the support of infrared communication using the Infrared Data Association (IrDA) protocol the IrSock API has been added.

Windows CE can access network resources from remote. The Remote Access Service (RAS) supports upload and download of files. The PPP protocol is used for connecting to the server.

Modem communication to a remote host using a telephone line can be achieved with the separate Telephony API (TAPI), leveraging dialing and managing telephone connections. Windows CE only supports outbound calls.

7.1.6
Peripherals and Device Drivers

Drivers for manifold peripherals are necessary: battery, display, serial, IR, touch panel – all need to interact with the operating system. Windows CE differs between four kinds of drivers:

- Native drivers are usually applied to built-in features of a device, like the power supply or keyboard. These low-level drivers are linked with the kernel and reside in the binary image of Windows CE on the ROM.
- Stream Interface drivers are loaded as standalone DLLs. They support all stream-based peripherals, e.g. when connected with the serial port. The driver interfaces offer functions to open and close the data stream.
- Drivers based on the Network Driver Interface specification (NDIS) allow the network protocols to be implemented independently from the hardware drivers itself.
- Universal Serial Bus (USB) drivers support the external bus architecture for connecting peripherals.

Typically, native drivers are integrated by the device manufacturer who builds his own Windows CE configuration. The other types of drivers are often user-installable and allow to add support for temporarily attached peripherals, e.g. a printer or a modem.

Similar to other Windows platforms, Windows CE uses the concept of Cryptographic Service Providers (CSP):

CSPs allow to separate a generic cryptographic application programming interface (CAPI) from the actual implementation. Applications can call the well-defined CAPI-functions. Various CSPs offer different implementations of the same CAPI – some having stronger or weaker encryption, others having specific algorithms. A CSP implementation could also integrate hardware security features, e.g. a cryptographic chip.

7.1.7
Platform-Builder – Creating a Custom OS

Configurations

When applying Windows CE on a particular device, the hardware manufacturer needs to configure and build a customized version of the operating system. Depending on the capabilities of his device, different operating system components will be included. By selecting only those building blocks, which are actually needed, the hardware manufacturer can minimize memory footprint and optimize performance of his device.

A minimum operating system configuration includes the core OS kernel and the essential parts of the GWES (messaging and power management). This minimum version has a footprint of about 256 kilobytes.

A full-fledged handheld device featuring all available components would need about 2 MB of ROM and 512 kilobytes of RAM. P/PC, H/PC, H/PC Pro and Auto/PC are the default Windows CE 2.11 versions, pre-configured by Microsoft. PocketPC is currently the only configuration for Windows CE 3.0. For debugging and evaluation purposes, there is another configuration for an x86 based PC.

Platform Builder

To ease the creation of a particular Windows CE device, Microsoft provides the Windows CE Platform Builder. The Platform Builder includes all available modules and components of the operating system in binary form. Tools are supplied to custom configure Windows CE and to create the kernel and drivers according to the constraints and requirements of each target device. Kernel debugger, cross-compilers, operating system loaders, sample device drivers, and some application code are also part of this product. Currently supported target CPUs are from AMD, ARM, Hitachi, Intel, Motorola, NEC, Philips, and Toshiba.

Underneath the Windows CE components, the manufacturer must provide a hardware abstraction layer (HAL) specific to his device hardware. This layer of adaptation code between the operating system kernel and the processor implements the interrupt service routine handlers, the real-time clock, and an interval timer. Beside the HAL, other native drivers must also be supplied by the hardware manufacturer for built-in peripherals (Figure 7.5).

Software Development Kit

After the customized operating system has been generated on a development workstation, a binary image is loaded onto the device (Figure 7.6). Additionally, a specific Software Development Kit (SDK) for the particular device is generated. The SDK defines all available application interfaces in this configuration, which a developer is allowed to use. The application developer can load the SDK into the common Toolkit for Visual C++ or Visual Basic Environment.

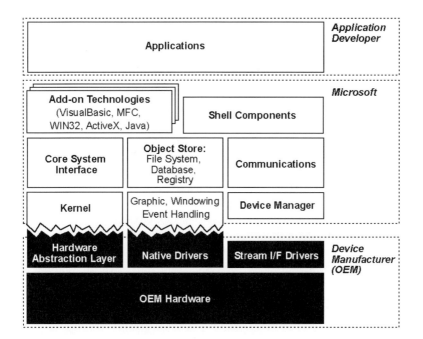

Figure 7.5:
Integrating the
Windows CE
Kernek with the
Hardware
Platform.

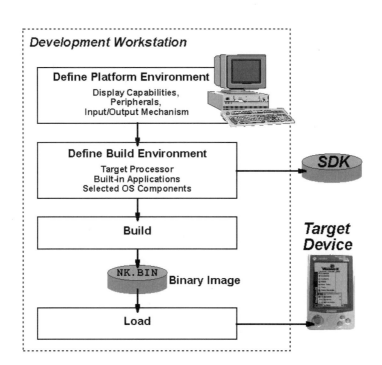

Figure 7.6:
How to Build a
Customized
Windows CE
Version

7.1.8
Developing Applications

Visual Studio Application programming is done on a separate workstation running Windows NT and using an integrated development environment based on Microsoft Visual Studio. The Windows CE Toolkit and the Windows CE Services supply a comprehensive set of tools:

- Cross-compilers are available for all supported Windows CE processors.
- A simulator allows testing on the developing system, before the code is actually downloaded onto the device itself.
- Resources like window layouts, test strings, or bitmaps can be designed within a Resource Editor and linked to the application.
- There are powerful debugging tools included: Remote Spy, Registry Editor, Zoom, Process Viewer, and Heap Walker are already familiar to Windows developers. Developers can set breakpoints, step through the code, view register values, manipulate the system registry, browse the object store, observe processes and threads in the kernel, read the memory spaces. A scripting utility is used to record and execute automated regression tests. Since it is not possible, to execute such sophisticated debugging software directly on the device itself, these tools must connect from remote to the device. Windows CE supports serial, parallel, and Ethernet transports for remote monitoring and manipulation of the target system.

Different Windows CE configurations differ in the supported set of functionality. Microsoft provides default Platform Software Development Kits (SDK) for each of the common standard configurations: Palm-Size PC Edition, Handheld PC Edition, and Auto PC Edition for Windows CE version 2.11 as well as the new PocketPC Edition for version 3.0. The Visual Studio project properties include the information which specific device the code is developed for. Once the target processor has been selected within a project, the development tool does not allow to change that setting! Targeting another processor or device requires explicit and significant porting efforts!

Win32 API The SDK exposes the supported subset of the C/C++ Win32 programming API, which is the consistent development interface across all Windows operating systems. Although the programming of these APIs is quite similar, the various subsets for different targets are incompatible! Compared to other Windows versions, CE is constrained regarding functionality. Only half the functions of the Win32 API exposed by Windows NT are provided. New features like touch screen support or the database functions have been added to meet the requirements of small mobile devices. With these modifications a relatively small footprint is achieved and it is possible to focus on the needs of the targeted class of devices.

To make programming more efficient, the Microsoft Foundation Classes *MFC Classes*
(MFC) were ported to Windows CE. MFC is a very popular C++ class library
for developing graphical user interfaces and manipulating objects. Compared to
the basic Win32 API, MFC offers more high-level functions. MFC classes are
convenient wrappers around the basic Win32 API functions, leveraging application development. The original set of MFC classes was modified for the Windows CE version. For example, there are some new classes for programming
primitive displays. Currently the usage of the MFC classes is restricted to the H/
PC and P/PC SDK. The class libraries take 200 to 600 additional kilobytes of
memory on the target device.

Performance considerations make C/C++ the preferred choice for application *Visual Basic*
development. As an alternative, Visual Basic can be used too. Since the Visual
Basic runtime environment itself takes about 300 kB of RAM, not all Windows
CE configurations support Visual Basic. It is included for instance within P/PC
and H/PC, but not in Auto/PC.

With the new Windows CE version 3.0 "PocketPC", the DirectX API is supported. DirectX is a real-time multimedia library for the creation of photo-real-
istic graphics and high quality sound. This makes handheld computers a promising platform for computer games.

7.2
Palm OS

Palm OS has been developed by Palm, Inc. Although it is a proprietary operating system, it has a share of 70% of the handheld device market.

The latest version is Palm OS 4.0. While 3.1, 3.2, and 3.3 were specific to a *Version 3.5*
particular Palm device model, OS 3.5 has been designed as one common operating system for all Palm devices. Version 3.5 includes color support, some new
features in the user interface and several changes in the programming interfaces
to achieve more consistency and clarity. A hardware abstraction layer eases
porting Palm OS to new processor platforms.

Version 4.0 provides enhanced security features (like locking of handhelds, *Version 4.0*
assignment of passwords and data encryption), 65,000 color support and enhanced wireless Internet and e-mail access.

Figure 7.7 gives an overview of the Palm OS architecture, which is explained
in the following sections.

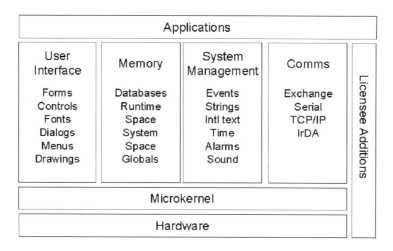

7.2.1
Memory Management

All storage of a Palm device resides on so called memory *cards*. A card is a logical unit of RAM, ROM, or both. It may represent a physical memory module, but it can also be a defined area of memory. Each memory card has a maximum address space of theoretically 256 MB!

The entire available memory is divided into multiple heaps:

- one single dynamic heap
- and multiple storage heaps.

The concept of differentiating between storage and dynamic memory is similar to Windows CE. Palm's dynamic memory is equivalent to Windows CE's program memory.

Dynamic heap The dynamic heap is used for memory allocation, global variables and application stacks (Figure 7.8). The dynamic heap has a fixed size, regardless of how much is used or needed by the applications. The total dynamic memory area of OS 3.x is limited to 96 kB. System globals occupy about 2,5 kB of this amount, the application stack takes 3 kB. If used, the TCP/IP stack requires 32 kB and reduces the available heap once more. The space, which is left the application's memory allocations, stack variables and global variables is less than 36 kB! Thus, it is necessary to keep the memory footprint of an application as small as possible. Especially the optimization of dynamic heap space is very important. Another important point when optimizing memory usage is to avoid the generation of large structures on the stack, to avoid the use of global variables, and to avoid copies of databases in the dynamic memory.

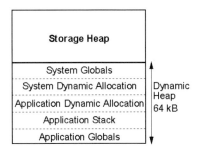

The storage heaps hold data and applications. The operating system and built-in applications are placed in the ROM. The user's data and installed applications are stored within the RAM, which is always powered to allow persistent storage. All information is accessed in place, instead of reading parts of a file into the memory buffer. Since version 3.0 of the operating system, storage heaps are not limited in size.

Storage heap

The Memory Manager allocates, disposes, resizes, and locks chunks, which can be either be located within storage or dynamic memory. Both kinds of memory are managed separately. In order to avoid destroying data accidentally, storage memory is write-protected by the operating system and can only be accessed through the Memory Manager.

Memory Manager

The Memory Manager allocates memory in chunks. Currently each single data chunk has a maximum size of 64 kB. Pointers are non-relocatable memory chunks, while handles are relocatable. The latter is automatically rearranged by the Memory Manager to avoid defragmentation of the heap. When accessing the content of a handle, the memory must be temporarily locked.

Palm OS does not use a traditional file system. Storage is structured into databases instead. Each database comprises multiple records and is managed by the Database Manager. A database holds a set of information about the database itself and about the records which are referenced by the database. The database information fields contain the name, the attributes, and the application-specific version number of the database. A modification number is also maintained by the database. It contains an access counter for the database, which allows an application to check whether the database has been accessed by another process since the last database operation of the application. The application and sort info fields are optional fields, which hold application specific data and a local ID of a sort table. Each record has a Record ID and Record Attributes. The Record Attributes indicate whether the record is deleted (delete bit), updated (dirty bit), locked (busy bit), or protected (secret bit).

*Database
Manager*

Records of a database are chunks within the storage memory (and therefore also limited to 64 kB each). Internally, the Database Manager uses the Memory Manager's functions when manipulating records. From the Memory Managers point of view, a database is just a list of references to correlated chunks, which are distributed discontinuously in the physical memory (although they must reside within the same memory card). An advantage of this concept is that the entire database does not have to be moved within the memory space during record operations (like deleting, adding, etc.).

Figure 7.9:
Palm OS Memory

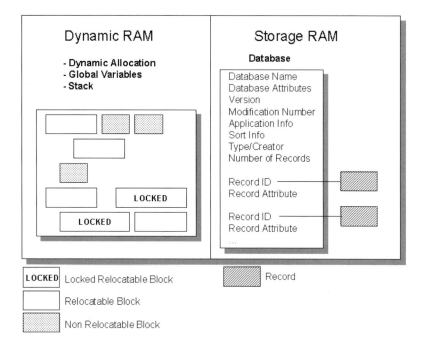

7.2.2
Events

Only one application can run at a time. There is no threading mechanism available for Palm OS. Instead, it is possible to sub-launch other applications. For instance a "search"-application can invoke other applications, asking them to perform a query on their databases. After the search is completed, the control is passed back to the calling application.

Palm OS is event driven. A main event loop for applications dispatches events to handlers. Events can be

- user interface actions, e.g. a tap on the touch screen,
- system notifications, e.g. a timer alarm,
- or application specific events, e.g. a "search"-request.

Programmers should avoid code blocks which need processing power while waiting for an event. For example, it is recommended to use poll intervals instead of continual polling.

7.2.3
User Interface

A very obvious difference between a desktop computer and a palm computing device is the screen size. Up to now, palm devices have only a 160 x 160 pixel screen. It is therefore very important to design the screen layout very carefully. For rapid entry and retrieval of data, switching between windows should be minimized to perform certain tasks with a Palm application. The screen layout should be self-explaining and reduced to the absolutely necessary elements. Different views or representations of data should be easily accessible. Important commands must be reachable with only a few pen taps. When using a Palm device, it is very helpful that different applications have a consistent "look and feel". This reduces the time needed to get accustomed to an application.

Palm OS supports the typical components of a graphical user interface: forms, menus, tables, dialogs, buttons, and scrollbars can be implemented with the help of an user interface API.

7.2.4
Communication and Networking

Palm OS supports a number of communication standards like serial, infrared, and TCP/IP communication.

7.2.4.1
Infrared Communication

The physical infrared layer of a Palm device consists of hardware components for synchronous and asynchronous infrared communication. The asynchronous serial communication supports speeds up to 115 kb/s; the synchronous serial layer is capable of transmitting up to 1.152 Mb/s. The fast IR synchronous layer allows a speed up to 4 Mb/s.

Applications can access infrared communication in two ways:

- Using the Exchange Manager, which provides a transparent high-level interface for infrared communication.
- Using an infrared library, which provides a low-level interface to the Palm OS infrared communication capabilities.

The Exchange Manager enables applications to communicate without using conduits or HotSync. The Exchange Manager API facilitates the exchange of byte streams containing the data itself and some additional information about the data content, like the application's identifier, an optional file name, and a MIME data type. When an application wants to receive data, it has to register it-

Infrared Exchange Manager

self with the Exchange Manager, specifying which kind of data it is able to receive. An application which wants to send data has to supply an exchange socket structure, which contains information about the connection and the data type that has to be exchanged.

The communication between an application and the Exchange Manager takes place using a set of "launch codes". When the Exchange Manager receives a message, it sends a launch code to all applications, which can take appropriate actions on that event if necessary.

This first launch code (sysAppLaunchCmdExgAskUser) asks the application whether a dialog box should be displayed which asks the user if the incoming data should be accepted or not.

If the application decides to receive data, the Exchange Manager sends the next launch code (sysAppLaunchExgReceiveData), which tells the application that it should receive the data now.

To receive the data, the application can use the Exchange Manager functions ExgAccept, ExgReceive, and ExgDisconnect.

Infrared Low Level API   The low-level API to the infrared (IR) communication capabilities of Palm OS is provided via a shared library. It allows applications to directly use the IR capabilities of Palm OS, without accessing the Exchange Manager layer.

The library API is compliant with the IrDA standard and supports both mandatory and two of the optional protocol layers.

- The IrDA Link Access Protocol (IrLAP) is a mandatory part of the IrDA standard. It provides a reliable connection between devices for the data transfer.
- The IrDA Link Management Protocol (IrLMP) is also mandatory and provides a kind of session handling for the IrLAP and thus manages multiple channels over an IrLAP connection.
- The optional Tiny TP layer contains a light-weight transfer protocol for the higher layers in the IrDA stack and a flow control on IrLMP with a segmentation and reassembly service.
- The Object Exchange Protocol (IrOBEX) is also an optional layer of IrDA. It provides HTTP-like object exchange services. These services can be accessed via the Exchange Manager.

The full IrDA protocol stack is described in more detail in Chapter 12.

7.2.4.2
Serial Communication

The serial communication between a Palm and other devices takes place via the cradle port. Like the architecture of the infrared communication, the architecture of the serial communication has multiple layers. The base layer consists of the Serial Manager, which is the interface to the serial port and thus to the RS232 signals.

There are three major function blocks, which base on the Serial Manager (Figure 7.10):

- The Connection Management Protocol (CMP) is used to establish connections using baud rate arbitration and exchange of version numbers by the communication software.
- The Modem Manager facilitates modem dialing.
- The Serial Link Protocol (SLP) is the base layer for the Packet Assembly/ Disassembly Protocol, which is used by the Desktop Link Protocol. The Packet Assembly/Disassembly Protocol sends and receives buffered data. The Desktop Link Protocol is used for remote access to Palm OS subsystems and thus is used for any kind of data exchange between Palm OS and external applications. SLP implements packet send and receive functionality.

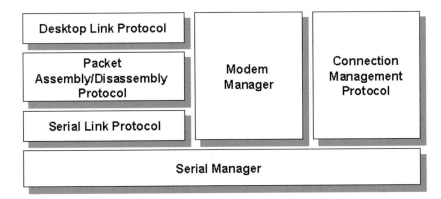

Figure 7.10: Palm OS Serial Communication Model

7.2.4.3
TCP/IP Communication

Palm OS provides two libraries to access TCP/IP and Internet networks (Figure 7.11).

- The Net Library provides a socket API to access TCP and UDP services. The TCP/IP stack on the Palm is part of the Net Protocol Stack, which runs in a separate operating system task. The access to this protocol stack is provided by an interface, which is part of the Net Library. Additionally, the Berkeley UNIX sockets API is supported.
- The Internet Library uses the Net Library to build higher level Internet protocols, like HTTP and HTTPS.

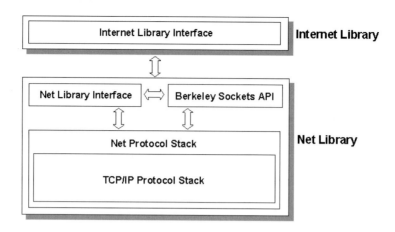

7.2.5
Conduits

Conduits are desktop programs which are used to perform the HotSync for a special kind of application or data (like the Address Book, Mail, etc.). There are three possible tasks which can be performed by conduits:

- Copying data and applications from the desktop computer to the Palm OS device or vice versa.
- Performing a two-way synchronization in order to update the data on the Palm OS device and/or the desktop computer.
- Sending transactions from/to a backend application.

Conduits have to be registered with the HotSync Manager first, before they can be used. Thus, a conduit needs a mechanism for registering the application it is responsible for. A conduit programmed in C needs C entry points for the registration of the conduit's name, the version number and a start point to enter the conduit. In order to provide the user with information about the synchronization process, a conduit has to provide log messages.

Figure 7.12 shows the architecture on which the Conduit technology is based.

148 ■   *7 Operating Systems*
    ■
    ■

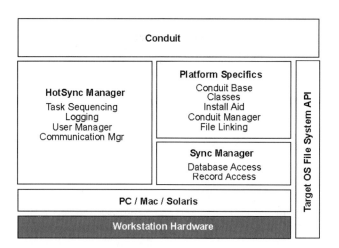

7.2.6
Developing Applications

Theoretically, there are three programming languages available for palm developers: C, C++ and Java. Nevertheless, the usage of C++ and Java is still very restricted due to the limited dynamic heap size. This is why, in most cases, C is the language of choice, when creating own Palm applications. Conduits, which run on a complementary workstation are usually written in C++ or Java.

Palm provides two packages for developing applications:

- The Software Development Kit (SDK) includes the available application programming interfaces (API) for developing Palm OS applications. There are functions for user interfaces, system management, and communication.
- The Conduit Development Kit (CDK) supports the implementation of conduits, which enable the exchange and synchronization of data between a desktop application and an application running on a device. The CDK is available for Windows, Mac, and Solaris.

Each Palm OS application must have assigned an unique identification, called *CreatorID*, which can be obtained from Palm.

Beyond these development kits, programming is supported by several development environments and tools. The following section, will concentrate on the two most common C/C++ development tools for Palm OS:

- Metrowerk's CodeWarrior
- GNU C compiler for Palm OS

CodeWarrior   The CodeWarrior development environment for Palm OS allows programmers to write ANSI C/C++ code for Palm OS platforms on most of the MS Windows platforms and on Macintosh systems. The CodeWarrior is an Integrated Development Environment (IDE) and contains a source code editor, a class browser, a compiler, a linker, and a debugger.

These tools are common for all CodeWarrior IDEs and are not specific to the Palm OS environment. However, CodeWarrior for Palm OS also contains some additional tools, which are specific to the Palm OS platform:

■ PalmRez – a plug-in tool that converts the file generated from the project into an executable file, which can be executed on the Palm OS device.
■ The C/C++ compiler and linker generate code for the Motorola 68K processor family.
■ The Palm OS Emulator allows the developer to test applications on the desktop computer. The debugger transfers applications between a desktop computer and a Palm OS device or emulator for debugging.

Figure 7.13:
CodeWarrior
Source Editor

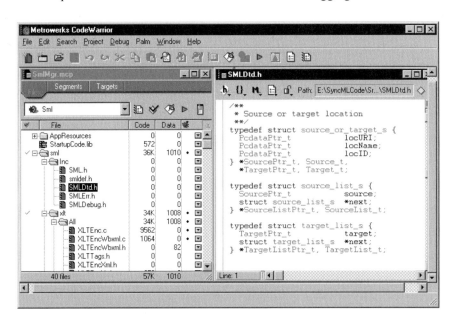

GNU   In the early days of Palm OS devices, the development of applications for these devices was bound to CodeWarrior development on Macintosh systems. This was one of the reasons for extending the GNU C Compiler to create Palm OS binaries. The result was a collection of tools to create C/C++ Palm OS applications on Windows and Unix systems:

■ The GNU PalmPilot SDK contains a C compiler generating Motorola 68K binary code, a source debugger, a resource compiler, a displaying tool, and a hardware level Palm emulator .

- The GNU resource compiler generates Palm resources on the base of textual resource descriptions. The generated resources can be displayed by the GNU displaying tool.

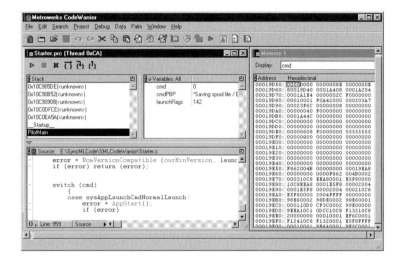

Figure 7.14:
CodeWarrior
Debugger

7.3
Symbian OS

Symbian OS, formerly know as EPOC, is a real-time, multitasking, pre-emptive, 32-bit operating system written using C++ with an object-oriented design.

The roots of Symbian and EPOC can be found in Psion's software division. In June 1998, Psion, together with Ericsson, Motorola, and Nokia, founded Symbian as an independent joint venture to develop an industry platform for wireless information devices. Matsushita (Panasonic), Sony Ericsson, and Siemens joint Symbian later. The Symbian OS operating system is based on over 20 years of experience with handheld devices and provides a stable base for handheld applications.

Currently PDAs from Psion as well as the following mobile phones use Symbian OS: Ericsson R380, R380e, and R380 World as well as Nokia 3650, 7650, 9200 Communicator, and Sony Ericsson P800.

The current version of Symbian OS, Release 7, was released in February 2002. New features in Version 7 are support for SyncML, 2.5GSM/GRPS as well as 3G phones, EMS, MMS, IPv6, ISEC and as MIDP Java. Email and messaging support, software for the synchronization of data between the device and a PC, encryption and certificate management, WTLS, SSL, and support for various audio/image formats are also included. Symbian OS Version 7 runs on x86 PCs (in the emulator), on ARM4 (StrongARM SA1), ARMv4T (ARM710T, ARM720T, ARM920T, ARM922T, ARM925T), ARMv5T (XScale, ARM 1020T), ARMv5TJ (ARM926EJ) processors.

7.3.1
Operating System Architecture

The core of the Symbian OS operating system consists of the following major components: Base, Telephony, Security, Communications infrastructure, Application framework, Personal area network, Application engines, Messageing, and Java (Figure 7.15).

Figure 7.15:
Symbian OS v7
Architecture

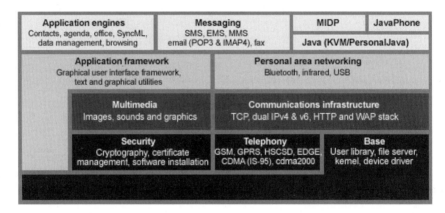

Diagram courtesy of Symbian

Base

The *Base layer* contains the runtime system, kernel, files server and device drivers. It provides a bootstrap loader, access to and monitoring of file systems, an API for implementing new file services, a hardware abstraction layer, a kernel with scheduler, tick interrupt, memory management, and device driver management.

Telephony

The *Telephony layer* provides support for a wide variety of networks including GSM, GPRS, EDGE, CDMA (IS-95), 3GGP2 cdma2000, and 3GPP WCDMA.

Security

The *Security layer* supports the DES, 3DES, RC2, RC4, RC5, RSA, DSA, DH, MD5, SHA1 and HMAC cryptographic algorithms. It can also handle cryptographic tokens and manage certificates.

Communication Infrastructure

The *Communication infrastructure layer* provides the infrastructure for networking and communicating. It consists of basic networking support (TCP, UDP, PPP, DNS, SSL, FTP, IPv6, Telnet, and Ethernet), an HTTP 1.1 stack, as well as a WAP 1.2.1 stack (with WSP, WTP, WTLS, WDP, and WAP push).

Application Framework

The *Application framework layer* allows the development of event-driven GUI applications using C++ and Java. It contains a widget architecture with libraries for data, graphics, and text.

Personal Area Networking

The *Personal area networking layer* provides Bluetooth, InfraRed, and USB connectivity. The Bluetooth stack is compliant with Version 1.1 of the Bluetooth specification and allows to connect the device to a large number of other devices supporting Bluetooth, like other PDAs, laptops, or mobile phones. Infrared is currently the most widely used way to connect a PDA or mobile phone to another device without using cables. Symbian OS supports the IrOBEX protocol version 1.2. Also a USB 1.1 client implementation is included.

The *Application engine layer* provides the base infrastructure to build great applications on top of it. It includes components for calendaring, address book, spreadsheets, database, word processor, and text converters. On top of this it comes with built-in SyncML support to make it really easy to synchronize with a large variety of servers providing all sorts of data (SyncML is explained in greater detail later in this book). The Web Engine provides everything one knows from a full-functional browser.

Application Engine

With the *Messaging layer* one can do email and fax as well as SMS, EMS (Enhanced Messaging Services operates over 3GPP bearers and can include pictures as well as sound), and MMS (Multimedia Messaging Service which operates over GPRS, supports also WAP Push and provides a number of additional messaging features like notification, priority, and delivery reports).

Messaging

Symbian OS implements the JavaPhone specification, which is an enhancement to the PersonalJava specification with APIs required by smart phones, like for telephony, calendaring, messaging.

Java

7.3.2
Application Architecture

Developers designing and implementing applications for the Symbian OS platform should layer their software as shown in Figure 7.16 to achieve greatest flexibility and portability. This layering corresponds to the Symbian OS operating system architecture, which is layered in the same way.

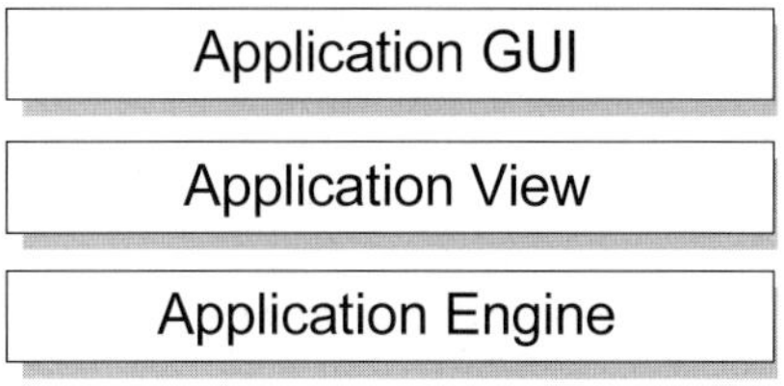

Figure 7.16: Layering an Symbian OS Application

The *Application Engine* contains the basic logic of the application. It should be independent of any representation of this information to the end user. The engine could also be viewed as the model of the application.

The *Application View* provides a simple graphical representation of the application data, such as the array of cells in a spread-sheet. One advantage of having an application view is that the GUI framework could differ from one Symbian OS device to another, but the basic graphics support is the same on all Symbian OS devices and so there is no need to rewrite the view part, but maybe only the GUI part.

The *Application GUI* part is based on a application. It drives the engine and uses the views to present a GUI to the end user. Applications can use parts of application framework to make the handling of features, like hotkeys, menus or toolbars, easier for the developer.

7.3.3
Developing Applications

Symbian OS was designed and developed using C++, and all native APIs are implemented in C++. The only way to get access to all the components of the Symbian OS operating system is to select C++ for the development of an application. It is also possible to develop applications using C with the Symbian C Standard Library, a standard POSIX-like set of APIs implemented on top of the native services.

Symbian OS Release also fully supports PersonalJava 1.1.1a, is the reference platform for PhoneJava 1.0 and was the first platform on which the JavaPhone API was fully supported.

A third supported language environments is OPL, a Basic-like language originating from Psion's past.

The development process depends on the programming language you decide to use. Since only C++ gives access to all components of the operating system, the rest of this section focuses on application development using C++. The Developer section on Symbian's website would be the right place to start for developers interested in using one of the other languages.

The following tools are required to write and test an Symbian OS application using Windows:

- The Symbian C++ Software Development Kit (SDK) includes two Symbian OS Emulator for Windows (one that runs with Metrowerks CodeWarrior for Symbian and one for Visual C++), and a tailored build of the GNU C++ compiler which is used to build the binaries before they are downloaded to the Symbian OS device. Online documentation, a tutorial, and many other useful tools can also be found in the SDK. This SDK is available for free from the Symbian Developer Network website at http://www.symbian.com/developer.
- Metrowerks CodeWarrior for Symbian or Microsoft Visual C++ is the environment used to test and debug the Symbian OS application on a Windows development workstation.

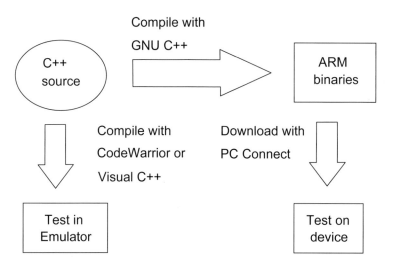

Similar to Palm OS based applications, each application targeting Symbian OS must have a unique UID for file identification and association. The range from 0x01000000 to 0x0fffffff is reserved for development purposes and can be used by every developer for testing. But before the application is released, a unique UID must be requested from Symbian.

Application ID

7.3.4
A Sample Application

The goal of this small example is to write the text "Pervasive Computing" to the console of an Symbian OS device.

First, some header files are included, which are needed in this example. Every program should have an E32Main() function, which is the entry point into every Symbian OS application and can be compared with the main() function in a PC application.

An application can use a console to display text without having its own GUI. This console is created using Console::NewL(...).

```
#include <e32base.h>
#include <e32cons.h>
GLDEF_C TInt E32Main() // main function called by E32 {
    // here comes the code that performs the action
    }
```

Finally, after doing all this setup and initialization, we can use the `printf(...)` method of the Console object to write the text we want to display to the console. The _LIT macro is used to create objects for constant literal text.

```
// create constant literal text objects
_LIT(KTxtPvC,"Pervasive Computing");
_LIT(KTxtPressAnyKey," [press any key to end application]");
// create a new cleanup stack
CTrapCleanup* cleanup=CTrapCleanup::New();
// create a new console
console=Console::NewL(KTxtExampleCode,TSize(KConsFullScreen,
        KConsFullScreen));
// push the object on the cleanup stack
CleanupStack::PushL(console);
// write text to the console
console->Printf(KTxtPvC);
console->Printf(KTxtPressAnyKey);
console->Getch();
// get the console object and close the console
CleanupStack::PopAndDestroy();
__ASSERT_ALWAYS(!error,User::Panic(KTxtEPOC32EX,error));
// destroy the cleanup object
delete cleanup;
// end the application
return 0;
```

Without the `Getch()`, the console would close immediately after the text was written, which would probably be too fast for you to see if the text was written to the console correctly. `Getch()` just waits till the next key is pressed and returns it. In our case, we don't care which key was pressed and just ignore the returned result.

Each application should have a cleanup stack to clean up memory left over by a function which was exited, may be due to an exception or a trap. With `CTrapCleanup::New()` a cleanup stack is created, `PushL()` is used to push an object on the cleanup stack and, `PopAndDestroy()` can be used to get the top object from the stack and destroy it.

7.4
Java Card

The Java Card platform is a smart card operating system, allowing the applications on a smart card to be written in Java. This brings the platform independence of Java to on-card software development, which used to be very proprietary for each card operating system manufacturer. In addition, it provides a good basis for multi-application cards, which support more than one application at a time.

The on-card executables are referred as *Card applets* and consist of a Java Card specific byte code, which is interpreted by the Java Card Runtime Environment. This runtime environment controls the execution and makes sure that different applets do not interfere. The goal is that Java Card applets run in any Java Card. This goal is not fully achieved yet because current implementations still differ slightly from each other.

The Java Card specification is owned by Sun, extensions and improvements to it are discussed and contributed in the Java Card Forum [JCF00]. This section is based on version 2.1.1 of the Java Card specification, which was released in May 2000.

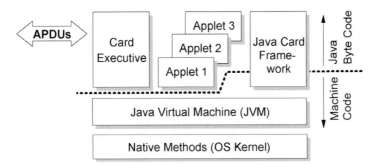

Figure 7.18: Java Card Software Stack

Strictly speaking, Java Card is a platform but not a smart card operating system. This platform is carried by a native operating system kernel, which is not directly accessible to the applications. The applications are written against the interfaces of this platform in the same way, in which they would use operating system services otherwise. The software stack of a Java Card is shown in Figure 7.18.

The Java Card Runtime Environment (JCRE) has following interfaces:

- The Card Executive manages the card and is the communication link between the card applet and the off-card code.
- The Java Virtual Machine (JVM) executes the byte code of the applet and of the library functions it uses.
- The Java Card Framework provides the library functions. These comprise the standard Java Card API.

The operating system kernel and the Java virtual machine (JVM) are native code, the layers above it are Java Card byte code.

7.4.1
Java Card Framework

The Java Card Framework consists of four packages. These packages and their contents are listed in Table 7.1.

Package	Contents
java.lang	A subset of the Java programming language. This package contains the fundamental classes Object, Throwable, Exception, RuntimeException, and several specialized exceptions. The fundamental classes differ slightly from their corresponding classes in JDK's java.lang.
javacard.framework	Classes and interfaces for the core functionality of a Java Card applet. This package provides smart card specific interfaces like ISO7816 and PIN, and classes like Applet, AID, and APDU. In addition, it contains the class JCSystem, which corresponds to JDK's java.lang.System.
javacard.security	Classes and interfaces for the Java Card security framework. This package provides interfaces for all kinds of keys, and the classes like KeyBuilder, MessageDigest, RandomData, and Signature.
javacardx.crypto	Extension package containing security classes and interfaces for export-controlled functionality. Interface KeyEncryption and class Cipher.

7.4.2
Lifetime of On-Card Applets and Objects

An important distinction between a Java environment on a workstation and the Java Card environment is the lifetime of the applets and of the objects owned by the applets.

Lifetime of the applets
The applets installed in the card have a theoretically infinite lifetime. Only when explicitly de-installed they will terminate. Otherwise, they will stay alive as long as the card is usable.

Lifetime of the applet's objects
The objects created and owned by an applet are allocated in the EEPROM by default and thus are persistent. The specification defines a way to create transient arrays but does not define a way to make an object transient. The Java Card system class JCSystem provides makeTransient...Array() methods for Boolean, Byte, Short, and Object. Transient arrays are alive until the card is powered down or the applet is deselected.

When an applet is installed on the card, its installation method `public static install(...)` is called. This method should allocate all objects the applet will use, thus making sure during installation that sufficient space will be available for the applet later. Otherwise, an out-of-memory failure should surface already during installation. If the installation was successful, the applet must call one of its inherited `register()` methods to get its application identifier (AID) registered with the Java Card runtime environment.

After an applet has been installed, it can be used by an off-card application. For communicating and working with an applet, the off-card application must select it by sending an appropriate select command using the AID of the applet. The applet that became selected is called the "active" applet. All APDUs that the application is sending to a card are passed to the active applet. This applet remains active until another applet is selected.

Selecting the active applet

Such change of the active applet causes the `deselect()` of the currently active applet being called, followed by `select()` of the selected applet being called.

Once an applet is active, all APDUs sent to it are causing a call to its method `process(APDU)` with the APDU forwarded as argument.

process(APDU)

7.4.3
Developing a Card Applet

The process for developing a card applet is shown in Figure 7.19. There are variations in the tools and in the development environment provided for the Java Cards of the various suppliers. Therefore, each development setup might differ in details.

The code development starts with the creation of a Java source file. It should not come as a surprise that the Java language subset that can be used to write a card applet has several limitations compared to the Java language for PC's or network computers. Some of the differences between the Java Card specification [JC00] and the specification of full Java are:

- Dynamic class loading is not supported.
- Garbage collection is not mandatory.
- There is no SecurityManager class – security policies are implemented by the JVM directly.
- The JVM does not support multiple threads.
- Objects cannot be cloned.
- `Java.lang` is significantly different, for example there is no `String` class available.
- The basic types char, double, float and long are not available, the `int` keyword and 32-bit integers are only optionally supported.

The source files can be compiled into regular Java byte code with a standard Java compiler. Instead of the JDK class libraries, the Java Card Framework is included. The compiled "class" files can be tested in the Java Card simulation environment on a regular Java workstation.

In the next step, the byte code converter verifies the "class" file and optimizes them for the limited resources of a smart card. They are statically linked and converted into a Java Card specific format. The resulting "cap" files can be tested in the Java Card emulator on the development computer.

Finally, all "cap" files comprising an on-card application are downloaded onto the card, where they can be executed. To prevent loading unauthorized applets on a card, this download can be secured by signing the code and letting the card verify the signatures.

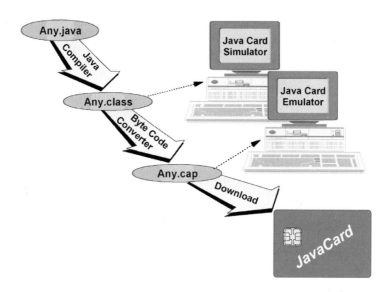

Figure 7.19: The Process for Developing a Card Applet

7.4.4
A Sample Applet

The following simple example applet takes two parameters and calculates the results based on the formula, which is defined in the source code of the applet.

The smart card this applet is developed for is the IBM "JCOP10" Java Card, which reflects the Java Card 2.1.1 specification. This card also contains the support for the Open Platform 2.0.1 specification.

Now we start creating the applet. We call it "Compute". To make it a Java Card applet, we derive it from the class `Applet` in the Java Card Framework:

Card applet definition   Remember that before we can communicate with the applet to use its services, we need to select it. The off-card part of the application must send a SELECT APDU to the Java Card runtime environment, which then calls the card applet's method `select()` and subsequently its method `process(APDU)` with the SELECT APDU passed to it.

Here we encounter another class from the Java Card Framework, the class `APDU`. The APDU was passed to the applet from the off-card part of the application.

In our applet, we do not need to perform any special functions during selection. We merely perform some checks on the APDU and return successful, when the APDU is as expected. Because you will see such APDU checking in other methods, we do not show it here and move on to the `process(APDU)` method instead.

After our applet was successfully selected, it receives all subsequent APDUs in its method `process(APDU)`.

The responsibility of the `process(APDU)` is to dispatch incoming APDUs to the appropriate methods of the applet. To determine, where to forward the call to, we examine the instruction byte of the APDU. In a more complex applet with more work to do, one could use the class byte in addition to the instruction byte to determine which function the applet should perform. In our very simple applet, the instruction byte is good enough.

On any error, we exit by throwing an `ISOException` with the appropriate constant defined in the class `ISO7816`.

```
public void process(APDU apdu) throws ISOException {
  byte[] buffer = apdu.getBuffer();


  // Dispatch commands depending on
  // the instruction bytes.
  switch ((byte)(buffer[ISO7816.OFFSET_INS])
  {
    case (byte) 0x0020 :
                byte p1 = buffer[ISO.OFFSET_P1];
                byte p2 = buffer[ISO.OFFSET_P2];


                // ENTER sample algorithm on p1 and p2 here
                buffer[0] = (byte)(2*p1 + p2);


                apdu.setOutgoingAndSend((short) 0, (short)1);
      return;
          case (byte) 0x00A4 :
                if(! selectingApplet() )
          ISOException.throwit(
```

```
                                ISO7816.SW_APPLET_SELECT_FAILED);
        return;
      default :
        ISOException.throwIt(ISO7816.SW_INS_NOT_SUPPORTED);
        break;
    }
}
```

If all went well, we pass back an APDU with data contained in the ap-
duBuffer and the return code to indicate successful execution. To do this we
use the method setOutgoingAndSend(...) of class APDU:

```
// Send the response: The content of the buffer is
// returned
  apdu.setOutgoingAndSend((short) 0, (short) 1);
```

Now we have seen all run-time functions of our Compute applet. The bit that
remains to be done is the initial installation of our applet. We provide a method
install(...). It invokes the constructor of our applet (which we made protected) and then registers the applet to the Java Card runtime:

```
/**
 * Install the applet on the Java Card.
 *
 * @param All parameters are urrently ignored.
 *
 * Note: In Java Card 2.1 the function signature has
 * three parameters instead of this single parameter.
 */
public static void install( byte [] bArray,
                            short bOffset, byte bLength) {
  Compute me = new Compute()
  me.register();
}
```

If we would need any memory for the lifetime of the applet or any special
setup for our applet, we would need to provide a constructor. In our case we
could just use the default constructor.

7.5
Further Readings

Dallas' IButton

http://www.ibutton.com

The IButton is a security module in form of a ring. Its programming interface is very close to the Java Card specification.

On this web site, you can order your IButton including accessories on-line. The 1-Wire for Java™, the latest development tool for the IButton supporting OCF, is also available for download.

Embedded Linux

http://www.emlinux.com

EmLinux is a provider of embedded Linux operating systems. This link covers a white paper as well as some example applications.

Gemplus

http://www.gemplus.com

Gemplus is one of the leading providers of smart cards. This Web Site contains information about smart cards in general and about products from Gemplus of course. You could also order smart cards like Gemplus' GemXpresso online.

Gemplus Developers

http://www.gemplus.com/developers

This site contains the latest information for developers, like a forum, latest drivers, and trends. It also provides access to OpenCard CardTerminals and Card-Services for readers and cards from Gemplus.

Java Card Forum

http://www.javacardforum.org

The Java Card Forum is the group working on the Java Card specification.

On this web site of the Java Card Forum, you can find the list of its members, its charter, information on the membership, minutes of recent meetings, and technical documents.

Linux Devices

http://www.linuxdevices.com

A vast of information related to using a Linux operating system for Internet appliances and mobile devices.

Palm Computing Tools

http://www.palmos.com/dev/tech/tools

Palm Computing provides links to SDKs, Palm OS Emulators, and a Conduit development kit on this Web page.

Psion

http://www.psion.com

Psion provides the largest variety of EPOC powered devices. Their web site provides information about Psion devices, support, and free downloads of various utilities.

Purple Software

http://www.purplesoft.com

Purple Software is a software provider for the EPOC operating system. Examples are chess, a WAP browser, a database, and backgammon, just to name a few. In addition to this, some utilities are also available for free.

QNX

http://www.qnx.com

QNX is a provider of real-time operating systems applied for various pervasive devices.

Symbian

http://www.Symbian.com

The Symbian web site is a good starting point for every EPOC developer. It provides useful technical papers, which give an insight in the architecture of the EPOC operating system, as well as EPOC event information.

Symbian Developer Network

http://www.Symbian.com/developer

The Symbian Developer Network provides the developer with a free SDK to start developing for EPOC, which also include an on-line manual and a large collection of samples.

Windows CE

http://www.microsoft.com/windows/embedded

The starting point, when searching for Windows CE related resources. There are plenty of technical papers about programming environment, and OS characteristics
http://www.microsoft.com/mobile

An overview of Windows CE devices. This site includes reviews, features, and comparisons of handheld and PocketPC devices. There are also developer and enterprise resources.

Microsoft Debuts PocketPC For Handhelds

http://www.byte.com/clumn/BYT20000315S0009

An overview of the new PocketPC from Microsoft [RUL00].

Windows for Smart Cards

http://www.microsoft.com/smartcard

The homepage of Microsoft's smart cards engagement.

8 Client Middleware

8.1
Overview

The operating system of a device provides the basic layer of a client software stack. On top of that a infrastructure of middleware and application components needs to be provided. The more flexible those components are, the easier it can be leveraged for powerful end-user applications. A generic of a client software stack is shown in Figure 8.1.

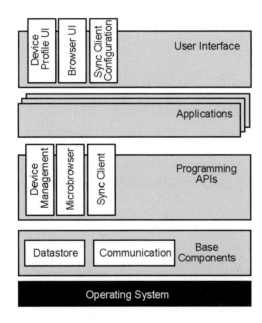

Figure 8.1: A generic client software stack

The following list gives an overview of typical middleware components and their characteristics:

A fundamental component which is used by almost every application is the local data store. In many cases this is a proprietary database with its own structure and API. But there are also more powerful solutions available, for instance a relational databases with a set of SQL commands to access the stored information. DB2e, a relational database for devices is described in section section 8.5.

Data storage

Communication *management*	Communication with networks is essential and almost any device provides a native implementation of some network protocol, e.g. simple TCP/IP, HTTP or WAP. More sophisticated communication components support transaction based communication, e.g. MQe which is described in section 8.4.
Device *management*	Device management components on devices are gaining more and more importance. Most mobile devices are now capable of installing additional applications and services during their lifetime. These applications needs to be managed and maintained. There are technical problems to solve, such as to install and maintain software without requiring the user to have technical skills. But there are business related tasks to solve as well. Such tasks include billing of licenses, subscription management. To achieve this, a local device management agent is installed on the device and interacts with a server side device management system. More details can be found in chapter 19.
Programming *APIs*	A lot of extensible libraries can be leveraged to build sophisticated applications on devices. In this chapter, two of them will be described in a little more detail:

- The first example explains two Java APIs and a markup language for developing applications targeting phones and television sets.
- Another example covers the two leading frameworks for smart card programming.

Admittedly, this selection is quite random. But they give software developers an exemplary introduction to programming interfaces, which are applied in the pervasive landscape

Connectivity	Being connected to a server infrastructure is essential to access backend data and applications from a device. Devices have connectivity components which applications can leverage for that purpose. Two basic device usage models can be distinguished: the always and the intermittently connected model:

- When intermittently connected to a server the user is mostly disconnected from a network and executes applications locally on his device. The data of these local applications is stored on the device itself and is synchronized with the corresponding data stored on a server at convenience. The synchronization process represents the interaction with the enterprise applications on the backend servers. It ensures that the device and server copy of a set of data is consistent. Typical applications built on top of synchronization are PIM applications, where the user carries a copy of his address book and calendar with him on a mobile device, whereas the original data set is stored on a corporate mail server, such as Exchange or Domino. But there are also very specific vertical applications, e.g. for sales force automation, where corporate data like orders or customer status are synchronized between a server and the mobile device of a sales man. Palm or PocketPC devices typically operate in this offline mode and require synchronization software. Palm´s synchronization application is called HotSync, PocketPC is equipped with ActiveSync. Besides those proprietary implementations, SyncML is a standard which defines a common protocol independent of a particular device and achieves interoperability when synchronizing various mobile devices with data servers. A client side implementation of this standard is a typical component of the architecture described in this chapter. The topic synchronization and especially its server side aspects are covered in more detail in Chapter 20.

- Working always connected does not require any synchronization. Data is accessed directly on the server. The business logic of the application is executed on the server itself and offered to the device as services, e.g. provided by a wireless portal. The client device only renders the accessed data, which is most often prepared as HTML web pages and displayed by a local micro browser component. A user would access those services with his browser. E.g. he would lookup his appointments in the server side Domino calendar, accept invitations, or work with his emails. The mobile workforce would read and manage their business data online. Such web services which are accessed by mobile devices are explained in more detail in chapter 18. WAP phones use this always connected usage model. A RIM two-way pager is another typical example for a device operated in always connected mode.

Instant Messaging is a tremendously growing business. Such application allow users to check the availability of colleagues and discuss things quickly in an online chat dialog. Extending instant messaging to devices enables the online chat regardless of the way a user is connected to the network. He can chat with desktop users as well as with other mobile users running an instant messaging application on a PDA or phone. Unifying the instant messaging by including various ways of communication and various kinds of client devices makes it easier to collaborate with travelling colleagues. For instant messaging and beyond, a new standard called SIP (Session Initiation Protocol) is emerging as the protocol of choice for setting up conferencing, telephony, multimedia and other types of communication sessions on the Internet. SIP aims at interoperability across various wired and wireless networks as well as devices. SIP is defined by the IETF organization. Lotus Sametime Everyplace is one implementation of an instant messaging component for mobile devices.

Instant Messaging and Awareness

In many if not in most cases, the applications built on top of those components are provided by the manufacturer. Especially the PIM application which can be found on almost every mobile device, are typically a core part of a purchased mobile device. But a growing number of a 3rd party vendors offer their own applications built on top of the device software stack. Those include improved and more sophisticated PIM applications, various tools and utilities and very specific business applications.

Applications

8.2
Programming APIs

8.2.1
JavaPhone API

The JavaPhone API is an extension to the Java platform. The API is supported by Ericsson, Motorola, Nokia, Psion, Texas Instruments, and Symbian. It aims at unifying a set of functions, on which applications for phones can be built on. Using the JavaPhone API, network operators and content providers can deliver value-added services and distribute applications to any telephony device supporting that platform.

JavaPhone packages

The JavaPhone API comprises of the following packages:

- The Direct Telephony Control API includes the basic calling functionality. For this purpose the JavaPhone API uses selected classes of the common Java Telephony API (JTAPI). As an enhancement of JTAPI, a specific device can publish a customized set of optional Java features.
- The Address Book and Calendar APIs allow applications to take advantage of PIM functionality.
- The User Profile API publishes information about the user. A simple address book entry is used for that purpose.
- The Network Datagram API encapsulates message delivery and helps applications like Short Message Text (SMS) to be independent from a particular physical network or bearer.
- The Power Monitoring API observes the power level of the system, enabling the applications to take actions, for example when battery power drains.
- The Power Management API entitles the application to control the power consumption of the system actively. The system can be set into a sleep mode for example.
- The Application Installation API provides functionality to install and remove applications during the life-time of a device. This allows service providers to deploy updates remotely or enable users to download the latest applications to their phones.
- The Communication API includes support for the serial port, allowing to attach peripherals like smart card readers, printers, scanners, or fax machines.
- The Secure Socket Layers API offers classes for establishing a secure connection between the phone and a server.

Profiles

Out of these packages, JavaPhone specifies two overlapping sets of libraries: The functionality of these libraries is defined by a wireless Smart Phone profile and an Internet Screenphone profile.

- The Internet Screenphone profile applies a mandatory subset of JTAPI. As further mandatory classes this profile comprises the address book API and the user profile API. Optional APIs cover calendar functionality, power management, application installation, SSL connection, and serial port communication.
- The Smart Phone profile uses a different subset of JTAPI to meet the specific needs of mobile devices. Additionally, it includes the mandatory power monitoring API and the datagram API. The optional APIs are the same as listed for the screenphone profile above, except that the calendar API belongs to the mandatory features.

In order to meet the requirements of a wide variety of phones, two different Java Platforms can be applied for both profiles: *Java Platforms*

- Devices which execute applets downloaded from a network should base on PersonalJava. They must provide a set of common Java classes in order to be able to run a broad variety of applets. Based on those core classes, new applications can extend the functionality of a phone during their life-time. A screenphone is a typical example for this class of devices.
- Devices which have very strict memory limitations and do not support applet downloading can apply EmbeddedJava instead. They only need to provide those Java classes, which are needed for tasks known at device conception. Since the functionality cannot be extended, the footprint is optimized, reducing the memory requirement of the device. A simple cellular phone belongs to this class of devices.

Figure 8.2 shows the architecture of a JavaPhone based application.

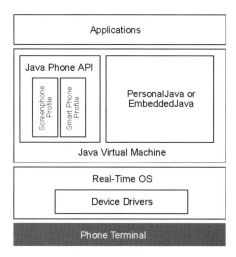

Figure 8.2:
JavaPhone

8.2.2
JavaTV API

The JavaTV API provides Java classes for implementing transparent overlays on video, graphics layering, and interactive control elements. Due to the platform independent nature of Java, applications will run on any set-top box, satellite receiver or digital TV, supporting Java. This enables numerous technology and content providers to offer interactive television, regardless of the specific hardware setup and transmission media an individual uses.

Hardware manufacturers, software vendors, and telecommunication service providers are engaged in the JavaTV activities. Among the supporters are Sony, Toshiba, Matsushita, Philips, Hongkong Telecom, and OpenTV.

The layers of the JavaTV application stack are shown in Figure 8.3.

Figure 8.3:
JavaTV API

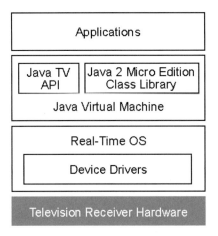

A real-time operating system is usually running on top of a common digital television receiver hardware. A Java Virtual Machine and the core Java classes have been implemented for several operating systems, which encapsulate all hardware specific features.

For example, the common java.net package is re-implemented for cable television networks in order to tunnel IP packets through a digital broadcast stream. A special profile specifies the used language subset of the Java 2 Micro Edition. The common Java classes combined with the JavaTV package expose the programming interface to the application developer.

Xlet   Java applications running on a TV receiver are often referred to as Xlets. A television service in terms of JavaTV is a collection or container of service subcomponents, such as video or audio streams, Java applications or other data. Every service needs corresponding ServiceInformation, including program descriptions and schedule data, as well as a ServiceLocator. The ServiceLocator holds those parameters needed to extract the various content streams from a transmission signal. Among these parameters are transponder frequency and de-multiplexer information. JavaTV uses a database to store all service information. Several views of that database have been defined and can be accessed through a `ServiceInformation` Manager object:

■ Within the navigation view services, information can be selected. `ServiceCollections` gather service objects according to specified `ServiceFilters`. Authorization and authentication mechanisms can be applied for conditional access to particular services.
■ The guide view supports Electronic Program Guides. A `ProgramSchedule` object encapsulates available program information.
■ The transport view retrieves physical media information. A generic `TransportInformation` class can be subclassed for implementing MPEG-2 or IP transport.
■ The utility view is used for managing applications and services.

The Service and Selection API within JavaTV includes the `ServiceContext` class, encapsulating all information about a currently running service or application. Since multiple programs can run simultaneously, more than one instance of `ServiceContext` is possible. A `select()`-method allows to start processing a particular service.

Another important package is the Java Media Framework (JMF)JMF is used to easily create players for real-time media, such as video, audio or text based data. For example, there are rendering utilities and a user interface object that can display video streams. For implementing a `Player` object, a protocol handler and a media handler are needed. The protocol handler manages the data delivery mechanism – for example using HTTP over an IP connection or receiving MPEG-2 through a cable signal. `DataSource` is an abstract class, which can be extended for particular protocols. A media handler interprets a specific type of received content. All media handlers implement the provided `MediaHandler` interface.

Java Media Framework

The Broadcast Data API supports asynchronous data transmission without any timing constraints, as well as synchronous transmission, which needs to synchronize multiple content streams with the help of time stamps.

A so called `DataCarousel` is a pragmatic way to provide interactive television capabilities even using a one-way connection: various content is send continuously to the viewers, who can select pieces of the provided data individually. Since the users might zap into a ongoing show at any time, all offered information is retransmitted within an infinite loop. In order to access requested information, an application just has to wait a few seconds, until the corresponding content is repeated the next time.

Finally, the Application Lifecycle API manages the Java applications on a television receiver. Xlets are similar to applets within a usual browser. They must implement a init, start, pause, and destroy method and have a `XletContext` object.

8.2.3
WebTV

The Advanced Television Enhancement Forum's (ATEVF) Enhanced Content Specification describes interactive TV enhancements: "A collection of Web content displayed in conjunction with a TV broadcast" [ATVEF]. This specification defines a standardized way to deliver television enhancements over networks. It covers terrestrial, cable, satellite, and Internet based broadcasting, using either analog or digital signals. Both, one-way and two-way communication is supported. Along with the specification goes a reference implementation for WebTV based Internet Receivers and Personal Computers running Windows 98.

Software within a set-top box or PC interprets the received data and presents the information as web sites on the television screen. An icon signals if interactive content is available for the current program. With a remote control or keyboard, the viewer can now switch between traditional TV pictures and interactive mode. In interactive mode browsing capabilities are offered. Beside viewing graphics, menus, and links, even TV video streams can be overlaid or placed within a web page.

The ATVF specification consists of three conceptual elements:

- "triggers" for delivering events between linked pages,
- content formatting using the HTML language, and
- "announcements" for starting enhanced TV content.

Triggers   For synchronizing regular television and interactive information content, a trigger concept is used. Triggers are real-time events. They invoke links to web pages, which include some TV enhancements. Trigger definitions must include the URL of the referred HTML document. Optional information assigned to a trigger is readable description text (name), expiration date (expires), and a executable script. After a trigger is invoked the linked page is loaded.

In order to optimize the response time for interactive requests, it is inevitable to preload some content, which is likely requested by the viewer, and to make it locally available on the client television system. Local copies of web content also reduces the peak traffic on the server (for instance, the load on a server is high, if all viewers would access the questionnaire at the same moment during a popular quiz show). This is taken into account by introducing the Local Identifier URL scheme. With a "lid:" URL resources can be retrieved relatively to a given namespace directly on the client system.

The following example shows the definition of a local link and an Internet link:

```
<lid://myTV.com/surprise.html> [name:Get a surprise]
<http://myTV.com/freeTV.html> [expires:20000818]
```

In order to handle the trigger event that corresponding web page must include a trigger receiver object, which can look like this:

```
<object type="application/tve-trigger"
        id="myTriggerReceiver">
```

The enhanced television content is provided within a HTML document. Television receivers supporting this specification must be able to interpret common MIME types, such as HTML, plain text, CSS style sheets, "jpg" and "pnp" images, as well as BASIC audio.

Content formatting

"tv:" is a new URL scheme and links to a broadcasting stream. Embedding TV pictures into a standard HTML web site is similar to the way simple images are included. The following example defines a TV screen within the web page, which is zoomed to half of its size:

```
<object data="tv:" width="50%" height="50%">
<img src="tv:" width=320 height=240>
```

In order to overlay a web page and a TV screen, the video stream is defined as background:

```
<body style="background:url(tv:)">
```

Announcements are the third element of the specification. They are used to inform the viewer about available enhancement programs.

Announcements

The announcement protocol and format is specified by the IETF. An announcement contains information about the enhancements such as duration, author, and local memory requirements for the content download. It further specifies, where the trigger data and content of an enhancement can be fetched from. Each enhancement can be executed automatically or only on user confirmation.

Enhancements are broadcasted in parallel to the regular television picture. The described elements of the ATVF specification make the delivery of interactive television independent from the underlying transport media.

When sent via a NTSC analog television signal, the Vertical Blanking Interval is used for encoded trigger data. Due to the limited bandwidth the enhanced content itself cannot be transmitted this way. The viewer's television system must use a separate HTTP Internet connection to fetch the linked content. There is no explicit announcement process.

When the Internet protocol is used, the video stream including all interactive content, triggers, and referred resources is sent over the same network. The data is sent in multiple packages through a network to the destination address.

8.3
Smart Card Programming

One obstacle smart card application developers face is the very primitive communication interface between the card and the corresponding host system to which the card connects. Although the format of the exchanged APDUs is standardized [ISO99], the set of commands a particular card supports varies. Additionally, the smart card readers are controlled in a proprietary way. Each card acceptance device, or card reader, has its own communication protocol. This results in a dependency of smart card aware applications on specific card types and readers. It requires a big effort to adopt additional cards to existing software. Developing a large-scale smart card solution that involves many different manufacturers and users is quite difficult under these circumstances.

These incompatibilities and integration problems, as well as the demand for modern multi-application solutions led to a process of standardization. Two important standards for developing smart card aware applications have evolved in the past few years: The *OpenCard Framework* (*OCF*) and the *Interoperability Specification for ICCs and Personal Computer Systems* (*PC/SC*).

8.3.1
OpenCard Framework

OpenCard Framework (OCF) is a Java API for developing terminal side smart card applications, which has been defined by an industry consortium formed by companies like Bull, Gemplus, IBM, Sun, Visa, and others. One of the main objectives of OCF is to make the parts of a smart card solution, which are typically provided by different parties, independent of each other.

Interoperability   OCF integrates components from different providers and offers access to their functionality through a high-level interface. These standardized interfaces ensure interoperability, since applications do not depend on proprietary commands of a particular smart card operating system or smart card reader. Since the OCF reference implementation is written in Java, applications are platform independent and can even be used on Pervasive Computing devices, like mobiles or set top boxes.

Simplicity   Additionally, the components hide the complexity of the low-level communication with cards and readers. Application developers can use this functionality without detailed knowledge of protocols or APDUs. This significantly reduces the programming effort and the complexity of solutions.

OCF is supported by many traditional file system oriented cards, as well as Java Cards and many smart card readers. The current version of the OCF reference implementation is 1.2. The source code and documentation can be downloaded for free from the Internet location http://www.opencard.org. Besides the regular OCF, a small footprint version for embedded devices has been specified by the consortium, although the implementation itself is an IBM development.

The OCF is composed of two basic concepts (Figure 8.4):

- the Card Terminal layer and
- the Card Service layer.

The Card Terminal is an abstraction of a card reader. It consists of an interface and its implementation, which allow to access the reader and its slots. For instance, there are methods to detect if a card is currently inserted. *Card Terminal*

The exposed interface is independent of the physical reader device. If a new reader is used, it is only required to switch to another Card Terminal implementation. The application itself remains unchanged.

Card Services represent the capabilities of smart cards. Each Card Service has a corresponding interface, defining a set of available methods. One card can provide multiple Card Services, to represent its full functionality. Card Services are general-purpose components and make up the infrastructure, which an application can use to fulfil its smart card related tasks. Internally, these components would use basic services provided by the smart card operating system and the installed card-resident executables. *Card Services*

The following Card Services are specified by OCF:

- The `FileAccessCardService` provides a standard way to access files on an ISO file system card. It allows to read and write from files on an arbitrary smart card. There are methods to select directories and files.
- The `SignatureCardService` exposes methods to import, export, and verify cryptographic keys and to handle digital signatures.
- The `AppletAccessCardService` is capable of listing executable card-resident programs ("*applets*") on a smart card. An additional `AppletManagerCardService` defines a high-level interface through which applications can install and remove applets in a standardized way. These Card Services manages the coexistence of the various application functions and data on the same multi-application card.

Besides these very generic Card Services, it also possible to define services for a specific application context. For instance, a `VisaCashCardService` would interface the Visa Cash electronic purse card. To leverage this, a hierarchical usage of CardServices, which draw on each other is supported. The `VisaCashCardService`, which needs to access files can use the `FileSystemCardService` available for the card.

With programmable cards, the APDUs sent to the card are not determined by the card operating system, but by the on-card application. Here, a single Card Service developed by the application provider could be used regardless of the card provider. An example of this is the Java Card. The APDUs and data needed by the application are determined by the card-resident applet.

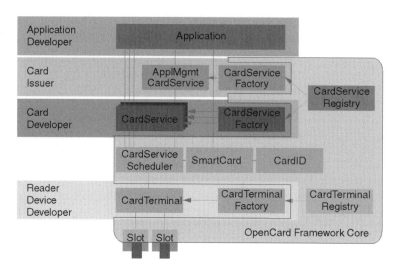

The class `SmartCard` represents an inserted card. A `SmartCard` object has a `CardID` that contains the information identifying the card type. To obtain Card Service modules for the card, an application uses the `SmartCard` object to request them.

Scheduler
A Card Service object communicates with the smart card through the `CardServiceScheduler`, which is also a class provided by the OpenCard Framework core. The `CardServiceScheduler` synchronizes concurrent accesses to one card from different applications. Consequently, there is exactly one `CardServiceScheduler` object for every `SmartCard`.

Registry and Factory
For maximum flexibility, OCF is able to instantiate appropriate Card Terminal and Card Service classes dynamically during runtime. The card's Answer-to-Reset (ATR) is used to determine which Card Service implementation is required for a particular card. If a Java Card is inserted into the reader, a different Card Service would be used, than when a file system card is detected. In order to support cards as flexible as possible, it is even possible to download required Card Service classes from a network when they are needed. This allows a local system to communicate with an unknown card without any installation or setup.

Instances of the required Card Terminals and Card Services are created by factory classes. Each Card Terminal needs a corresponding Card Terminal Factory.

Registry classes keep track of the available implementations. The system-wide `CardServiceRegistry` keeps track of the installed Card Service Factory objects. At the time a Card Service with a particular interface is requested, the `CardServiceRegistry` calls every registered Card Service factory until an appropriate Card Service implementation has been created. The new Card Service object is connected to the `SmartCard` object with which it will be used. While we can have many Card Service and Card Service Factory objects in a system, there is always only one single `CardServiceRegistry`.

Looking at the terminal layer, we find similar mechanisms: A `CardTerminalRegistry` lists all readers, which are available on a system.

The following sample code shows how to read the content of a file on an inserted smart card using a `FileAccessCardService`:

A sample

```
...
SmartCard.start();
// wait until a card is inserted
SmartCard sc = SmartCard.waitForCard();


// get an appropriate Card Service implementation
FileAccessCardService facs = (FileAccessCardService)
        sc.getCardService(FileAccessCardServie.class, true);


// access the files
CardFile root = new CardFile (facs);
CardFile file = new CardFile (root, ":C009");


// read the data
byte[] data = facs.read(file.getPath(),0,file.getLength());
...
```

8.3.2
PC/SC

The "Interoperability Specification for ICCs and Personal Computing Systems" is commonly know as PC/SC. The specification covers the entire range from physical characteristics of smart cards up to application programming guidelines. It has been defined by the PC/SC Workgroup, formed by Bull, Gemplus, Hewlett-Packard, Intel, Microsoft, Schlumberger, Siemens, Sun, and Toshiba.

The principle goal of PC/SC is quite similar to OCF: A standardized programming interface which allows to access smart cards in an application independent way. Smart card and reader interoperability facilitates the application development. But there is also a fundamental difference: While OCF uses Java to achieve portability, PC/SC focuses on Microsoft's Windows platforms. Although the PC/SC specification is actually independent from a particular operating system, the implementation of PC/SC revision 1.0 has been developed by Microsoft for the various Windows operating systems. PC/SC is already integrated in Windows 2000 as well in Windows CE 3.0. It can be installed on Windows 9x and Windows NT systems.

Linux is a possible candidate for an "unofficial" additional platform: The Movement for the Use of Smart Card in a Linux Environment (MUSCLE) is porting PC/SC to Linux.

Compliance testing
A significant achievement of PC/SC, was to establish a compliance testing program for smart cards and readers. In spite of existing standards like ISO 7816 [ISO 7816] and EMV [EMV3.0] interoperability was a severe threat to the smart card industry: an arbitrary real-life smart card reader would not necessarily work with any card. This problem has been solved: Every PC/SC compliant reader type has been tested to work properly with a specified set of compliance test cards. The compliance test program is executed by Microsoft.

PC/SC uses a particular terminology, which should be familiar to understand the specification: smart cards are referred as *Integrated Circuit Cards* (ICCs). Card readers are called *Interface Devices* (IFDs).

The PC/SC architecture comprises three basic concepts:

IFD Handler
The *IFD Handler* is the equivalent to OCF's Card Terminal. An IFD Handler encapsulates the reader specific protocols and exposes a common programming interface, which is device independent. One IFD Handler for each reader supporting PC/SC must be implemented.

ICC Service Provider
The *ICC Service Provider* offers the functionality of smart cards in a card independent manner. Similar to OCF's Card Services, there are several pre-defined Service Providers, exposing APIs for general purpose tasks, such as file access or authentication. It is also possible to define own industry or application specific Service Providers.

If cryptographic functions are needed for the access to the card, these functions are localized in the optional Crypto Service Provider (CSP). Specified interfaces include key generation, key management, key import/export, digital signature, hashing, and bulk encryption services.

Resource Manager
The central component of PC/SC is the *ICC Resource Manager*. The Resource Manager keeps track of attached smart card readers and off-card resources, like IFD Handlers and ICC Service Providers. The Resource Manager detects card insertion and removal. A card recognition mechanism analyses the card's Answer to Reset (ATR) and maps this information to corresponding Service Provider implementations for each requested interface. For this purpose a mapping table is maintained. The Resource Manager is a privileged component. It offers the same degree of security as the base operating system. To achieve this, the Resource Manager is an integral part of the Windows 2000 operating system.

Figure 8.5 shows the PC/SC architecture.

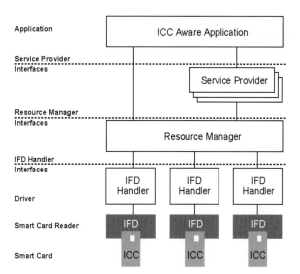

The following list, gives a brief overview of the revision 1 specification: *Revision 1.0*

- Part 1 gives a high-level introduction and an architectural overview.
- Part 2 specifies physical characteristics and lower transport protocol levels of ICCs as well as reader hardware.
- Part 3 and 4 specify the requirements and characteristics of an IFD subsystem including IFD Handler.
- Part 5 includes the ICC Resource Manager specification.
- Part 6 defines ICC Service Providers, including the Crypto Service Provider interfaces.
- Part 7 advises application developers on how to use the interface of the components and concepts of PC/SC.
- Part 8 gives recommendations how to handle identification, authentication, and how to achieve secure storage and confidentiality in a smart card solution.

The revision 2.0 of PC/SC is currently in progress and will provide several   *Revision 2.0*
enhancements:

- The update will support readers with extended capabilities, such as integrated PIN pad, display, biometrics, and multiple slots. The abstraction of such reader capabilities is represented by an IFD Service Provider
- In revision 1.0, the mapping between the ATR the and corresponding Service Providers is done statically by the Resource Manager. Additionally, revision 2.0 introduces a new plug-and-play mechanism: Each smart card stores a Global Unique Identifier (GUID), which links the card to a new

type of Service Provider, called *Application Domain Service Provider Locator* (*ADSP-L*). The ADSP-L assigned to an inserted card is capable of listing all application specific interfaces as well as their corresponding implementations. This mechanism gives more flexibility when dealing with multi-application cards.

■ The specification will be extended to cover contactless smart cards.

8.4
Messaging Components

Exchanging messages is a recurring task when dealing with any device, which is connected to a network or just to another device. Applications need to be enabled to transmit information using a high-level interface and without having to take care of communication issues themselves. Once a message is entrusted to a messaging software, delivery is assured from an application point of view. The messaging software deals with network interfaces, supports multiple communication bearers and protocols, allows intermittent connections, assures delivery of messages, and handles recovery after system problems.

8.4.1
WebSphere MQ Everyplace

This section describes IBM's WebSphere MQ Everyplace (MQe) as one example of a messaging component for lightweight devices. MQe is part of the MQ-Series product family, which provides a scalable messaging infrastructure. MQe is currently available for different Windows Platforms, AIX, Solaris, Palm OS, EPOC, and Windows CE.

Transferring messages   The MQSeries messaging services base on *queue managers*. A queue manager is able to manage multiple *queues*. Each queue stores *messages*. For transferring a message into a queue of a remote application, the originating application puts the message into a queue of its local queue manager. The local queue manager transmits the message to the remote queue manager over channels. The essence of messaging is to decouple the sending application from the receiving application, queuing and routing messages, if necessary.

8.4.1.1
MQSeries System Configurations

MQSeries supports several different configurations for "traditional" workstation and server environments:

The simplest configuration is one (standalone) server with a queue manager running. One or more applications run on that server, exchanging messages via queues. The server provides assured messaging for applications, synchronous local queue access, and asynchronous delivery to remote queues. It supports one or more local queue managers, messaging channels (to attach different queue managers), and client channels (to attach clients and gateways).

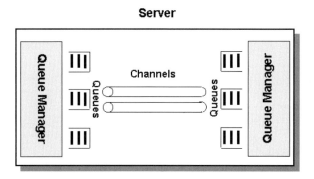

Figure 8.6: Standalone MQSeries Server

Another configuration is *client-server*. In this configuration the queue manager only exists on the server, but each client has access to it via a *client channel*. The client channel is a bi-directional communication. Applications can run on the client, accessing server queues. One advantage of the client-server configuration is that the client-messaging infrastructure is lightweight and dependent on the server queue manager. The downside is that clients and their associated server operate synchronously and require the client channels always to be available.

MQSeries clients require synchronous client channel connections to the attached servers. They provide synchronous server queue access and asynchronous message delivery to remote queues.

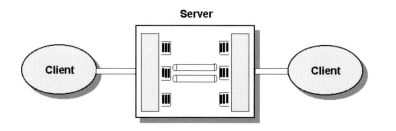

Figure 8.7: Client-Server Configuration

A further and more complex possibility to configure MQSeries is the distributed client-server configuration. In this case the servers communicate via message channels.

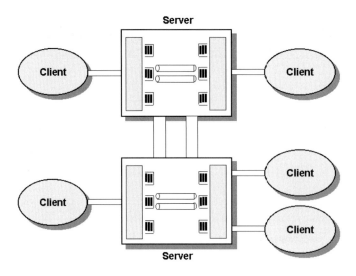

8.4.1.2
MQe System Configurations

While the previous configurations are applicable for all products of the MQ-Series family, the following setup is used for MQe only:

Devices running MQe have synchronous access to local and remote queues and asynchronous delivery to remote queues. Devices can support one or more local queue managers and dynamic channels to attach other devices or a gateway module. The gateway module connects devices to other devices or systems running the regular MQSeries software.

It is possible to configure devices with a full queue manager in opposite to MQ clients. Thus, devices are able to communicate asynchronously. They are also able to communicate with remote queues. Via peer-to-peer messaging, they can communicate directly with other devices. Devices as well as gateways support "dynamic channels", which are bi-directional and support synchronous messaging as well as asynchronous messaging.

Figure 8.9:
Peer-to-Peer
Devices

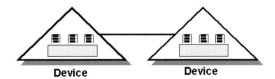

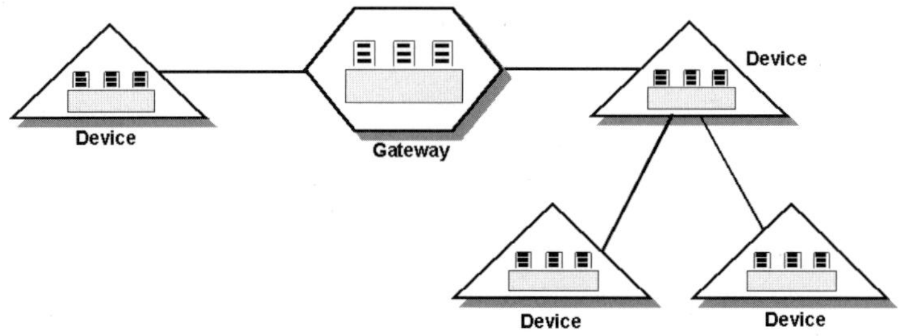

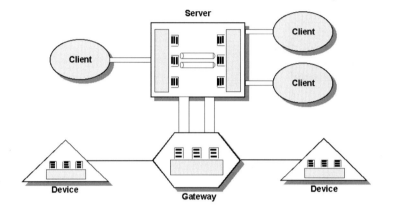

Figure 8.10:
Device Cluster

Figure 8.11:
Integration of
MQSeries
and MQe
Components

The gateway can also support client channels in order to communicate with servers. Gateways have queue managers and can therefore be used to run local messaging applications.

Figure 8.11 shows a combination of MQe and "traditional" MQSeries elements: Devices are connected to a server and its clients via a MQe gateway.

8.5
Database Components

Storing data on devices is another task middleware components need to leverage. Applications targeting multiple devices, need high-level interfaces, which are as independent as possible from a particular hardware and software environment. Besides just storing and retrieving data on the device itself, database components on handheld devices also need the technology to synchronize local information with corresponding server databases. For instance, mobile professionals use handheld computers to keep local copies of the data stored on the company's backend systems.

8.5.1
DB2 Everyplace

DB2 Everyplace is a relational database that has been designed to reside on small devices. It gives enterprises the ability to deliver and synchronize DB2 database content to small devices. Supported platforms are Palm OS, Windows CE, Symbian V6, Java2 Micro Edition, QNX Neutrino, Linux, and Win 32 Platforms.

Figure 8.12:
DB2 Everyplace
Integration
Scenario

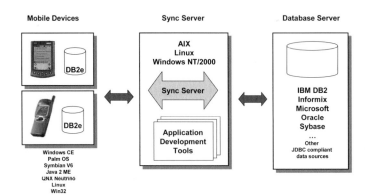

Figure 8.12 shows the three key components of a DB2 Everyplace solution:

- DB2 Everyplace database
- Synchronization Server:
  This server runs on a Windows NT/2000, AIX or Linux machine and exchanges the data between the DB2 Everyplace database on a device and a corresponding back-end database.
- Back-end database server.

The DB2 Everyplace is a relation al database engine that serves as a persistent storage for enterprise data on mobile devices. The size of a DB2 Everyplace can be as small as 100 KB to 150 KB, but it could manage up to 120 MB. Currently, the published limit is 16 million pages at 512 bytes per page. The engine size is 100-150 KB and the engine can be shared by multiple applications (not running simultaneously). The integrity mechanism guaranties that data is not lost even if the handheld device is powered off or dropped off during processing. If the processing is not complete, the session cancellation processes take care of the data integrity.

The DB2 Synchronization Server is a client-server program that manages two way data synchronization between a source and a target database. Two way synchronization means propagating the changes made by mobile professionals in their local databases to the enterprise database, and making available to the mobile professionals, the changes that are made to the enterprise database by some other user or process. There are two parts for the Sync Server. The server program and the client program. The server part is called Sync Server and the client part is known as Sync Client. In a three-tier architecture, the mid-tier server hosts the Sync Server and the Sync Client is installed on the mobile devices.

8.6
Further Readings

Advanced Television Enhancement Forum

http://www.atvef.com

The homepage of the Advanced Television Enhancement Forum (ATVEF).

IBM DB2 Everyplace

http://www-3.ibm.com/software/data/db2/everyplace/

Further information and news related to DB2 Everywhere can be found at this site.

IBM WebSphere MQ Everyplace

http://www-3.ibm.com/software/ts/mqseries/everyplace/

A web page featuring more information about supported platforms and software requirements.

JavaPhone

http://java.sun.com/products/javaphone

The starting point for Internet research on the Java Phone initiative.

JavaTV

http://java.sun.com/products/javatv

This site hosts technical papers, links to supporting companies, and several user scenarios describing the benefits of interactive television.

Microsoft's ATVEF Site

http://www.microsoft.com/atvef/TVE-public-1-1r26.htm

A technical description of the ATVEF specification.

Microsoft's PC/SC Site

http://www.microsoft.com/security/tech/smartcards/default.asp

Using this link, the developer kits for the Microsoft PC/SC implementation can be obtained. There is also information on the hardware compliance test program available.

MUSCLE

http://www.linuxnet.com/smartcard/index.html

The homepage of the PC/SC Linux project provides API specification and documentation as well as developer kits and hardware drivers.

OpenCard Consortium

http://www.opencard.org

The OCF reference implementation is available from this site. A programmers guide and several presentations are additional sources for further reading.

PC/SC

http://www.pcscworkgroup.com

The PC/SC specifications and information about the workgroup can be obtained using this link. There is also a list of compatible products. A white paper describes the enhancements of the revision 2.0 in more detail.

Smart Card Application Development Using Java

A detailed book about the OpenCard Framework and its usage [HAN99].

Television Futures – JavaTV Technology Looms Large

http://developer.java.sun.com/developer/technicalArticles/Media/javatv

This technical article explains the JavaTV API in more detail.

9 Security

Security issues are critical for the success of Pervasive Computing. Because Pervasive Computing and mobile e-business may provide millions of people with the power to move trillions of dollars in goods or money by a few mouse clicks, the security of e-business transactions is a top priority.

Cryptography can be used to assure security in a lot of e-business scenarios. It can be used to enable the secure spending of money on the net, for secure authentication of users, or for generating digital signatures for electronic contracts, just to name a few examples.

In this chapter, we first give an overview of the various needs for security. We start by explaining some of the new challenges that appear when business is moved from traditional stores to mobile devices connected to the Internet. Then we present you the basic concepts and technologies of cryptography. [RSA00] provides a detailed coverage of all major security mechanisms used to protect communication in the Internet. The last parts of this chapter describe how cryptography is used.

9.1
The Importance of Security

In the last few years, the Internet was discovered as a huge market with billions of customers around the world. The Internet already enables companies to sell to customers around the world with minimal investment. This is frequently called e-commerce. We consider electronic commerce a subset of e-business[1]. With Pervasive Computing, as well as more and more devices and appliances getting connected to the Internet, security becomes more and more important.

Together with the advantages, Pervasive Computing brings new challenges that didn't exist before.

A merchant must know the identity of the customer and the recipient of a message, a command, or an order should know the identity of the sender. For some kinds of business, it is not sufficient that the customer authenticates himself by the use of a password. Or imagine you can control heating at home over the Internet, in this case you better make sure that only you, or other authorized

Authentication

---

[1] The term "e-business" is more general than e-commerce. It includes also applications like supply chain management and other business-to-business flows.

persons, can turn on or off the heat and not anybody else surfing around in the Internet. In these cases, an electronic version of today's identity or credit card is required. This challenge is met using cryptographic methods to authenticate persons or messages.

Integrity   The recipient of a document should be able to recognize if a document or message was altered during transmission. It wouldn't be good if anybody could increase or decrease the number of shares in a stock order sent to an e-broker by changing the message content. At least the e-broker should be able to check if the message was altered or not.

Privacy   The exchange of data between two individuals, for example the merchant and the customer, should in most cases be kept secret. No unauthorized party should be able to read or copy such a communication. This challenge is met using encryption.

Cryptography can help to address all these challenges. It can be used to authenticate persons and transactions, to get secure access to data or services, and to protect the privacy of communication.

9.2
Cryptographic Patterns and Methods

Cryptographic algorithms are used to encrypt information in a form that cannot be read or altered by third parties. The sender of the information encrypts the data using a key, the recipient of the data decrypts the data back into a usable form by applying a second cryptographic operation also using a key. Cryptographic algorithms can be divided into two groups: symmetric and asymmetric algorithms.

9.2.1
Symmetric Cryptographic Algorithms

Symmetric cryptographic algorithms, also known as secret key algorithms, are characterized by the fact that the sender and the receiver use the same key to encrypt and decrypt the data.

Cryptography   Compared to asymmetric algorithms, symmetric cryptography is fast and it can be used to encrypt and decrypt a large amount of data. To keep the communication secret, only the sender and the receiver of the information should know the key that was used to encrypt the data. If someone is exchanging data with a lot of other parties, he should maintain a separate key for each of them. This could become a complicated task if the network is quite large.

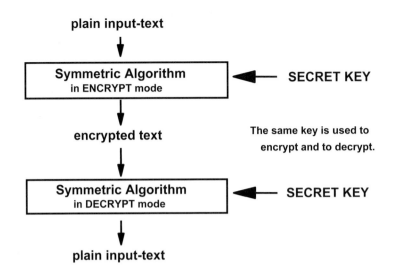

The same key is used to
encrypt and to decrypt.

Symmetric cryptographic algorithms can be divided into two groups, based on the way the data is processed:

- block-cipher and
- stream-cipher algorithms.

Cipher is another word for encrypt. Block-cipher algorithms split the data into blocks with fixed length. The last block is padded, if necessary. Today, a block length of 64 bit is usually used. Systems based on stream-cipher algorithms encrypt each byte separately.

Today, only block-cipher algorithms are standardized in the industry; thus they are the ones that are used in most situations.

9.2.1.1
Data Encryption Standard (DES)

The Data Encryption Standard (DES) defines the Data Encryption Algorithm (DEA). Usually, DES and DEA are used interchangeable.

DEA is a symmetric block-cipher algorithm developed by IBM in 1974. The US government declared DEA an official standard in 1977 and it is also defined in the ANSI standard X9.32. A lot of experts have studied DES extensively in searching for ways to break it, but no practical and fast way was found so far.

Brute-force attacks or exhaustive key searches can be used to break almost every cryptographic algorithm, by simply trying each key until the correct one is found. On the average half of the possible keys have to be tested to have success. At the time DES was developed, it was considered secure against brute-force attacks. Building systems that could find the correct key using this technique in a reasonable time was too expensive at that time. Computing power

Brute-force attacks

has tremendously increased since the mid-seventies and therefore DES protects the data today no longer than for a few hours.

Figure 9.2:
DES in ECB
Mode

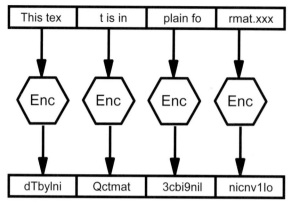

Enc = Encryption

Figure 9.3:
DES in CBC
Mode

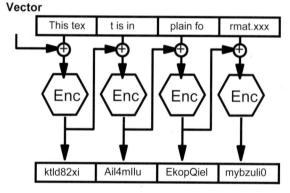

Enc = Encryption

DES uses a key length of 64 bits containing 8 parity bits; therefore the actual key size is only 56 bits. The DES block length is 64 bits. It operates in two modes: electronic codebook mode (ECB) and cipher block chaining mode (CBC).

ECB encrypts each block separately. Therefore the blocks can be encrypted or decrypted in parallel, which leads to a better performance.

In the CBC mode, each block is combined with the encrypted previous block using an XOR operation, before this block is encrypted. This has the advantage that the same plain text block results in another encrypted block, depending on the text that was encrypted before. To start the encryption, the first block is combined with the Initial Chaining Vector.

9.2.1.2
Triple DES

Triple DES makes encryption with DES more secure by applying three DES operations to the same plain text. The following four variations have been used in practice:

- DES-EEE3: The input data is encrypted three times using the same key each time.
- DES-EEE2: It is similar to DES-EEE3, but during the second encryption operation, a different key is used.
- DES-EDE3: The data is first encrypted, then decrypted, and finally encrypted again, using a different key each time.
- DES-EDE2: It is similar to DES-EDE3, but now the key for the first and the third encryption operation is the same.

9.2.1.3
Advanced Encryption Standard (AES) / Rijandael

As DES was getting closer to the end of its secure lifetime, the National Institute of Standards and Technology (NIST) started looking for a successor of DES in January of 1997. On October 2, 2000 NIST selected Rijndael as the algorithm of choice for AES and published it as FIPS (Federal Information Processing Standards) Publication 197.

The two Belgium cryptographers Joan Daemen and Vincent Rijmen designed Rijandeal as a block cipher with variable block and key length. AES currently standardizes Rijandael with key lengths of 128, 192, or 256 bits and block length of 128, 192, or 256 bits. A combination of every key length with every block length is possible.

The strengths of Rijandael are its good performance as well as its low memory requirements, which makes it suitable for embedded or pervasive devices.

Currently the scientific community expects that an AES key could be detected with around 2^{100} operations. Currently nobody is able to build a cluster of computers that would be able to execute this enormous amount of operations for example in one year [COU02].

9.2.1.4
RC2, RC4, RC5

These algorithms were developed by Dr. Ronald L. Rivest for RSA Data Security and RC stands for "Rivest's Cipher".

RC2 is a block-cipher algorithm, which works with a variable key length. Therefore the strength of the encryption and the performance of the encryption operation can be varied by the key length used. With a shorter key, the performance is better, but the encryption is less strong. RC2 has a better performance compared to DES, but the algorithm is the property of the company RSA Data Security and was not published. This has the consequence that it wasn't possible to analyze and test the algorithm as thoroughly as it was done for DES. Therefore the risks and problems of the algorithm are not known.

RC4 is a stream-cipher algorithm with variable key length that needs about 8 to 16 operations for each output bit. RC5 is, as RC2, a block-cipher algorithm with variable key and block length.

9.2.2
Asymmetric Cryptographic Algorithms

Asymmetric cryptographic algorithms, also know as public key algorithms, were developed to solve the key distribution problem that every user of symmetric cryptography has. In 1976, Whitfield Diffie and Martin Hellman developed the Diffie-Hellman algorithm, the base for today's public key systems.

The main areas of use for asymmetric cryptography are

- the distribution of keys,
- the generation of digital signatures, and of course
- the encryption and decryption of information.

According to the concept of Diffie-Hellman, everybody has two keys, a public and a private one. The public key is accessible by the public and can be requested from a Trusted Third Party. This Trusted Third Party guarantees that a specific public key really belongs to that specific person. The private key stays with the owner of the key and should be kept in secret. A smart card is for example an ideal device for securely storing private keys. It can generate the signatures or the encrypted data on the card in a way that the private key never has to leave the card.

Figure 9.4:
Public Key
Cryptography

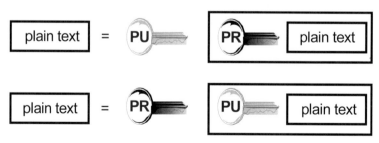

PU = public key;   PR = private key

In a public key system, information that was encrypted using a person's public key can only be decrypted using the same person's private key.

With the development of the public key algorithm, Diffie and Hellman primarily wanted to solve the problem of distributing keys. Symmetric cryptographic algorithms are today about 10 to 1000 times faster than asymmetric algorithms. This depends, of course, on the compared algorithms and if they are implemented in software or hardware. To combine the easy key distribution of asymmetric systems and the performance of symmetric algorithms, a lot of systems are today using asymmetric algorithms to distribute symmetric key (for example session keys) which are then used to encrypt the data exchanged between these two systems. *Distributing keys*

If person A wants to send person B some information, which should be readable only by person B, then person A uses person B's public key to encrypt the information. Because the information can be decrypted only using person B's private key, which should only be accessible by person B, the information sent by person A can be read only by person B. *Transmitting encrypted data*

If the sender now uses his private key to encrypt the data, the receiver needs the sender's public key to decrypt the information. Due to the fact that the public key is publicly available, it can't be used to ensure a private communication between two individuals but the receiver could verify, if the data was really sent by a specific individual. Using the private key to encrypt data is also known as signing data. *Digital Signature*

9.2.2.1
Rivest Shamir Adleman (RSA)

The RSA algorithm is the most widely used asymmetric cryptographic algorithm today. Ron Rivest, Adi Shamir, and Leonard Adleman developed it in 1977. RSA Data Security acquired the patent, which protected the implementation of RSA till September 20th 2000.

RSA is this popular because the algorithm is quite easy to understand and to implement. The only disadvantage is the slow performance compared to symmetric algorithms, as we mentioned before. In practice, most systems use RSA for the generation of digital signature and for the distribution of symmetric keys.

Today, a key length of 1024 bit is used in most systems. RSA is also referenced as one of the possible cryptographic algorithms in several standards, like S/MIME, IPSec, TLS (the successor of SSL), and PKCS.

9.2.2.2
Digital Signature Algorithm (DSA)

The National Institute for Standards and Technology (NIST) published the Digital Signature Algorithm (DSA) with a variable key length of 512 to 1024 bits, as part of the Digital Signature Standard (DSS). The US Government is using DSS for digital authentication.

One of the characteristics of DSA is that the generation of a signature is faster than its validation. In contrast, RSA can validate signatures faster than generating them.

9.2.2.3
Elliptic Curve Cryptography (ECC)

Victor Miller [Mil86] and Neal Koblitz [Kob87] were the first to propose the elliptic curve algorithm in the mid-80s. ECC provides the same security as other public key algorithm with a shorter key length. In pervasive devices with limited memory and computing power, ECC could be an alternative to other public key systems, like e.g. RSA. Some of the most effective implementations of ECC are patented by a company called Certicom. That makes the widespread adoption of ECC somewhat more difficult. With the fast growing number of pervasive devices, ECC will also be more widely used. ANSI is currently working on standardizing ECC as part of ANSI X9.62 for digital signatures and ANSI X9.63 for key agreement.

9.2.2.4
MD2, MD4, and MD5

Rivest developed MD2, MD2, and MD5 for RSA Data Security. MD2 is optimized for 8-bit computing platforms, MD4 and MD5 for 32-bit processors. MD5 is a more secure version of MD4, which makes it on the other hand a little bit slower.

MD5 first splits the message into 512 bit blocks and generates in three steps a 128 bit hash.

9.2.2.5
Secure Hash Algorithms (SHA & SHA-1)

The Secure Hash Algorithm (SHA) was standardized in the Secure Hash Standard (SHS) and published by the US government as a "federal information processing standard". SHA-1 is an improved version of SHA.

The algorithm is used to generate a 160-bit Message Authentication Code (MAC) from a message that should not be longer than 2^{64} bit. Compared to MD-5, the algorithm is a little bit slower, but due to its longer MAC, it is more secure against brute-force attacks.

9.2.3
How Secure Is an Algorithm?

The security of a cryptographic algorithm of course depends on the algorithm itself, but generally the security of an algorithm increases with the length of the key used.

A brute-force attack can always be used to break an encryption. To break DES with a key length of 56 bit, somebody would have to test up to 2^{56} possible keys before the right one is found. In 1998, it took for example 56 hours before a DES cracking machine found the key used. Today it would only take a few hours to do so.

The key length used to encrypt data should be increased regularly to keep up with increasing computing power. DES, with only 56 bit is in the meantime considered as too weak to protect important data. The US government for example has been using Triple DES, instead of DES, since November 1998 and will switch to AES in 2000. A key length of at least 128 bit should be used today, where as 192 bit or 256 bit are recommended.

Asymmetric algorithms today should have a key length of at least 1024 bit to be considered secure.

9.3
Cryptographic Tools

There are several ways cryptography is used to secure operations and data. The following section highlights the most important ones.

9.3.1
Hash

A hash function is a one-way function that generates a fixed-length string, the hash, out of a given input. A one-way function is a function that is hard to invert. Due to this characteristic, the hash of a document is also sometimes called the message digest or digital fingerprint. A hash is often attached to a document which is transferred to the recipient. The recipient then uses the same hash function to generate a hash himself. If this hash and the received hash are not identical, then the data was changed during transmission.

9.3.2
Message Authentication Code (MAC)

A MAC is an authentication tag or checksum computed by applying a secret key to a message. The MAC is always verified using the same key. The generation of a MAC can be based on a hash function, on a stream-cipher or on a block-cipher algorithm. In the Internet, MACs are often generated using the MD5 algorithm.

9.3.3
Digital Signature

The signature on a contract or a letter shows and guarantees to the recipient the identity of the sender. Today, more and more data, orders, or emails are transmitted electronically, but most of the recipients today just have to trust that the data is coming from the person listed as the sender. It is quite simple to fake an email by attaching a wrong sender address.

Digital Signatures enable the recipient to verify the identity of the sender and the origin as well as the integrity of the document.

Figure 9.5:
Digital Signature

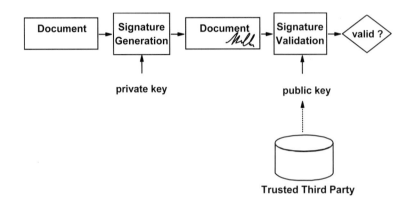

Digital Signatures are based on asymmetric cryptographic algorithms. The documents are signed with the private key of the sender. The recipient can take the sender's public key, which is provided to him by a Trusted Third Party, and validate the integrity of the document he received.

The following two types of Digital Signatures are used today:

- a Digital Signature which encrypts the whole message, and
- a Digital Signature based on a Message Authentication Code that is attached at the end of the document. The document itself is transmitted unencrypted.

The second possibility is faster than the first, because it does not have to encrypt the complete message. If the document itself should also be encrypted to make sure that an unauthorized person doesn't read the content, the document is often encrypted using a symmetric algorithm. The key used is encrypted using a public key algorithm and attached to the message. As mentioned before, a symmetric algorithm is 10 to 1000 times faster than an asymmetric one. In the above-mentioned combination of symmetric and asymmetric cryptographic functions, the large amount of data is encrypted with the symmetric algorithm.

Only the symmetric key, which is usually quite short compared to the data, is encrypted with the slower asymmetric algorithm.

The recipient first decrypts the attached symmetric key using his private key and then decrypts the document with the symmetric key.

9.3.4
Certificate

A certificate is a document distributed by a Trusted Third Party that binds a public key to a specific person. The Trusted Third Party guarantees that the information contained in the certificate is valid and correct.

Certificates are standardized by X.509. They should at least contain the following information:

- the digital signature of the Trusted Third Party,
- the name of the person owning the public key, and
- the public key itself.

9.4
Secure Socket Layer (SSL)

Netscape developed SSL in 1995 to provide security and privacy on the Internet. Today, most web servers and browsers support SSL. A user recognizes an SSL-session at the "https://" instead of "http://" before the URL.

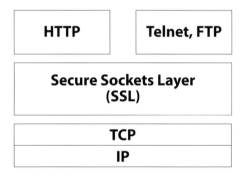

Figure 9.6: Secure Socket Layer

SSL sits on top of TCP and below the application layer. This way, SSL is not only able to secure an HTTP connection, it can also be used for other services on the Internet, like telnet or ftp.

The SSL protocol itself consists of two parts:

- the handshake protocol: it is used to setup a session, the two partners authenticate each other and the security features for the data transfer are negotiated.
- the record protocol: it is used to transfer the actual data between the two communication partners.

SSL supports a variety of different cryptographic algorithms. The handshake protocol is used to find an algorithm, which is supported by both, the client and the server. Due to the US export restrictions for cryptography, two versions of SSL are defined. The one with strong encryption is only used in the US, the exportable version has unfortunately shorter key length and therefore weaker encryption.

Table 9.1:
SSL Security

SSL	US-Version	Exportable version
data encryption	RC2 - 128 bit key RC4 - 128 bit key IDEA - 128 bit key DES - 64 (56) bit key Triple-DES - 192 (168) bit key	RC2 - 40 bit key RC4 - 40 bit key DES - 40 bit key
data integrity	MAC based on MD5 and SHA	MAC based on MD5 and SHA
key exchange	RSA - 1024 bit, based on PKCS 1, Diffie-Hellmann and Fortezza	RSA - 512 bit, based on PKCS 1, Deffie-Hellmann and Fortezza
certificates	X.509	X.509
digital signature	RSA- 1024 bit with MD5 SHA DSS	RSA - 512 bit with MD5 SHA DSS
hash function	MD2, MD5	MD2, MD5

The next version of SSL will have a new name: Transport Layer Security (TLS).

9.5
Further Readings

AES Homepage

http://csrc.nist.gov/encryption/aes

This site is the homepage of the Advance Encryption Standard hosted by the National Institute of Standards and Technology. It contains detailed design documentations on Rijndael, information about the selection process that took place, links to several related sites, as well as the AES standard itself (FIPS 197).

Cryptography FAQ index

http://www.faqs.org/faqs/cryptography-faq

This site provides an index of cryptography related FAQs.

National Institute of Standards and Technology

http://www.nist.gov

This National Institute of Standards and Technology (NIST) is also responsible for the cryptographic strategy of the US government. NIST provides also a log list of publications related to security and cryptography.

PKI Page

http://www.pki-page.info

On this web page you can find excellent links to information on public key infrastructure (PKI) and cryptography.

RSA Security

http://www.rsa.com

RSA is one of the leading companies providing cryptographic solutions and utilities. RSA also owns some patents on several cryptography algorithms. On RSA's homepage you can find information about cryptography in general as well as cryptographic code and information about events organized by RSA.

Part III
Connecting the World

Communication and interoperability between devices, networks, and back-end systems is an ubiquitous task when developing pervasive applications. Phones interact with handheld computers, home appliances transmit information through a power line, and navigation systems plug into in-vehicle networks (Figure III.1). Many similar examples have already been described in this book.

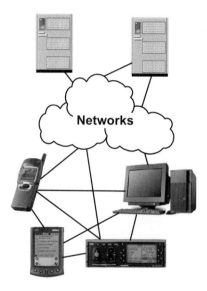

Figure III.1:
Boundless
Connectivity

This part explains the technologies and standards, which are applied to connect the different devices. Three aspects will be covered:

- The various physical layers over which information is transmitted,
- the communication protocols, and
- the mechanism to achieve interoperability on the application level.

Physical layers include infrared, phone line, twisted pair cable, radio, and power line, just to mention a few. Especially the various kinds of wireless communication technologies play an important role, since they are suitable for mobile devices and require only minimum installation efforts by the end-user.

Many of the common protocols base on Internet related standards. Among them are the Hypertext Transport Protocol (HTTP) and the Wireless Application Protocol (WAP). Both are used when devices communicate directly with server systems. Infrared Data Association (IrDA) or Bluetooth define protocols for local communication between devices. The applied protocols need to be robust and fault tolerant since mobile computing environments are extremely dynamic. Network connections can come and go, the network infrastructure, bandwidth and peripherals can change, and users might roam from one service provider to another.

The use of protocol independent data formats is an important way to achieve interoperability of applications. The Extensible Markup Language (XML) has evolved as the de-facto standard to exchanging content. But besides simple data exchange, it gets increasingly important to access network resources without knowing the details of local system configurations in advance. Some network resources are physical devices, like printers. Other resources are software based services, such as security key management, or application libraries. The number of network based services, which will become available in the future, is expected to grow rapidly as the Internet itself becomes even more pervasive.

10 Internet Protocols and Formats

This chapter provides an overview of standard protocols and formats used for data communication on the Internet.

10.1
Hypertext Transfer Protocol (HTTP)

The Hypertext Transfer Protocol (HTTP) was one of the first World Wide Web technologies. It is used by the global information initiative since 1990.

It is an application-level protocol for distributed systems. The HTTP protocol follows the request/response paradigm, like many other protocols in the networking area, but it allows more than just simple data retrieval. A client sends a request to the server, followed by a message containing request modifiers, client information, and possible body content. The server responds with a status line containing the protocol version and an status code, followed by the server information, meta information, and possible entity-body content.

Usually HTTP takes place over TCP/IP (Transmission Control Protocol/Internet Protocol), however HTTP is not dependant on TCP/IP.

An HTTP request consists of the HTTP method (GET, HEAD, POST, etc.), the Universal Resource Identifier (URI), the protocol version and an optional supplemental information. The method is executed on the object named by the URI. In the simplest case, a single connection is enough to establish the communication. The connection is established by the client prior to the request and terminated by the server after the response is transmitted.

HTTP methods

An HTTP response consists of a status line, which indicates success or failure of the request, a description of the information in the response (called meta information) and the actual information request.

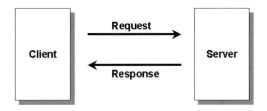

Figure 10.1:
HTTP
Connection

In the client-server communication chain intermediaries can be present. Usual intermediaries are gateways, tunnels, or proxies.

- A proxy is an agent for forwarding a client's request. The proxy rewrites parts of the received message and forwards it to the server.
- A gateway acts as a receiving part in the request chain. If necessary, the gateway translates the request to a protocol, the receiving server can understand.
- Tunnels are usually used when the communication takes place across a firewall. The tunnel acts like a relay between connections and doesn't change the message.

Figure 10.2: Client-Server Communication Chain

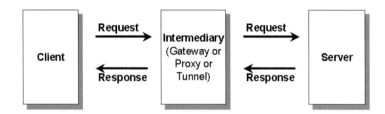

In order to identify a network resource, HTTP uses a Uniform Resource Identifier (URI). The resource is identified by its name, location, or other characteristics.

MIME   To identify pieces of information which cannot be handled by the browser itself, Multipurpose Internet Mail Extension (MIME) types are used by the Web browsers. The Web browsers can be configured to map special MIME types to particular functions, like plugins, multi media programs, etc.

MIME content-types are sent by the HTTP server in header messages before it delivers the actual files to the browser. To do this, the server usually uses a configuration file which maps the filename extensions to the MIME types. In case a server does not send the MIME type information, the Web browser has to relate on its configuration list, where typical file extensions are mapped to the appropriate type of data they contain.

10.2
Hypertext Markup Language (HTML)

To publish documents over the Web, which have to be processed by different browsers, a standardized language is needed. HTML is an universal representation which can be used by any Web browser.

Such a language is the Hypertext Markup Language (HTML). HTML first became popular during the 1990s with the explosive growth of the Web. The first definitive version however was HTML 2.0 in 1995. HTML 3.0 proposed many upgrades of several features but was never implemented because of the missing consensus in standards discussions. HTML 4 includes the support of style sheets, scripting, frames, embedding objects, and some of the browser proprietary extensions. HTML not only provides the means to build documents with elements known from other word processors. It also enables authors to retrieve online information via hypertext links, to design forms for conducting transactions with remote services, and to include different kinds of multi media information into documents.

Thus, HTML can specify the logical organization in a document. It is not intended to act like a WYSIWIG word processor because HTML has to be processed by many different Web browsers.

HTML instructions are called tags. Tags are surrounded by angle brackets (e.g. <BODY>). Tags can be differentiated in "Empty Tags", "Opening Tags" and "Closing Tags". *Tags*

"Closing Tags" contain a leading slash character. Most tags mark blocks of a document for a special purpose. The above <BODY> tag for example marks the beginning of the body section of a HTML document. The tag </BODY> marks the end of the body section of that document. Elements (like <HR>) do not affect a block of a HTML document and have no "Closing Tag". <HR> for example would draw a line across the page.

Tags can contain attributes within their brackets. Attributes are used to influence the tag behavior by passing parameters to the interpreter handling the tag. Attributes are placed between the tag name and the right angle bracket of the tag. Tag name and attribute name are separated by a blank. The attribute name and its value are separated by a "=" character. *Tag attributes*

The HTML document is structured into two parts, the HTML HEAD and the HTML BODY. Both parts are contained within HTML tags. The HEAD contains general information about the document, like its title. The BODY contains information, which will be displayed in the browser. A typical HTML document may look like this: *HTML Document Tags*

```
<HTML>
<HEAD>
<TITLE>Test Document</TITLE>
</HEAD>
<BODY bgcolor="#FFFFFF">
<h3>Hello World</h3>
<p><strong>Description</strong></p>
<p>This is an example HTML page.</p>
.

.

.

</BODY>
```

HTML HEAD   The HTML `HEAD` contains meta information about the document, like the author name, the document name etc.

One element in the document header is the title of the document. Another important element is `<LINK>`, which can be used for interconnecting documents.

HTML BODY   The `BODY` contains the content of the document. This can be ASCII text or some tags for structuring the document.

Some examples for structure elements are the headline tags `<H1>` to `<H6>`, which provide different levels of headlines, the `<P>` tag to start a new paragraph, the `<HR>` tag to draw a horizontal line, etc.

There are two major possibilities for Web browser to determine the type of the document it receives:

If it gets a document from an HTTP server, the server tells it the type of the document. In other cases, the Web browser identifies the document type by the file extension (e.g. ".html", ".gif" etc.).

10.3
Extensible Markup Language (XML)

The Extensible Markup Language (XML) is a standard format for interchanging structured documents. It is a subset of the Standard Generalized Markup Language (SGML), which is defined in ISO 8879.

User Defined Tags   Unlike HTML, XML enables the creation of user-defined tags. Traditional HTML tags are extended by user-defined elements. Like in HTML, the appropriate tags mark the start and end of each logical block. The World Wide Web Consortium (W3C) published XML as an open specification.

Document Type Definitions (DTD)   Document Type Definitions (DTDs) are an important element of XML. They define the role of text elements in a formal model. The DTD can be used to check if a XML document contains valid tags and if the tags occur in the right place within the document. It also specifies the attributes that belong to an element and the valid values of these attributes.

Thus, DTDs have two main functions:

■ They specify which document structures can be used by the author of an XML document.
■ They specify which document structures have to be handled by parsers that process that defined kind of XML document.

In SGML, the use of DTDs is required. However, XML does not require the use of a DTD. A parser that processes a document without a DTD has to extract the relevant information from the document itself.

The statement that XML is "extensible" can be misunderstood. XML defines a syntax, in the form of a number of rules in order to define document structures. These rules are defined in the XML specification and cannot be extended. "Extensible" means that special instances of tag languages can be built on the fundamental rules of XML.

The basic concept of XML is the composition of documents using a series of entities. The entities themselves are composed of one or more logical elements. The elements can contain attributes that describe the way in which the elements have to be processed.

There are some basic rules to determine if a XML document can be called "valid":

Properties of "valid" XML documents

- The XML document contains a header information, which consists of the XML version, the encoding information, and an information whether other files are referenced.
- The document contains a DTD with the markup declarations in the document itself or as a link to an external DTD document.
- The XML document contains a root element with the same name as the DTD name.

If only the last point is present, the document still can be called "well formed", which means that the structure of the document still behaves according the XML rules.

A "valid" XML document may look like this:

```
<?XML version="1.0" STANDALONE="YES"?>


<address>


  <name> Henry King </name>

  <street> Wherever St., 20 </street>

  <zip> 12345 </zip>

  <city> Anywhere City </city>


</address>
```

The corresponding DTD looks like this:

```
<!DOCTYPE address [
<!ELEMENT address (name, street, zip, city)>
<!ELEMENT name    (#PCDATA)>
<!ELEMENT street  (#PCDATA)>
<!ELEMENT zip     (#PCDATA)>
<!ELEMENT city    (#PCDATA)>
]>
```

Like in HTML, start and end tags are use to delimit each logical element (`<tag name>` content `</tag name>`).

The DTD tells the computer, that an address consists of a sequence of elements `<name>`, `<street>`, `<zip>` and `<city>`. If elements names are followed by a "?" (like `<!ELEMENT address (name?, street?, zip?, city?)>`), they are optional. Elements followed by a "*" are optional but can be repeated multiple times.

A "+" after the element indicates, that the element must occur at least once and also can be repeated multiple times (like `<!ELEMENT address (name, street*, zip*, city+)>`).

Leaf nodes of a document that can contain character data are identified by "#PCDATA", which means "parsed character data". In order to process elements in a particular way, attributes can be specified to apply certain properties to the elements. One special type of attribute is the unique identifier. It is used to provide cross references between two points of a document. These identifiers can be referenced by the use of attributes.

Text entities   Text entities are a further technique used by XML. Text entities are used to include commonly used text or characters outside the standard character set of the document. A text entity definition could look like this: `<!ENTITY city "New York">`. The entity reference `&city` can then be used in place of the string `"New York"`.

Other possibilities to use text entities are the incorporation of text or illustrations stored in another XML file or the use of non standard characters.

XSL   A further technology often used when working with XML is the Extensible Stylesheet Language (XSL).

XSL provides a style sheet syntax, which can be used to specify how a XML document is presented. XSL uses XML element types and is intended to be independent from any output format.

This means that a style sheet could be used for different output formats (like HTML, RTF etc.).

At the moment we can find a lot of emerging standards based on XML and HTML. One of them is XHTML.

XHTML   The World Wide Web Consortium (W3C) describes XHTML (Extensible Hypertext Markup Language) as "a reformulation of HTML 4 as an application of the Extensible Markup Language (XML)". XHTML is more or less a follow-on version of HTML 4; all markup tags and attributes of HTML 4 will be supported in XHTML 1.0.

However XHTML provides a few advantages in extensibility and portability. At a first glance, it looks like HTML, but it can be extended by anyone that uses it. Thus, new tags or attributes can be defined assuming some program that can understand and act on them. In XHTML, specific extensions are planned for some fields like mathematics and multimedia.

Extensibility doesn't necessarily mean that the Web pages become more complicated. XHTML also provides a means to simplify certain Web pages by a defined tag and attribute subset to meet the needs of small devices.

Another new XML based technology is VoiceXML. VoiceXML is set up for automated speech technologies, like digitized audio, speech recognition, etc. in order to bring the advantages of web-based development and content delivery to interactive voice response applications.

VoiceXML

10.4
XForms

The paradigm of separating purpose from presentation is not new in application development. However, the realization of this paradigm is up to the software designer and the current design of Web forms doesn't support separating purpose and presentation. Other limitations of Web forms are the limited accessibility features, the poor integration with XML and device dependencies.

XForms is W3C's name for a specification of Web forms that can be used with a wide variety of platforms. It aims to meld XML and forms in order to provide XML and Schema integration, universal accessibility, device independence and strong separation of purpose from presentation.

The XForm Model – a device-independent XML form definition – is able to work with a variety of standard and proprietary user interfaces. The XForms specification adopts the XML Schema data-types mechanism to provide additional data collection parameters. This, combined with form-specific properties, makes up the XForm Model.

XForm Model

A further concept of XForms is the XForms User Interface which provides a standard set of visual controls. These controls are targeted to replace today's XHTML form controls and deal with interactive data entry and display. XForms offer text controls, like `<input>`, `<textarea>` and `<secret>`, which behave essentially as in HTML, when attached to a string data-type. More interesting possibilities become available which richer data-types. List controls like `<selectOne>` and `<selectMany>` are also supported. hese controls provide a higher level of abstraction than HTML. Forms become usable acreoss a wide variety of devices, since each device can tailor the controls for their own requirements. XForms also presents some new controls, like `<output>` which provides the display of form data within ordinary inline text,or `<range>` which provides a continuous selection of a value within two extremes.

XForms User Interface

Forms are collecting data. Thus, an other, and maybe the most important concept of XForms is "instance data", which is an internal representation of the data mapped to the familiar form controls.

Instance Data

The XForms Submit Protocol provides a channel for instance data to flow to and from the XForms processor. This means, the XForm Submit Protocol defines the way how data is received and sent by XForms (Figure 10.3).

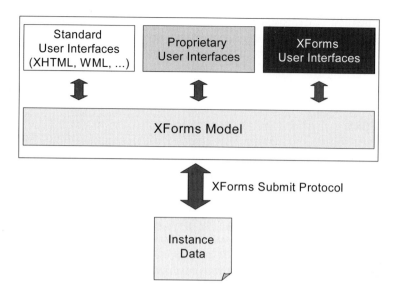

XForm processing is similar to XSLT processing but combines input and output into the same tree. In a first step, the instance data from an input source is parsed into memory. The input source can either be inline or an XML document on a server. The processing of instance data involves interacting with the user and recording of any changes in the data. Upon submit, the instance data is serialized, typically as XML, and sent to the server.

10.5
Further Readings

HTML

http://www.cc.ukans.edu/~acs/docs/other/HTML_quick.shtml

An HTML quick reference.

The Internet Engineering Task Force

http://www.ietf.org

Links to all kinds of Internet related documents.

World Wide Web Consortium

http://www.w3.org

Links to HTTP, XML, HTML, WAP, VoiceXML, XHTML and other standard protocol documents.

XML

http://www.xml.org

Resource for news, education, and information about the application of XML in industrial and commercial settings.

XForms

http://www.w3.org/TR/xforms/

W3C XForms Working Draft.

11 Mobile Internet

Compared to a wired connection like ADSL or cable modem, the bandwidth of today's wireless connections is reduced and wireless networks usually tend to be less stable and to have more latency.

Browsing multimedia HTML pages on the Internet from a wireless device like a GPRS mobile phone is getting better, but still not the same as from a 2 Mbps cable modem from home or the office. Receiving content via HTTP in HTML format with a data-rate of 53.6 Kbps (with GRPS) and displaying pages that were originally created for large high-resolution screens on a tiny display with only 96 x 65 pixels, is not perfect.

Two standards were established to bring the benefits of the Internet to the mobile world: WAP and i-mode.

In June 1997, Ericsson, Motorola, Nokia, and Openwave (formerly know as Unwired Planet and Phone.com) founded the Wireless Application Protocol (WAP) Forum as an industry group for the purpose of extending the existing Internet standards for the use with wireless communication. By June of 2002, the WAP Forum and the Open Mobile Architecture Initiative consolidated to the Open Mobile Alliance with members from all parts of the industry, including network operators, device manufacturers, service providers, and software vendors.

The WAP Specification Version 1.1 was released in the summer of 1999, and the first WAP devices and services were available in the fourth quarter of 1999. Most of the devices today support WAP Version 1.1, but the WAP Forum is currently working on Version 2 of the specification which acknowledges the increase in performance of wireless networks from 9.6 Kbps with GSM, over 53.6 Kbps with GRPS to UTMS with 2 Mbps and the new possibilities this gives to make WAP a real user experience with colored images, sound, and video.

As WAP is based on GSM it was introduced by a large number of GSM carriers all over the world. Around the year 2000, together with the Internet boom, were has a lot of hype also around WAP, which got back to normal levels as the Internet economic back got to more realistic expectations.

In parallel to this i-mode was developed and successfully introduced by NTT DoCoMo, the largest carrier in Japan. Early 2002, o2 in Germany was the first carrier in Europe also offering i-mode, followed by carriers in the Netherlands and Belgium. This chapter will begin with a closer look at WAP 1.1, as this is today support by most, continued by an outlook on WAP 2.0, followed by an introduction on iMode.

11.1
The WAP 1.1 Architecture

The Wireless Application Protocol (WAP) Version 1.1 defines specifications for the communication with wireless devices, like mobile phones or personal digital assistants (PDAs). The specifications of the protocol are based on existing Internet and network technologies and extend or optimize them for the use in a wireless environment.

Figure 11.1:
WAP
Infrastructure

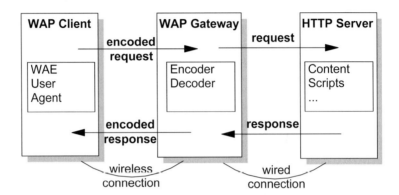

Figure 11.1 shows the usual infrastructure for a connection from a wireless WAP-enabled device to a standard Internet HTTP server. WAP protocols are used between the WAP client and the WAP gateway. Between the WAP gateway and the HTTP server, TCP/IP and HTTP are used. It is the responsibility of the WAP gateway to translate requests from the WAP protocol stack to the WWW protocol stack (TCP/IP and HTTP) and to decode requests sent from the WAP client to the server. The responses from the server are encoded by the gateway into a compact binary format, which the client is able to interpret.

The WAE (Wireless Application Environment) User Agent uses WML (Wireless Markup Language) as the format to display the content. The WAP gateway transforms the HTML documents received from the server into WML in case the server does not provides WML.

WAP 1.1 uses a layered model similar to the ISO OSI Reference Model. On top of the bearers, which are not in the scope of WAP, the WAP Forum has defined protocols for the following layers (see Figure 11.2):

- Transport Layer: Wireless Datagram Protocol (WDP);
- Security Layer: Wireless Transport Layer Security (WTLS);
- Transaction Layer: Wireless Transaction Protocol (WTP);
- Session Layer: Wireless Session Protocol (WSP);
- Application Layer: Wireless Application Environment (WAE).

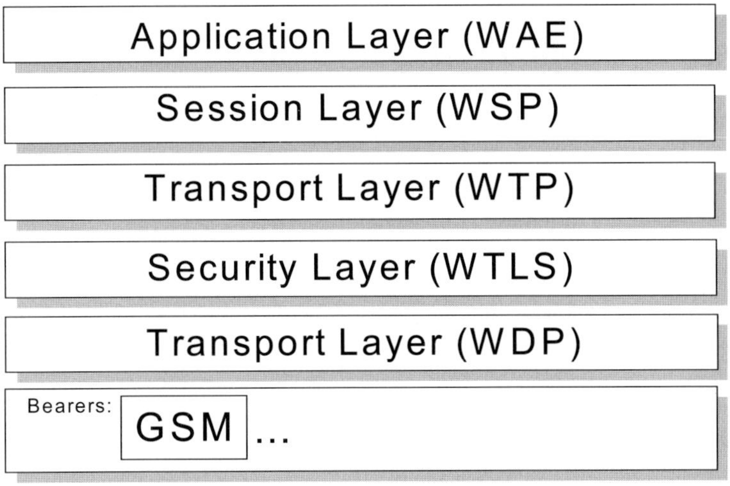

Application Layer (WAE)
Session Layer (WSP)
Transport Layer (WTP)
Security Layer (WTLS)
Transport Layer (WDP)

Bearers: | GSM | ...

Figure 11.2:
WAP Architecture

WAP is designed to be independent of the underlying network layer and to operate over a variety of carrier services, including Short Message, packet data, and circuit switched data. It handles the differences between the carriers with respect to error handling, throughput, and delays.

Network layer

The *Wireless Datagram Protocol (WDP)* layer sits on top of the transport layer and offers a consistent interface to the security layer. To do so, WDP has to handle the differences between the carriers and to define which carrier services are supported by WAP. This ensures the independence of the upper WAP layers.

Wireless Datagram Protocol

The *Wireless Transport Layer Security (WTLS)* specification is derived from the Transport Layer Security (TLS), formerly known as Secure Socket Layer (SSL), and is optimized for use on low bandwidth communication channels. Applications have the possibility of using WTLS depending on the security requirements and capabilities of the underlying network layer.

Wireless Transport Layer Security

The *Wireless Transaction Protocol (WTP)* provides a transaction-oriented interface to the upper layers. WTP supports unreliable one-way requests as well as reliable one- and two-way requests.

Wireless Transaction Protocol

The *Wireless Session Protocol (WSP)* provides a connection-oriented service on top of WTP. In addition, it provides a second connection-less service that is directly based on WDP. WSP currently supports services for browsing, like HTTP 1.1 functionality and semantics in a compact format for wireless connections, long-lived session state, session suspend and resume with session migration, and protocol negotiation. Data push will come with WAP 1.2.

Wireless Session Protocol

The *Wireless Application Environment (WAE)* enables the operators and service providers to develop interoperable applications for all WAP-compatible environments. WAE is based on the World Wide Web (WWW) as well as on mobile telephony technology. WAE is, from an application provider's and developer's point of view, the most interesting part of WAP and the following section will discuss it in detail.

Wireless Application Environment

11.2
Wireless Application Environment 1.1

The following specifications are part of the WAE:

- Wireless Markup Language (WML) – a markup language optimized for wireless communication channels and based on HTML as well as Unwired Planet's Hand Held Markup Language (HDML).
- WAP Binary XML Format (WBXML) – a specification for the binary encoding and transfer of XML documents in a WAP environment with the goal of reducing the actual amount of data transferred over the wireless connection.
- WMLScript – a scripting language based on JavaScript, which allows the execution of commands on the client to reduce the number of necessary turn-arounds over the network. In the WMLScript Standard Libraries Specification, WAE also defines a set of standard functions available on the client.
- Wireless Telephony Application (WTA) – a collection of telephony specific features for call and feature control mechanisms. The corresponding interfaces are defined in the Wireless Telephony Application Interface (WTAI) specification.
- Content formats – the data formats supported by a WAP environment, like calendar entries, images, and address book records.

Figure 11.3:
Distribution of
Components in a
WAP System

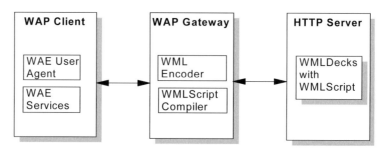

The interaction between a client and a server is described in Figure 11.3. The client contains a so-called WAE User Agent, like for example a micro browser, which is used to display the content and to enter information requested by the application. The User Agent is capable of displaying documents received in WML and executing procedures written using WMLScript.

A WML document is called a WML deck, which can consist of several cards. A card should contain the amount of information, which fits on the display of a device. A card could also contain WMLScripts to perform actions on the client.

To request data from a server, the user navigates through menus, which ulti-mately results into a WSP request to the WAP Gateway. The WAP Gateway re-ceives the request, transforms it into an HTTP request, and forwards it to the HTTP server.

The server performs the requested action and returns the deck to the WAP Gateway. If the server is not able to generate WML directly, it could also reply sending plain text to the gateway. In this case, the gateway transforms the plain text into WML decks. Now the gateway binary encodes the deck into WBXML to reduce the number of bytes to be transferred to the client.

WML Encoder

WMLScript contained in the decks is also compiled into byte code. First, this reduces the number of bytes to transfer and secondly takes workload away from the client. Performing work on the gateway instead of on the client also allows mobile devices with less computing power to be compliant with WAP.

WMLScript compiler

The client receives the binary encoded WML and optionally the WMLScript byte code, displays the cards and executes the scripts. The contained scripts might be used to perform input validation or other actions to reduce the number of turn-arounds between the client and the server to perform a transaction.

HTTP only supports pull-mode, which means that the server reacts only on requests sent by the client by returning information. WAP 1.2 also defines a push-mode, which allows the server to inform the client in case an important event occurred, like the drop of some stocks or an update in a database. Today, such functions are usually performed using SMS.

Push / Pull

We will take a closer look at WML, WBXML, and WMLScript in the fol-lowing sections.

Together with a request, the WAE User Agent sends some information about its capabilities in the WSP/HTTP header. With this data, the gateway and the server are able to transform and send the reply to the client in an appropriate format. In the `uaprof` parameter, the User Agent can add a URI to a profile describing its capabil-ities.

```
Accept:
  application/x-wap.wmlc;uaprof=http://www.supplier.com/dev,
  application/x-wap.wmlscriptc;uaprof=
                  http=//www.supplier.com/dev,
  text/x-vcard,
  text/x-vcal
```

Table 11.1 lists the header parameters that describe the capabilities and fea-tures of a WAE User Agent installed on a client. A complete description could be found in [WAESpec].

User Agent Characteristic	Description
Character set / encoding	supported character sets / encodings .
Language	supported language.
Media Type	supported contents / encodings.
WML Version	supported WML version, currently 1.0 or 1.1.
WMLScript Version	supported WMLScript version, currently 1.0 or 1.1.
Standard Libraries Supported	supported WMLScript standard libraries.
WTA Version	supported WTA version, currently 1.0.
WTAI Basic Version	supported WTAI Basic library version, currently 1.0.
WTAI Public Version	supported version of WTAI Public libraries, currently 1.0.
List WTAI Net-Spec version	supported WTAI net-spec libraries. Possible value are currently: 1.0GSM, 1.0IS-136, 1.0IS-95, 1.0PDC.

Content formats   The Wireless Application Environment supports the following content formats:

- Binary encoded WML as defined in the WBXML Specification;
- Binary encoded WMLScript as defined in the WMLScript Specification;
- Electronic Business Card Format (vCard 2.1) as maintained by the Internet Mail Consortium (IMC). Details can be found in [VCARD];
- Electronic Calendar and Scheduling Exchange Format (vCalendar 1.0) as maintained by the Internet Mail Consortium. Details can be found in [VCAL];
- Standard image formats, like TIFF, JPEG, and GIF as defined in the Wireless Session Protocol Specification [WSPSpec];
- WBMP WirelessBitMaP, as defined by the WAP Forum;
- Multipart messages and
- WTA events.

The Wireless Bitmap format is a generic bitmap format consisting of header information and the type-specific content. The header information provides the type, the width, and height of the bitmap as well as the WBMP version. Currently WAP has defined type 0, which is a black and white bitmap without compression. In the future, the WBMP format could be extended by adding new types to it.

11.2.1
Wireless Markup Language

The WAP Forum has defined WML as an Markup Language optimized for mobile devices. WML is based on the Extensible Markup Language (XML) as described in Chapter 10 of this book. It is also very similar to HTML, therefore writing WML documents is not very difficult.

The following section introduces WML with a simple example, which gathers some information from the user and sends it to a server for processing. In this sample, the user can query price information for a flight between two airports.

Without a device with a WAP browser, a WAP emulator available for desktop computers can be used to test applications. Even with a WAP-capable device, it might still be useful to test WAP pages on an emulator, which speeds up testing and saves on-air time.

Nokia offers the Nokia WAP Toolkit for download at the Nokia WAP Developer Forum. It contains an editor, a simulator, a micro browser in the design of an actual Nokia phone, a WML and WMLScript compiler, and a debugger. In addition to that the developer finds some samples as well as documentation. This is all one should need for the beginning.

Compared to the Nokia WAP toolkit, a smaller and simpler environment is for example the WinWap browser for Windows, which is available from Slot-Trot Software [SlotTrot].

Having obtained the required tools, it is now time to start writing a first small sample WML application. This application should allow the user to query the costs of flights between two cities. To do this, the user has to enter some personal information, like name and age, select the departure and the arrival airport from a list of available airports, and click a confirm button to transmit the data to the processing host. In this section, we will develop the required WML document to collect the information from the user.

WML is based on XML, and therefore every WML document must define the DTD that contains the definition of the elements that could be used in WML. The WAP Forum has defined a WML DTD and every WML document should point to this one. To do so, the following two lines must be added at the beginning of every WML document:

```
<?xml version="1.0"?>
<!DOCTYPE wml PUBLIC "-//WAPFORUM//DTD WML 1.1//EN"
  "http://www.wapforum.org/DTD/wml_1.1.xml">
```

WML documents are organized in decks and cards. Each .wml file contains one deck. A deck contains a collection of cards. A card is a container of text and input elements, which is flexible enough to be rendered by the WAE User Agent. Due to the different sizes and display capabilities of WAP-compatible devices, a card could be displayed quite differently on each device.

We start with a "Welcome" type of page. Every deck starts with the `<wml>` tag and ends with a `</wml>` tag. The content of a card is surrounded between a `<card>` and a `</card>` tag. Text could be displayed like in an HTML page. A paragraph starts with `<p>`. One difference between HTML and XML is that in XML, to have a valid document, each tag must be closed by an explicit end tag. For a paragraph it is the `</p>` tag. In HTML, the `</p>` is not required.

```
<wml>
  <card>
    <p>
      Welcome to WAP Flight price !
    </p>
  </card>
</wml>
```

On a mobile phone, the page could look as depicted in Figure 11.4.

Figure 11.4:
WML page on a
phone

We will now extend our application by asking the user to enter his name and his age. We will also put the name of our application ("WAP Flight price") into the title of the card.

In addition, we add a button to navigate to the next page by inserting a `<do>` tag. The `label` argument specifies the text the button should display. The `type` argument gives the User Agent a hint about the button's intended use.

Inside the `<do>` element, the `<go>` element is used to specify the next URL or card. In our case, we want to switch to the next card, the one with the ID card2.

Type	Description
accept	Positive acknowledgement
prev	back to previous card
help	display help
reset	clearing or resetting state
options	options for the current card
delete	remove item or choice
unknown	undefined, same as type=""

Table 11.2: Predefined Types inside the do *Element*

With the `input` element, the application can ask the user to enter data. The *Input* `name` attribute is used to specify the variable, which should contain the entered data. Either the type could be text; in this case, the entered data is displayed on the display or it could be password, in which case the data is not displayed.

```
<card id="card1" title="WAP Flight price">

   <p>
   <do type="accept" label="Goto next card" >
     <go href="#card2"/>
   </do>
   Please enter <br/>
   your Name: <input name="name" type="text"  format="*A"/>
   <br/>
   Your Age: <input name="age" type="text" format="3N"/>
   </p>
 </card>
```

The following code snippet shows you how to display the content of a varia- *Variables* ble. A variable is referenced by adding a $-sign in front of a variable name.

```
...
   Good morning  $(name) !
   Please enter your <br/>
   departure airport:
   <select name="dep">
     <option value="ORD">Chicago O'Hare</option>
     <option value="LAX">Los Angeles International</option>
     <option value="STR">Stuttgart (Germany)</option>
   </select>
...
```

In case the user should select one or more items out of a list of items, then the select element could be used. Inside the select element, the option argument is used to list the different items the user can select from. After the user made his selection, the variable dep contains the value specified in the option the user selected.

The user has now entered all information our server needs to process the request. The next and final step now is to send the information to the server. To do so, we create a separate card in which the user should confirm this action.

```
<card id="card3">
  <do type="accept" label="Start processing">
    <go href="/process" method="post">
      <postfield name="name" value="$(lastname)"/>
      <postfield name="age" value="$(age)"/>
      <postfield name="dep" value="$(dep)"/>
      <postfield name="dest" value="$(dest)"/>
    </go>
  </do>
  <p>
    Please confirm the data to start processing
  </p>
</card>
```

Using the go element, we specify the next WML deck that should be loaded. The postfield element inside the go element concatenates the variables names and values and adds it to the URL. Assuming the user has entered Hugo as the name, 57 as the age and wants the price information for a flight from Los Angeles to Stuttgart, then the complete URL will look as follows: /process?name=Hugo&age=57&dep=LAX&dest=STR. Now it's up to the server to process the data and return the right information to the user.

Besides the few elements we used in the sample, WML has of course a lot more to offer. There are for example tables to present information in a structured way. There are also templates that could be used to define a do element, which is common for all cards of a deck.

The complete list of all features and elements is part of the WML Specification [WMLSpec], which is available at the WAP Forum's homepage. Alternatively, the documentation that comes with the Nokia WAP Emulator is a very good source also.

11.2.2
WAP Binary XML Content Format 1.1

WAP Binary XML Content Format (WBXML) is used to reduce the amount of data to be transferred by binary encoding WML documents. During the encoding, the document is tokenized. This means that each tag is represented by a number and instead of transferring four bytes for a `card` tag, only one byte is transferred for the token. The same happens with attributes and entities.

All this is done by the WAP Gateway and the User Agent in the client and no developer of an application using WML has to bother with this.

In case you want to know how the encoding works in detail, then we would suggest you have a look at the WAP Forum's WBXML Specification [WBXML]. It is available for download at WAP Forum's homepage.

11.2.3
WMLScript 1.1

WMLScript was defined to enable the execution of scripts on WAP devices. The goal of using scripts on WAP devices is to reduce the number of turn-arounds between the client and the server, thus improving the performance of a WAP application. WMLScript could for example be used to validate data entered by the user before transmitting it over the network and to execute functions locally on the client. It could also be used to access local functions of a phone or an organizer, like retrieving an entry from the local address book or generating and sending a message. Updating the device configuration or the device software is also possible.

The rational for scripting

There is always a trade-off between placing functions on the client or on the server. Using the computing power of the server allows WAP devices to be simpler and reduces costs but increases the network traffic and the time to process a function. Executing functions on the client requires on the other hand more powerful clients, which increases production price but improves the performance of WAP applications by executing functions locally instead of contacting the server.

One of the major extensions the WAP Forum has made is the definition of the WMLScript byte code and the byte code interpreter. The WAP Gateway compiles the WMLScript into byte code before transmitting it to the client. This allows faster execution on the client and due to the compact byte code format, fewer bytes must be transferred to the client.

11.2.3.1
A WMLScript Sample

We would like to use an example to explain WMLScript to you. It should simply compute the result of a formula.

WMLScripts are stored in files with the extension ".wmls". We call our file `compute.wmls`.

The script takes three parameters. The first parameter (`varName`) should contain the name of the variable which the calling card uses to reference the result, the second and third parameter are the input for our formula.

Variables   extern function compute(varName,p1,p2) {

The script uses the variable `result` to store the computed result temporarily. Variables are not typed. A variable is defined by the keyword `var` in front of the name of the variable. The scope of the variable is the remainder of the function in which it is declared.

```
extern function compute(varName,p1,p2)   {

 var result;


  // Compute the formula
  result = 2*p1*p2;


  // Return the results to the browser

     WMLBrowser.setVar(varName,result);
     WMLBrowser.refresh();
}
```

WMLScript also knows `if`, `while`, and `for` statements and each statement ends with a semicolon (;).

To display the results on the WAP device in the current card, we are using the two external functions `setVar` and `refresh`, which are both part of the WML-Browser library. The WMLBrowser library is contained in the WMLScript Standard Libraries, which we will cover in the next section. The function `setVar` writes the result of our calculations into the variable the card uses to display the result. With `refresh`, the browser is forced to update its context and to display the new content of the variable.

```
<?xml version="1.0"?>

<!DOCTYPE  wml  PUBLIC  "-//WAPFORUM//DTD  WML  1.1//EN"  "http://
www.wapforum.org/DTD/wml_1.1.xml">


<wml>
```

```
<card id="card1" title="Compute">
  <p>
    Please enter <br/>
    Parameter 1: <input name="p1" type="text" format="N"/>
    <br/>
    Parameter 2: <input name="p2" type="text" format="N"/>
    <br/> = <u>$(result)</u>
  </p>
  <do type="accept" label="Compute">
    <go href="compute.wmls#compute('result',
      '$(p1)','$(p2)')"/>
  </do>
</card>
</wml>
```

The source above shows the card, which is used to enter the two parameters and to call the script `compute` in the `compute.wmls` document. A WMLScript file can contain one or more functions. To call one function in a script file, the URL is used to specify the document, followed by a hash mark (#), followed by the name of the function. Parameters are passed to a function within brackets, separated with commas.

This is how this small script looks like after it was compile to WMLScript byte code. It now has a size of just 47 bytes compared to 449 bytes of the script source.

```
012D016A  00020001  01000763  6F6D7075  74650301  1A12000E  01220E02
220F030E  03370E00  0E030A01  04370A06  04373B
```

The complete WMLScript Specification [WMLScript] is available at the WAP Forum.

11.2.3.2
WMLScript Standard Libraries

The WMLScript Standard Libraries provide client-side functionality, which are accessible by WAP applications.

The following libraries are available:

■ Lang – contains a set of functions that are closely related to the WMLScript language, like creating a random number or converting a string to an integer or float.

- String – provides a set of operations on strings, like comparing two strings or replacing a part of it.
- URL – contains functions for handling absolute and relative URLs, like getting the port or the path or checking if a URL is valid.
- WMLBrowser – provides WMLScript with functions to access the WML context of the browser. These functions can be used to set or get variables of a WML card, go to another URL or to refresh the context, for example.
- Dialogs – this library provides typical user interface functions like prompting the user or sending an alert.
- Float – provides floating-point arithmetic functions.

Float is the only optional library and is only available on devices that support floating-point arithmetic operations.

More libraries are currently being defined by the WAP Forum. These libraries will, among others, allow access to the telephony functions of mobile devices.

A complete list with all the functions the different libraries provide can be found in the WMLScript Standard Libraries Specification [WMLSLibs].

11.3
WAP 2.0 Architecture

Version 2.0 extends WAP in a way that directly builds on existing Internet technologies such as IP, TCP, TLS, HTTP while at the same time still supporting WAP 1.0 devices and applications. It now also supports GPRS and 3G bearers and provides a more flexible and more attractive User Interface.

The WAP Application Environment, usually seen by the end user as the WAP browser, is now based on XHTML (eXtensible HyperText Markup Language). The WAP Forum has created the XHTML Mobile Profile which defines what parts of XHTML have to be support on WAP 2.0 devices.

Instead of a WAP Gateway, which had to transform WAP specific requests from a WAP device into formats used in the Internet, now only a WAP Proxy is needed which has to forward the requests to a web server on the Internet.

Figure 11.5:
WAP 2.0 Gateway

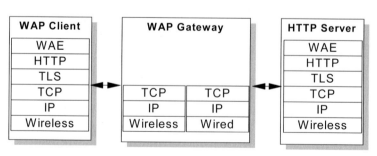

WAP 2.0 also introduces WAP Push, a way that a server can initiate a session with a the client. So far WAP sessions could only be instantiated by the client and work-arounds, like sending an SMS to a client which then starts a WAP session, were needed to provide this functionality.

WAP Push

WAP Push is especially important for applications providing customers with real-time information. Also there is no need to query a mailbox for incoming messages anymore; the server could push the new message to the device as soon as it arrived.

SyncML

With this version WAP defines a standard for data synchronization: SyncML. SyncML is the leading standard for data synchronization defined by the SyncML Initative, which is also mandated by standards like 3GPP. More details on SyncML can be found in Chapter 17.

MMS

Multimedia Messaging Service (MMS) is the "multimedia" version of SMS (short messaging service). It enables a user to quickly mail messages which can also contain images and sound besides text, of course.

These three are only the most important new features. In addition to them WAP 2.0 also brings:

- User Agent Profile (UAProf): A new service to better describe a device's capabilities to a server to allow the server better tailor its services for that device.
- Wireless Telephony Application (WTA): A tool to handle calls from with the WAP environment.
- External Functionality Interface (EFU): A way to plug-in enhancements and extensions in the WAP Application Environment (WAE).
- Persistent Storage Interface: A interface to store and organize data on a device.
- Provisioning: A way to pre-configure a device remotely by a service provider.
- Pictogram: Last but not least, a set of standard images, like ☺ or ☹, to be used within an application or message.

11.3.1
XHTML Mobile Profile (XHTMLMP)

The W3C defined XHTML modules to allow applications and devices which are not capable of supporting the full set of elements to work with a well defined support.

The WAP Forum defined a Mobile Profile for XHTML as the base from mobile devices supporting WAP.

```
<!DOCTYPE html
PUBLIC "-//WAPFORUM//DTD XHTML Mobile 1.0//EN"
"http://www.wapforum.org/DTD/xhtml-mobile10.dtd">
<html>
  <head>
    <title>Sample XHTMLMP document</title>
  </head>
  <body>
   <p>Sample paragraph</p>
  </body>
</html>
```

XHTMLMP consists of XHTML Basic and some additional XHTML modules. XHTML Basic contains the following elements: Strcuture, Text, Hypertext, List, Basic Forms, Basic Tables, Image, Object, Metainformation, Link, and Base. In addition to this WAP supports: Style Sheet, Style Attribute and parts of Forms, Legacy, and Presentation.

In the above sample the DOCTYPE element for a WAP XHTMLMP document must point to the DTD as defined by the WAP Forum. This is the only obvious difference between an XHTMLMP document and a "usual" XHTML document on the Internet, which makes it very easy to develop content for both Internet as well as WAP devices.

11.4
i-mode

In parallel to the efforts to establish WAP in Europe, NTT DoCoMo, a subsidiary of Japan's telecommunication giant NTT, introduced i-mode in February 1999 in Japan.

NTT DoCoMo based i-mode on existing Internet technologies like HTTP, HTML, and TCP/IP. I-mode content is developed in cHTML, which is a subset of HTML plus a few functions needed in a mobile phone environment to control the phone and for example make calls from within an application.

Figure 11.6:
i-mode phone
TS21i

Image courtesy of KPN Mobile

Right from the beginning i-mode was able to handle colored pictures, games, and sound. The penetration of mobile phones was already very high in Japan and the customers were willing to exchange their handsets for new ones very quickly.

This all together made i-mode a big success in Japan with about 35 million i-mode subscribers in Japan. This is over a quarter of Japan's total population.

In 1999 i-mode started with a network speed of 9.6 Kbps and the latest hand-sets are capable of downloading data with 28.8 Kbps. As with this tremendous success NTT DoCoMo's wireless network got close to its limit, a third genera-tion wireless network called FOMA was introduced in fall of 2001. FOMA pro-vides a typical download speed of 200 Kbps with a maximum of 386 Kbps.

Network speed

As i-mode is using cHTML (compact HTML), a derivative of HTML, devel-oping content for i-mode is pretty straight forward. It is like developing HTML pages, except that the developer should have the smaller screen in might as well as the lower connection speed.

Development

Emulators and editors for i-mode are available to make the development and testing easier. WapProfit (http.wapprofit.com) for example provides an i-mode emulator as well as an i-mode editor, even with a 30-day trial period.

Every i-mode phone has an i-mode button which brings the user to NTT DoCoMo's i-mode menu. In this menu the user finds a wide variety of content providers, which all must have a contract with NTT DoCoMo to get listed in the menu and must guarantee a certain level of quality. These i-mode partners are considered as official i-mode sites and are directly connected to the i-mode server. Every i-mode phone can also access the Internet and therefore access every site available on the Internet that provides content in cHTML for i-mode phones. These sites on the Internet are considered as unofficial i-mode sites.

11.5
Further Readings

Nokia WAP Developer Forum

http://www.forum.nokia.com/developers/wap/wap.html

Nokia provides a developer with a variety of tools and useful information, like the Nokia Mobile Internet Toolkit. It contains a set of tools needed for a quick start, like an emulator which also support WAP 2.0 with XHTML and CSS.

Open Mobile Alliance

http://www.openmobilealliance.org

The Open Mobile Alliance is the leading organization in the mobile world. The Open Mobile Architecture Initiative, the SycML Initiative, the WAP Forum, the Wireless Village, and the Multimedia Messaging Services Interoperability Process make up the Open Mobile Alliance.

OpenWAP.org

http://www.openwap.org/openwap/software.jsp

OpenWAP.org is a site focused especially on open source project around WAP. It provides links to open source implementations of a complete WAP stack and several WAP gateways.

Palo Wireless

http://www.palowireless.com

This "wireless resource center" is a good starting point to all sorts of information about wireless technology, like WAP, i-mode, SyncML, and many more.

WAPAG

http://www.wapag.com

This site allows to test of i-mode pages online with two i-mode phones: NEC21i and Toshiba21i

WAP.com

http://www.wap.com

WAP.com provides a collection of links to all kinds of WAP-related topics.

WAP Forum

http://www.wapforum.org

The WAP Forum is the industry initiative behind the Wireless Application Protocol, which has consolidated with the Open Mobile Architecture Initiative in June of 2002 to the Open Mobile Alliance. The WAP Forum defined WAP Version 1 and 2 still provides the specifications for download on their homepage.

wapprofit

http://www.wapprofit.com

WapProfit provides an i-mode emulators as well as editors for i-mode and WML.

12 Voice

As the access to the Internet becomes a fundamental necessity, alternative ways to go on-line without a computer are a requirement from mobile professionals. In addition, more companies are looking for new ways to extend the access to their data using voice technologies.

The last years brought an unbelievable convergence of communications and computers with the World Wide Web as its most important technology. With the simple access to the Internet, considerable amounts of information became available. Being connected to the Internet is a major interest as the pool of accessible information continues to grow and methods of information collection are improved.

Thus, the World Wide Web is causing us to look at telecommunications in a new light, where the major questions are:

- How can we deliver voice applications on standards-based technology?
- Is it useful and feasible to provide multi-vendor solutions that are both, practical and effective?

The primary method of accessing the Internet today is through the personal computer, which has certain advantages as well as some restrictions. However, computer-assisted access to the Internet is becoming less and less sufficient for professionals. Other chapters of this book have already illustrated possibilities for remote data access. This chapter shows some technology examples for access of data using voice interfaces.

12.1
Voice Technology Trends

Automatic speech recognition (ASR) has become very popular in the last few years. Earlier speech applications recognized only a small vocabulary, but the accuracy and vocabulary size of the automatic speech recognition machines has improved drastically. Today's systems partially support naturally spoken phrases even without previous training. Major vendors of speech recognition software are IBM, Nuance, Philips and Speechworks International. The heart of all modern high-performance speech recognition systems are statistical models that are able to characterize the sound properties of the language to be recognized. The process of speech recognition can be separated into several steps:

- Digitalization
- Representation
- Modeling

Digitalization

In the digitalization phase the incoming speech is detected by the telephony hardware. This stage must perform echo cancellation in order to allow the user to interrupt the prompt, i.e. to remove the echo of the outgoing prompt from the signal. To support speech detection in the presence of noise and residual echo, considerable computing performance is required. For this reason, the processing is mostly done by Digital Signal Processing cards.

Representation

In the representation phase, the input signal is converted into a spectral representation that shows the different aspects of the signal. These aspects of the acoustic signal are important for distinguishing different speech sounds.

Modeling

There are different ways of statistical modeling that can be applied to the speech sounds and are not described in detail within this book. Basic models are the segmental approach and the frame-based approach. The fundamental difference between these approaches is that the frame-based approach applies to uniform time segments whereas the segmental model takes into account the entire phonetic segment. Phonetic modeling is used to measure various properties of a speech signal and to apply probability distributions on each phonetic unit using statistical models. Training the speech technology product with previously collected speech data sets the parameters of the statistical models. This training allows the model to take account for the remaining variability in the speech signal.

Continuous Speech Processing

Continuous Speech Processing (CSP) optimizes the performance of host resources by preprocessing of voice data and enhances existing speech technologies with existing algorithms for large-scale systems.

CSP is a breakthrough in the support for large vocabulary, host-based speech recognition. Introduced by INTEL Dialogic, CSP technology allows establishing speech recognition applications more cost effectively.

Text-to-speech

As soon as information is accessed, it has to be communicated to the user. One way to do this is text-to-speech, or TTS. TTS is often used to speak e-mail and Web-based text to callers and will play a more important role in the future.

12.2
Voice on the Web

As the Web technology market exploded in the last decade and voice-based solutions came up using the rapidly advancing speech recognition technology, it was a logical step to combine these technologies.

12.2.1
Voice Infrastructure

Voice over IP (VoIP) gateways are servers that transform the analog audio signal into a digitized form suitable for transmission through an Internet Protocol (IP) network. Based on voice input, voice servers send HTTP requests to voice-enabled web applications. The applications return VoiceXML markup that the voice server converts into VoIP streams for transmission through the telephone network to users. VoiceXML is a base technology that links the world of voice technology with the Web space (VoiceXML will be covered in the next section). This information flow is illustrated in Figure 12.1.

*Figure 12.1:
Network
Components*

	Voice	VOIP	VoiceXML	
PSTN Network				
	Voice over IP Gateway	Voice Server	Web or other Application	

The voice over IP gateway processes the voice signal to produce a digital voice data stream. It then divides this digitized voice stream up into packets for sending through the Internet. VoIP also allows control information to be exchanged between the VoIP gateway and the voice server that allows calls to be controlled, set up, and closed. Voice over IP technology is important because it lets companies merge their data and voice networks.

Voice over IP gateway

The web application must be enabled to provide VoiceXML streams to the voice server. Either this can be done by changing the application code to directly handle VoiceXML, or technology can be used that automatically translate the usual application output to VoiceXML. One example of automatic translation technology would be transcoding, which is capable of translating HMTL markup into VoiceXML.

The voice server consists of several components as shown in Figure 12.2. It converts a VoiceXML stream from an application into digitized speech output and converts speech input into text.

The voice browser renders markup language input into speech just as a standard HTML browser renders markup for visual display. Instead of using a pointing device for input, the voice browser accepts text from a speech recognition engine. Based on the recognized text, it sends standard HTTP requests to the application in order to receive markup for a new voice "page". In this capacity, the voice browser operates in a manner similar to the way an HTML browser works when it reacts to mouse clicks. By speaking command phrases, the user can effectively click a link in order to move to a new VoiceXML page.

Voice browser

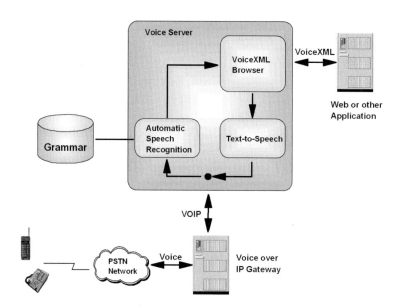

Figure 12.2: Voice Server

Automatic speech recognition
    The automatic speech recognition engine accepts a digitized voice stream as input and produces recognized text as output. A grammar is used to define the acceptable responses in a given context. For example, the allowed responses to the question "What fruit would you like delivered?" might be "oranges, grape-fruit, or tangerines".

Vocabulary
    The vocabulary to be recognized is an important variable that influences speech recognition quality and the training of a speech program for a particular user. A personal application program that recognizes speech for text processing requires a large vocabulary containing many words. Such programs typically need training for a specific user in order to achieve the required recognition quality. Programs that are to be used by many different users typically limit the allowed responses in order to achieve high-quality recognition without requiring training for each user.

Grammar
    Voice portals or other web applications are examples of such general-use programs. It would be unacceptable to require specific user training for such applications. The voice server recognition engines typically use grammars to exactly specify which words are to be recognized. This reduces the recognition vocabulary and improves recognition quality significantly.

Text to speech
    The text to speech component uses text input along with phoneme, pronunciation, frequency, and rate information to convert the text to spoken form. Phonemes correspond to individual sounds consisting of one or more letters that make up speech. For example, the "b" in "bat" and the "ou" in "house" are phonemes. The pronunciation information tells the text to speech component how the words are to be spoken, while the frequency information defines whether the synthesized voice will be higher or lower in tone. Female-sounding voices are typically about one octave higher in tone than male-sounding voices. The

rate information indicates how fast the words are to be rendered. By changing the rate, the text can be spoken slowly or more quickly.

12.2.2
VoiceXML

The "Voice Extensible Markup Language" is a standard that enables access to contents of the Internet via language and telephone. It was developed by a committee founded by AT&T, IBM, Lucent Technologies and Motorola. In this context, VoiceXML provides a specification language that is an application of the Extensible Markup Language (XML). To get to a working language-based Web application, a VoiceXML interpreter becomes necessary. This interpreter has to execute the VoiceXML code and to access the speech processing systems. Each VoiceXML document describes an interactive dialogue, which is executed by the VoiceXML interpreter. VoiceXML covers the following fields of application:

VoiceXML

- Telephone services
- Recording of spoken user inputs
- Speech recognition
- Synthetic speech generation
- Processing of audio files
- Recognition of dial tone impulses

A VoiceXML document consists of elements called "dialogs". We distinguish between two kinds of dialogs: "Forms" and "Menus". "Forms" are used for information description and for the reception of user information. "Menus" offer input options. A simple example of a VoiceXML document is shown in Figure 12.3.

VoiceXML document structure

```
<?xml version="1.0"?>
<vxml version="1.0">
  <form>
    <field name="name">
      <prompt>Please enter your Name?</prompt>
      <grammar src="name.grm "type="application/pvc"/>
    </field>
    <block>
      <submit next="http://www.pvcxt.com/info "/>
    </block>
  </form>
</vxml>
```

Figure 12.3: VoiceXML document

The field-element describes the input field. The processing of the following VoiceXML elements continues after a user input. The submit-element then transmits the input data to the info-Servlet. A special feature of VoiceXML is the definition of grammar files (like name.grm) for automatic speech recognition. The implementation of the VoiceXML interpreter has to be able to process the grammar files dynamically. VoiceXML does not define the format of the grammar file.

Selection menus

Selection menus are frequently used elements in telephone-based applications. The example shown in Figure 12.4 shows the use of the selection menus in VoiceXML.

VoiceXML is a well-defined and useful standard for combining speech and Internet technology. However, the quality of voice applications strongly depends on the systems that are used for speech processing.

Figure 12.4:
Selection menus
in VoiceXML

```
<menu>
  <prompt>
    Please enter your choice:
    <enumerate/>
  </prompt>
  <choice next="http://www.pvcxt.com/voice/banking.vxml">
    Banking</choice>
  <choice next="http://www.pvcxt.com/voice/news.vxml">
    News</choice>
  <noinput>
    Please make your selection:
    <enumerate/>
  </noinput>
</menu>
```

12.2.3
Voice Portals

Voice portals

As voice-based Internet technology is advancing rapidly, voice portals offer enormous opportunities for service providers. Voice portals can be reached by every telephone and combine the advantages of voice-based communication with Internet information-services – they can be defined as speech-enabled access to the World Wide Web. In other words, a voice portal is a way to provide Web access from a telephone.

The services of voice portals can range from a simple access of information to a communication with a virtual personal assistant. A voice portal can also provide a Web messaging front end that integrates Web access with services like voice mail, fax and e-mail. An example voice portal will be introduced in Chapter 18, Internet Portals.

12.3
Standardization

There are a number of organizations working to standardize web access through voice.

The VoiceXML Forum is an industry group founded by IBM, AT&T, Lucent, and Motorola to establish and promote VoiceXML as the standard markup language for voice applications. At present there are 57 promoter and 476 supporter member companies in addition to the four founding companies mentioned above. Promoter members include 724 Solutions, Aether Systems, Hitachi, SAP, and Intel. Supporter members include Bowstreet.com, Daimler-Chrysler AG, Nokia Corporation, Microsoft, and Ericsson.

VoiceXML Forum

The VoiceXML Forum is developing markup language specification that will be submitted to the World Wide Web Consortium (W3C) for standardization. In addition to specification work, the VoiceXML Forum provides educational materials about the specification in order to promote industry-wide usage. The resulting open systems will create a level playing field for all companies, and will lead to greater innovation and quicker adoption of voice technology.

The W3C organization supports a number of standardization efforts in the field of voice technology. The Voice Browser Workgroup within W3C drives this effort. The group is developing a suite of markup language specifications that will comprehensively cover interactive voice response applications.

World Wide Web Consortium

The markup languages under development include the Speech Recognition Grammar Specification (SRGS), the Speech Synthesis Markup Language (SSML), Semantic Interpretation for Speech Recognition, Call Control XML (CCXML), and VoiceXML.

Markup languages

SRGS defines syntax for describing words and phrases to be listened for by the speech recognizer. Grammars are important because they limit the potential user responses at a given point in the voice application. For example, an application asking the user what flavor of ice cream she would like delivered might limit the allowed responses to vanilla, chocolate, or strawberry.

SRGS

Another important use of SRGS is recognition of Dual Tone, Multiple Frequency (DTMF), or touch-tone dialer, signals. In some applications it may be convenient to allow either voice or touch-tone dialer input. Many people may find entering a personnel or social security number more convenient through a keypad than through speech.

DTMF

The Semantic Interpretation specification defines semantic interpretation tags that can be added to speech recognition grammars. These tags are designed to aid speech recognition engines recognize not just a list of words, but to identify the actual meaning intended by the speaker and representing that meaning in computer-readable form.

Semantic interpretation

SSML   SSML is a markup language designed to assist the generation of synthetic speech in web applications. Through SSML, applications can control aspects such as pronunciation, pitch, volume, and speaking rate. Another supported function is text normalization. Text normalization is used to resolve inconsistencies between the written and spoken forms of a language. For example, "$200" is usually pronounced as "two hundred dollars" rather than "dollars two hundred". SSML is based on the java Speech Markup Language (JSML) from Sun Microsystems.

CCXML   CCXML specifies elements that can be used by an application for telephony call control. It enables applications to support multi-party conference calls, to handle call transfer, to support multiple calls in parallel, and for handling asynchronous messages and events. Asynchronous events occur when a caller hangs up or a new caller dials in, for example.

VoiceXML, described more fully earlier in this document, leverages the other markup language specifications. It should be noted that there is a major difference between VoiceXML Versions 1.0 and 2.0. The markup languages listed above are companion specifications to VoiceXML Version 2.0. The older VoiceXML Version 1.0 specified some of this functionality directly, overlapping with the other markup language specifications.

12.4
Further Readings

IBM Corporation Main Site

http://www.ibm.com

This site provides a wealth of information about many computing topics.

IBM WebSphere Voice Zone

http://www7b.software.ibm.com/wsdd/zones/voice/

This is a one-stop information site for voice application developers using IBM voice technology. However, many resources are not IBM product specific and can be read by the interested general user.

IBM Voice Technology Page

http://www-3.ibm.com/software/pervasive/products/

voice/voice_technologies.shtml

Provides information about IBM voice technology.

VoiceXML Forum Web Site

http://www.voicexmlforum.org/

The VoiceXMl forum is dedicated to advance the use of the VoiceXML markup language. The forum web site provides tutorials, FAQs, specifications, and other materials of interest voice developers.

W3C Voice Browser Activity Site

http://www.w3.org/Voice/

This site, run by the World Wide Web Consortium (W3C), provides a wealth of information about voice browsers and VoiceXML.

13 Web Services

13.1
What are Web Services?

Today the Internet and private Intranets contain an enormous amount of data with is usually presented to a user via a webpage. These web pages are encoded using HTML.

```
<HTML
<HEAD>
<TITLE>Flights departing from Stuttgart Airport</TITLE>
</HEAD>
<BODY bgcolor="#FFFFFF">
<TABLE>
<TR>
<TD>DL 117
<TD>STR
<TD>ATL
<TD>11:05
<TD>15:30
</TR>
<TR>
<TD>LH 1105
<TD>STR
<TD>JFK
<TD>9:00
<TD>13:00
</TR>
...
</TABLE>
</BODY>
```

The example above shows a web page listing flights departing Stuttgart Airport at a given date with there current schedule. The data is arranged in a table and produces the following results for a user accessing that page using a web browser:

```
DL 117   STR ATL 11:05 15:30
LH 1105 STR JFK 9:00  13:00
```

This may not a very nicely designed web page, but it gives users the information they are requesting. Let's now assume a company wants to offer services to notify a customer about the delay of flights for example over SMS, pager or email. To do so this program would have to access this web page and parse the data provided here. This requires knowledge about the structure and layout of this web page, which might be different from airport to airport. Even worse, the application would have to be adjusted as soon as the airport makes a small change to the design of this web page and present the data in a different layout.

As seen in the example above, HTML does not differentiate between data and presentation. For an end user it is easy to interpret the information displayed on a page, but for an application it is hard to detect if some text is for example representing the departure time or the time of arrival.

Web services were introduced to overcome this problem and provide a "programming interface" to data on the web which allows the application to receive the data in a given, hopefully standardized format which allows easy processing.

13.2
Why should one use Web Services?

Web services provide a large potential for new or better integrated services on the Internet. Let's take the example of small loans where larger banks are currently starting to provide web sites which assess a customer's credit history without the involvement of a human being and at the end of the assessment offer the customer the possibility to sign a loan agreement, or not (depending on the results of the credit evaluation).

With a web service, the larger bank could offer this service to smaller banks which could easily integrate this in their web sites without requiring the investment of implementing this service completely on their own. For the customer it still looks like this service was completely offer by his local bank.

Or let's take google.com as another example. Google is now offering a web service as a new way to access their search engine. This allows Google's customer not only to start a search programmatically (which is already possible without web services by generating and posting the appropriate URL), but now an application can receive the results in a definite way without requiring saving and parsing the web page which was generated as a results of the search.

13.3
Web Services Architecture

In the web services world the following three components play an important role:

- Service provider: The service provider makes a web service available to another party to provide some service. The service provider does so by publishing a service to a service broker's registry.
- Service broker: The service broker helps the provider of the service and the consumer of a service to come together. Usually the service broker has as a central directory (like for example Yellow Pages) containing the available web services. The service provider publishes data about the web service he provides to the service broker's directory. The service requestor can search this directory using a specific protocol (this operation is called finding) which returns him the data that he needs to access the service directly.
- Service requestor: The service requestor is using a service provided by a service provider which he found using a service broker. The operation of a service requestor negotiating and accessing a web service provided by a service provider is called binding.

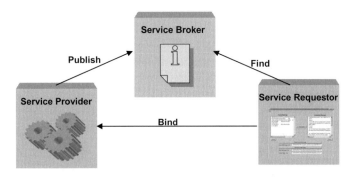

Figure 13.1:
Web Services
Architecture

In general the communication can happen over any protocol, but usually SOAP is used together with web services and UDDI is used as the central directory. SOAP is an XML-based protocol to access and execute remote objects. UDDI is a registry containing information about businesses and the web services they offer.

Any Web Services Protocol
Web Service Description Language (WSDL)
Universal Description, Discovery and Integration (UDDI)
Simple Object Access Protocol (SOAP)
Extensible Markup Language (XML)
HTML, TCP/IP

XML is described on greater detail in Chapter 10. A closer look at WSDL, UDDI, and SOAP is taken in the following sections of this chapter.

13.4
WSDL

The Web Service Description Language is used to describe web services and is based on XML. WSDL is based on a proposal from Ariba, IBM, and Microsoft and currently WSDL 1.2 is a W3C working draft.

A WSDL document has the following four elements:

- <portType> defines the offered service and which messages are involved. The communication can either be one-way, request-response, solicit-response (which means the sender will wait for the response) and notification.
- <message> defines each message involved the in communication.
- <types> defines the data types used.
- <binding> defines the protocol details and message formats for each port. The <binding> consists of a name and a type. The type defines the port for the service has to correspond with a <portType> definition.

The example below shows a snippet from a WSDL document which could describe the web service we introduced above in this chapter.

```
<message name="getFlightsRequest">

   <part name="term" type="xs:string"/>

</message>


<message name="getFlightsResponse">

   <part name="value" type="xs:string"/>
```

```
</message>
<portType name="flights">
  <operation name="getFlights">
      <input message="getFlightsRequest"/>
      <output message="getFlightsResponse"/>
  </operation>
</portType>
```

13.5
UDDI

The Universal Description, Discovery and Integration specification is describing the central database used to broker web services and allows businesses and services to be electronically published and retrieved.

UDDI.org develops and maintains the UDDI specification with the help of currently more than 200 members. Currently IBM, Microsoft, SAP, and NTT Com operate public business UDDI registries which could be used by anybody. Of course everybody could also setup his private UDDI registry, for example in his local network.

XML is used to send and retrieve data from a UDDI registry. UDDI.org developed a set of specifications which describes the protocol and XML data structures to be used. In addition to this an SOAP based API to access a UDDI registry was defined.

The following four types were defined to describe a web service:

■ Business information describes the general facts about the business offering one or more service. This could contain the name and address of the business, as well as details describing the experience of the business with the service that it offers. This data is stored in the "businessEntity" element.

■ Service information describes the service that a business has to offer, the corresponding data is store in the "businessService" element.

■ Binding information corresponds to service information and contains the relevant information to allow a service requestor to access ("bind") this web service at the service provider. This data resides in the "bindingTemplate" element.

■ Information about specifications for services points to specification describing the protocol and behavior that the web service expects. This information is contained in the tModel key. Web services can be offered for every possible service which could be made available over a network. Standards were and are developed to define a common behavior for each class of services. A closer look at "Web Services for Remote Portals" is taken later in this chapter as an example for a class of web services.

13.5.1
UDDI API

A SOAP based API was defined to access a UDDI registry. These APIs can be grouped in the

- Publisher's API which is used by service providers to publish a web service into the UDDI registry and the
- Inquiry API which is used by service requestor or consumer to find and bind a web service. The API consists of four APIs to search in the UDDI registry and four APIs to retrieve the details from it. A part of the Inquiry API can also be used to investigate web services invocation failures.

A service requestor has to take the following five steps to find and use a web service:

- Access a UDDI registry and find a business which provides the described service ("find_business"). The businessEntity elements are then retrieved with "get_businessDetail".
- Now that a business is selected, the next step is to search for a service ("find_service") and to retrieve more details about the service by requesting the businessService elements ("get_serviceDetail"). As these are sub-elements of the businessEntity element, an application could also simply retrieve a complete businessEntity element of a business including its businessService elements.
- The application now selects and saves the required part of the businessTemplate ("find_binding" and "get_bindingDetail").
- The application needs to select the appropriate protocol based on the data received about the standards that the selected web service complies to. The required data was accessed and retrieved from the UDDI registry using the "find_tModel" and "get_tModelDetail" APIs.
- Now the application has collected all details required to access the web service directly at the service provider and can bind the web service.

A service provider can save and delete each of the four UDDI keys (businessEntity, businessService, businessTemplate, and tModel key) with "save_business", "save_service", "save_binding", "save_tModel", and "delete_business", "delete_service", "delete_binding", "delete_tModel". In addition to this an API to query the information that was registered ("get_registeredInfo") and two APIs to logon and logoff from a UDDI registry ("get_authToken", "discard_authToken") are defined.

A Java class library to access UDDI directories is available as UDDI4J under the terms of the IBM Public License on http://www.uddi4j.org.

13.6
SOAP

The Simple Object Access Protocol is a simple way to access and execute objects over the network which was defined by the World Wide Web Consortium (W3C). Version 1.1 of SOAP was published in May of 2000 and the W3C is currently working on Version 1.2 which is due early 2003.

The example below shows a SOAP example embedded in an HTTP message. SOAP could be used over any transport, but the only one that the W3C defined a binding for is HTTP.

A SOAP message must at least consist of a SOAP envelope which is the bracket around the complete SOAP part of the message and a SOAP body. An optional SOAP header could be used for example for authentication, payment, session or transaction management. The SOAP header must come before the SOAP body, if used.

The body can also contain a fault element, which provides more details about errors which might have occurred while the previous message was processed.

```
POST /flights HTTP/1.1
Content-Type: text/xml; charset="utf-8"
Content-Length: nnnn
SOAPAction: "www.stuttgart-airport.com"


<SOAP-ENV:Envelope
  xmlns:SOAP-ENV="http://schemas.xmlsoap.org/soap/envelope/"
  SOAP-ENV:encodingStyle="http://schemas.xmlsoap.org/
    soap/encoding/">
  <SOAP-ENV:Body>
   <m:GetTodaysFlights xmlns:m="Some-URI">
          <query>All</query>
   </m:GetTodaysFlights>
  </SOAP-ENV:Body>
</SOAP-ENV:Envelope>
```

13.7
Web Services Security

There is no question that web services are the next step in the evolution of the Internet and they are already getting a lot of attention and success in the industry. But, one important part of the web services standards is missing: Security. So far every developer can choose between different proposals.

In April 2002 IBM, Verisign, and Microsoft have made a proposal how to secure web services and have given their proposal the name Web Services Security (WS-Security). Currently this is just a proposal and after detailed review by the industry, it is planned to submit this as a proposal for a new standard.

The proposal is based on a standard set of SOAP extension which would allow a common set of security mechanisms over all web services. WS-Security can work with various security mechanisms, like SSL, PKI, and Kerberos. WS-Security focuses on ensuring the integrity and confidentiality of web services transactions.

To complement and complete the set of standards required to secure web services, IBM, Verisign, and Microsoft suggest a set of additional specifications on top of WS-Security:

- WS-Policy to describe the capabilities of the security policies of the involved parties.
- WS-Trust to enable a framework for trust models.
- WS-Privacy for exchanging the privacy preferences of the involved parties.
- WS-SecureConversation to ensure the secure and authenticated exchange of messages.
- WS-Federation to manage and broker trust relationships.
- WS-Authorization to manage authorization data and authorization policies.

13.8
Web Services for Remote Portals (WSRP)

The Web Services for Remote Portals (WSRP) technical committee was formed in January 2002 at OASIS (Organization for the Advancement of Structured Information Standards) under the lead of IBM. The goal of this technical committee is to define a standard for interactive, user-facing web services that plug and play with portals.

The vision is that Portals can aggregate visual WSRP services which can be aware of the portal context. This context can be e.g. a user profile, a desired locale or markup-type or a user's device type. WSRP services enable businesses to provide content or applications in a form that does not require any manual adaptation. WSRP includes presentation, thus, service providers can determine how their content and applications are visualized to end-users and to which degree manipulation of the content may be allowed.

WSRP services can be published into service directories (UDDI) where they can be found by intermediary applications.

Making use of this mechanism, Portals can integrate content and applications from content providers (Figure 13.3).

The WSRP standard intends to define a web services interface description using WSDL and all the semantics that web services and consuming applications must comply with in order to be pluggable as well as the meta-information that has to be provided when publishing WSRP services into UDDI directories.

With this technology a set of interesting scenarios can be realized. One of these scenarios could be a Portal that shares Portlets with other Portals. An administrator could publish Portlets as WSRP services to UDDI. An administrator of another Portal will then be able to find and bind this WSRP service and enable it for the users of the Portal. In another scenario client applications can embed WSRP services through plug-in mechanisms like COM components or ActiveX controls.

Beside other tasks, the WSRP technical committee works on setting up an interface that enables the implementation of interactions between Portals and WSRP services. Figure 13.4 shows the lifecycle of interactions between a so called WSRP Consumer (e.g. a Client-Portal) and a WSRP Producer (service).

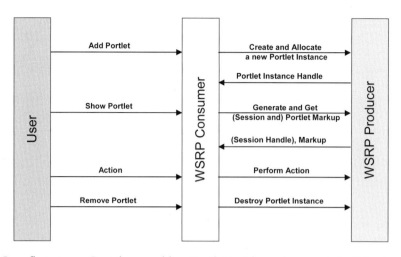

Figure 13.4:
Lifecycle of
WSRP
Interactions

In a first step, a Portal user adds a Portlet to his environment. In this phase, the Consumer invokes an interface method at the Producer side that allocates a new instance of the Portlet and returns the handle it. With this instance handle, the Consumer is able to request markup fragments from the Producer. These markup fragments are processed and displayed on the Consumer side. Before this step it might be necessary to create a session for the corresponding context. In a next step the user of the Portlet is able to perform actions (like clicking on buttons). These actions also invoke the corresponding interface methods at the Producer side. At the end of the Portlet life cycle the Portlet can be removed and the Producer's instance of the Portlet will be destroyed.

Like other Web Services, WSRP makes use of related standards, like different kinds of markup languages (HTML, WML, ...), UDDI, WSDL, and SOAP.

13.9
Further Readings

Apache Axis

http://xml.apache.org/axis

Apache provides an open source implementation of SOAP. Axis is the follow-on project of the Apache SOAP project.

Apache SOAP

http://xml.apache.org/soap

Apache provides an open source implementation of SOAP with is based on the IBM SOAP4J which was formerly available on the IBM Alphaworks website. The follow-on project is Apache Axis.

DeveloperWorks

http://www.ibm.com/developerworks

IBM provides this site to developers interested in the latest technology trends around Java, XML, web services, wireless, Linux, and open source. It contains tons of articles, tutorials, tools, source code, samples, news, forums, and so on for developers.

UDDI.org

http://www.uddi.org

UDDI.org is the organization maintaining the UDDI specification. On this website you find the specification, tutorials, as well as additional information about UDDI.

UDDI4J

http://www.uddi4j.org

UDDI4J is an open source projects to provide a Java class library for accessing a UDDI directory. Binary and source code, a bug tracker, documentation, access to the code library and a mailing list is available on this site. People who want to get involved in UDDI4J are more than welcome.

W3Schools

http://www.w3schools.com

This web site contains a broad set of free tutorial on all topics around the Web Wide Web including XML, SOAP, WAP, WSDL, .Net.

Web Services Description Language @ W3C

http://www.w3.org/TR/wsdl12/

This page contains the current working draft if WSDL 1.2 as the W3C plans to standardize it.

Web Services @ W3C

http://www.w3.org/2002/ws/

The Web Wide Web Consortium (W3C) is active in standardizing everything related to the Internet. This site lists all the current activities around web services in the W3C.

Web Services Architect

http://www.webservicesarchitect.com

This site provides articles, newsletters, and resources around web services.

Web Services Community Portal

http://www.webservices.org

This site provides all kind of news, like events, latest developments, announcements, directory containing web service vendors, links and more.

14 Connectivity

This chapter gives an overview of the relevant connectivity technologies for pervasive computing devices. Since analog technology plays a decreasing role in this application, only digital technology will be covered.

The topic of connectivity can be divided into three major themes. **Wireless Wide Area Networking** allows long distance communication through cellular radio. **Short-range Wireless** technology allows communication through radio waves or infrared beams up to a distance of a few tens of meters. Finally, **Home Networking** deals with communication between appliances and controls in a residential or small business environment.

14.1
Wireless Wide Area Networks

Digital cellular radio is the most interesting technology for pervasive device wide area connectivity. Cellular networks are widely deployed, to the extent of even allowing roaming between continents. Falling into the digital cellular communication category are a number of technologies that have differing characteristics and differing levels of deployment within particular geographic areas. Before getting to the topic of technology deployment, we will cover some basic ideas common to all digital cellular systems.

14.1.1
Cellular Basics

Cellular radio divides the geographic area into adjacent regions called cells, and places a radio transceiver, known as a base station, in each cell. A mobile phone communicates with the transceiver responsible for cell in which it is operating. This is illustrated in Figure 14.1.

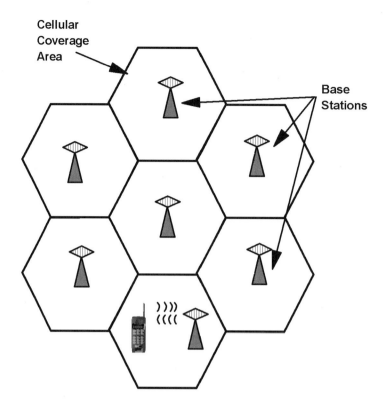

Figure 14.1: Cellular Area Coverage

Communication between the mobile equipment and the base station transceiver takes place using several channels that serve different purposes. When no telephone connection is active, the mobile equipment is still in constant communication with the base station transceiver through a control channel in order to monitor signal strength and provide status information to the base station. When a phone connection is active, a voice channel is activated in parallel with the control channel.

All transmission between the mobile equipment and the base station transceiver is in digital format. This is convenient for digital devices, but less so for a telephone. Since voice is an inherently analog signal, it must be converted to digital format for transmission. The voice signal is sampled and compressed using digital signal processing techniques.

The amount by which the voice data can be compressed depends upon the power of the Digital Signal Processor (DSP) built into the phone. This, in turn, determines the data rate required by the voice channel. The data rate available for voice transmission varies from system to system. The GSM system, for example, allows 13 kbps for the voice channel.

The DSP built into the mobile equipment also processes the radio signal received and acts upon commands sent through the control channel to the phone. For example, the base station transceiver monitors the signal strength of the mobile phone precisely, and can command the phone to increase its power level when the signal weakens. Since transmitter power level strongly affects battery life, the transmitter power level of the mobile phone is regulated by the base station very precisely.

14.1.1.1
Cells and Switching Centers

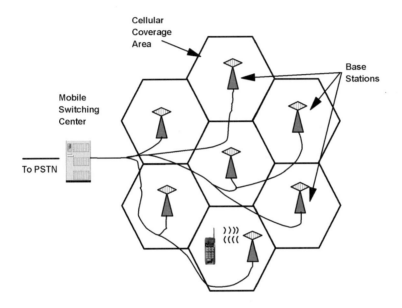

Figure 14.2:
The Mobile
Switching Center

The cellular system is organized hierarchically. The base station transceivers are connected to a mobile switching center (MSC). The mobile switching center is connected to the PSTN (Public Switched Telephone Network, also known as POTS – Plain Old Telephone System). The MSC is responsible for coordinating the work of a number of cells. When a mobile phone notifies a base station of its presence, the base station transmits the mobile equipment identity to the MSC.

When the mobile phone user wishes to place a call, the mobile phone signals the beginning of a call to the base station using a control channel. The base station sets up a voice channel circuit to handle the new call and relays the necessary descriptive information back to the mobile equipment, which then sets up the call.

Call placed to mobile phone

When a mobile phone number is dialed, the call is routed first to the MSC. The mobile switching center broadcasts a message to the cell in which the mobile phone was last located and possibly also to the surrounding cells informing the mobile phone that a call is pending.

Call from mobile phone

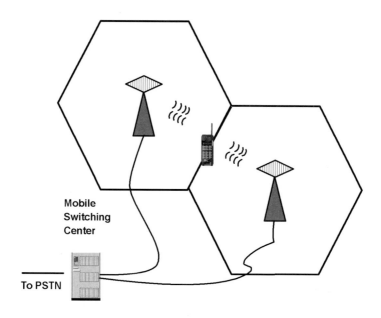

Mobile
Switching
Center

To PSTN

Figure 14.3 illustrates another situation that is handled by the mobile switching center. When a mobile phone moves from one cell to another, connection administration must be passed from one base station to another. This process is known as hand-off in the U.S. or handover in Europe. Since cell size at least within urban areas tends to be small in order to minimize the necessary mobile transmitter power level, hand-offs occur frequently as the mobile equipment moves from cell to cell.

Preparing for hand-off   One difficulty lies in recognizing when a hand-off should occur. The mobile equipment is in constant preparation for a hand-off. It prepares itself by listening to the available channels and building a list of the transceivers it can receive most strongly. This list is transmitted to its current base station.

The base station monitors the mobile transmitter signal strength. When the signal strength falls below a threshold level, the base station signals the mobile switching center to initiate a hand-off. The base station in the new cell is informed of its new charge, and the mobile phone is informed of the new channel information.

Hand-off with roaming   The situation is more complex when, say, the new cell is of the same technology but run by a different service provider. In this case, the MSC of the first provider will have to exchange billing and identification information with the MSC of the second provider (providing that there is a roaming agreement in place!).

There are a limited number of communication channels available to each cell. Furthermore, depending on the cellular technology used, it may be required that adjacent cells do not use the same communications channels. This implies that the cell size must be reduced in order to accommodate a larger number of mobile phones in one area, further increasing hand-off frequency.

14.1.1.2
Multiple Access

There are a number of methods for multiplexing data from many mobile devices into a given frequency spectrum. The three basic methods are Frequency Division Multiple Access (FDMA), Time Division Multiple Access (TDMA), and Code Division Multiple Access (CDMA).

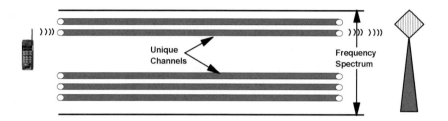

Figure 14.4:
Frequency
Division Multiple
Access

As illustrated in Figure 14.4, FDMA calls for dividing the available spectrum into unique carrier frequency channels. Each mobile device is assigned its own unique carrier frequency on which to operate. The frequencies are assigned using a control channel reserved for that purpose.

FDMA

Since analog mobile phones do not have any way of compressing the analog voice signal in the time domain, each analog phone require 100% access to a carrier frequency for operation. Analog cellular phone systems use FDMA to multiplex calls from multiple mobile phones into a given frequency spectrum.

Analog mobile phones

Digital mobile phones use sampling and digital signal processing techniques to compress the voice signal in the time domain – one second of voice signal can transmitted in much less than one second of real time.

Voice sampling & compression

A carrier frequency has a much higher data carrying capacity than required by a digitized audio signal. TDMA divides individual carrier frequencies into a series of time slots. Each device using the system is assigned a periodic time slot. The set of periodic time slots allocated to a particular mobile device can be thought of as a logical channel.

TDMA

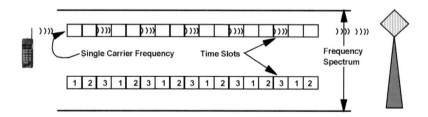

Figure 14.5:
Time Devision
Multiple Access

TDMA is often used in conjunction with FDMA. First, the available frequency spectrum is divided into carrier frequencies using FDMA, and then each carrier frequency is subdivided into time slots for use by the individual devices. A set of time slots on a single carrier frequency forms a logical channel.

CDMA   The Code Division Multiple Access multiplexing method is based on spread-spectrum technology. The data is divided into small packets and distributed in a predetermined pattern across the frequency spectrum. Many unique patterns for data distribution across the frequency spectrum are available. Each pattern is designated by a code known as a Pseudo-random Noise (PN) code. Each mobile device being serviced by a base station transceiver is assigned a unique PN code. This is illustrated in Figure 14.6.

Figure 14.6:
Code Division
Multiple Access

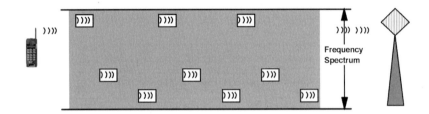

CDMA Chips   Neither entire packets nor entire bits are distributed across the frequency spectrum. Rather, information particles known as chips are distributed according to the PN code. Each bit is represented through the combination of a large number of chips. Since the representation of each bit is spread across the entire available spectrum in a manner that introduces redundancy in the signal, CDMA transmission is very reliable.

Pseudo-random   The PN code in CDMA corresponds roughly to a time slot number in TDMA
Noise (PN) code   or a carrier frequency in FDMA. Knowledge of the PN code allows the mobile phone to pick out its data from across the frequency spectrum. The set of all packets allocated for communication by a particular mobile phone makes up a logical channel.

Soft Hand-off   Since adjacent CDMA cells transmit and receive on the same frequencies, the same PN code can be used to identify the mobile phone in adjacent cells, allowing the mobile phone to communicate with two or more base stations simultaneously. This facilitates soft hand-off, which is a technique that results in higher-quality communication during the hand-off process.

14.1.1.3
Two-Way Communication

Half-duplex and   Old-fashioned walkie-talkies had a push-to-talk button that had to be held down
full-duplex   when speaking. The user could not speak and listen at the same time. This is known as half-duplex operation. Standard telephone systems allow the user to

listen to his or her partner and speak at the same time. This is known as full-duplex operation.

The half-duplex limitation is not acceptable for modern mobile phones. The user must have the feeling that he can talk and listen at the same time.

With analog cellular phones this could only be achieved through assigning separate frequencies for transmission and reception. The channel used by the mobile phone for reception from the base station is known as the forward channel, or downlink, and the channel used for transmission to the base station is known as the reverse channel, or uplink.

Downlinks and uplinks; forward channel and reverse channel

Digital systems can provide full-duplex operation in two different ways. Frequency Domain Duplex (FDD) mode calls for a separate carrier frequency to be assigned to the forward and reverse channels. Time Domain Duplex mode assigns different time slots on the same carrier frequency to the forward and return channels. TDD also has the effect of simplifying the logic in the mobile phone by removing the requirement for simultaneous transmission and reception.

FDD & TDD

Digital systems installed today use a mixture of FDD and TDD. They assign different frequencies to the forward and return channels and also stagger transmission and reception in time as illustrated in Figure 14.7. CDMA systems allocate separate frequency bands rather than single carrier frequencies to the forward and reverse channels.

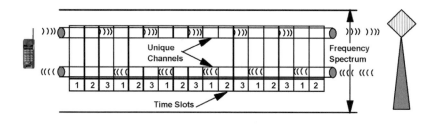

Unique Channels

Frequency Spectrum

1 2 3 1 2 3 1 2 3 1 2 3 1 2 3 1 2

Time Slots

Figure 14.7: Two-Way Communication

14.1.1.4
Short Message Service

Many cellular service providers offer Short Message Service, or SMS. This service allows mobile equipment to send and receive text messages of length limited to 160 single-byte characters or 70 double-byte characters (for Chinese and Arabic, for example). SMS can also be used as a transport medium for advanced protocols such as WAP (WAP is covered in Chapter 11 "Wireless Application Protocol").

SMS

The ability to receive short messages allows mobile phones to provide paging functionality in addition to voice service. Short message service is extremely popular. In December 2001 alone, 30 billion SMS messages were sent by the nearly 590 million GSM users worldwide. Worldwide traffic is expected to reach 100 billion messages per month by December 2003.

100 billion SMS messages

The end user usually pays for SMS messages on a per-message basis. The cost varies according to provider. In Europe, the per-message price is typically around _ 0.20, which works out to around 20 U.S. cents. Multiplying this amount by the projected monthly traffic indicates monthly revenue of about $20 billion for the service providers.

SMS transmission

Short messages are sent using store and forward technology. When a short message arrives, the mobile switching center first routes it to the short message service center, where it is stored while the MSC looks for the destination mobile phone. When the base station transceiver responsible for the destination phone has been located, the MSC retrieves the message from the short message service center and sends it to the mobile phone. When the mobile phone acknowledges that the message has been received, it is deleted from storage at the short message service center. This process is illustrated in Figure 14.8.

Figure 14.8:
Short Message
Mechanism

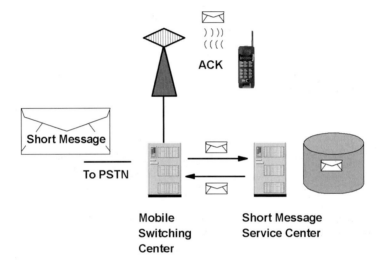

There are three types of short message delivery. With point-to-point delivery, the message is sent to a single mobile phone. The message is stored in the system until the phone acknowledges successful delivery.

With multi-point delivery, the message is delivered to several selected mobile phones. The MSC delivers the message to each mobile phone in turn and retains the message in the system until it has been delivered to all the addressees.

Broadcast delivery calls for the message to be sent to all mobile phones on the same broadcast channel. Broadcast messages are not acknowledged, so if a mobile phone is not available at the time of broadcast, it will miss the message.

14.1.1.5
Other Messaging Services

The SMS service described in the previous section is well-established and generating considerable revenue for mobile network operators. Next to SMS, other messaging standards have evolved to provide advanced functionality.

The Smart Messaging standard was created to allow provisioning of mobile phones through the network. Smart Messaging is typically used for downloading new ringing tones or operator logos into the phone. It can also be used for over the air provisioning of Internet access or WAP configuration information and bookmarks.

Smart Messaging

The Extended Messaging Standard (EMS) is a direct extension of the SMS standard. Multiple SMS messages can be concatenated to allow transport of larger blocks of information. EMS can be thought of as a stepping-stone to the Multimedia Messaging Service.

EMS

The Multimedia Messaging Service (MMS) can be viewed as an extension of the SMS service. Instead of sending only short lines of test, MMS will add the capability of sending still pictures, audio clips, and longer text messages. Over time, MMS will evolve, eventually allowing video clips to be exchanged between mobile devices as well as between mobile devices and personal computers.

Multimedia Messaging Service

Technically, MMS is conceived as store and forward system much like SMS. MMS messages originating either from a PC or from a mobile device are routed to an MMS Service center where they are stored while the target mobile device is located. However, unlike SMS, when the mobile device is located, a notification rather than the MMS message itself is transmitted to the device. The user can download the message at his convenience. After the device acknowledges receipt, the message is deleted from storage.

MMS transmission

If the mobile device is not immediately available (if it is turned off, for example), the message can be stored in the service center for longer periods. With SMS messages, the cost of storing the messages was low, since SMS messages are very short. However, with MMS messages containing image or even video data, the cost of storing data at the MMS service center could become significant, contributing to a higher per-message price.

MMS service is currently being rolled out in Europe. Prices for a still picture message are expected to range from 2-3 times the price of an SMS message.

Figure 14.9 shows the Nokia 6650 3G mobile phone. This phone has an integrated camera on the rear for taking still images at 640 x 480 pixel resolution as well as for recording videos at 128 x 96 bit resolution at up to 10 frames per second. This phone can operate in WCDMA 3G networks as well as in GSM networks through use of GPRS.

Photo courtesy of Nokia Corporation

14.1.1.6
Security

The telephone subscriber must be identified for billing purposes. This must be done in a secure manner to prevent fraud. In addition, privacy concerns call for secure methods of data transmission across the air interface. Cellular phone systems use cryptographic methods to implement these security features.

The first level of security concerns the mobile phone itself. Each mobile phone is assigned a unique identification code during manufacturing. When the mobile phone accesses a cellular system, it transmits its identification code to the mobile switching center. The mobile switching center to accesses its equipment database to check the equipment status. In general, the equipment status can take on the following values:

- **White-listed:** Access is allowed.
- **Black-listed:** Access is not allowed. This status can e set when the mobile phone has been reported as stolen, for example.
- **Grey-listed:** The mobile phone is under observation for possible problems.

Equipment identification

Status?

"White-listed"

Mobile
Switching
Center

Equipment
Database

The next level of security concerns the subscriber. Each subscriber is issued a unique security key. One copy of this key is stored at the cellular system authentication center and another copy is stored in the mobile phone itself.

SIM

The method used to store the secret key in the mobile phone is system dependent. The code may be preprogrammed into the phone at the factory, the user may key it into the phone, or it may be stored in a chip card module, known as a SIM (Subscriber Identity Module), which is inserted into the phone.

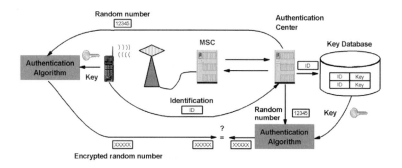

*Figure 14.11:
Authentication*

The cellular system uses an authentication process in order to verify the identity of the subscriber. This process is illustrated in Figure 14.11. Within the cellular system, authentication is carried out in the authentication center, which communicates with the mobile phone through the mobile service center.

*Challenge-
Response*

To begin the process, the authentication center sends an authentication request containing a random number (known in cryptographic circles as a "challenge") to the mobile phone. The mobile phone uses its secret key and a cryptographic algorithm to encrypt the random number. The mobile phone sends a response to the authentication center containing the encrypted random number along with identifying information.

The authentication center uses the identifying information to retrieve the secret key for the mobile phone from its key database. It uses this key to encrypt the same random number sent to the mobile phone using the same algorithm as the mobile phone.

The authentication center compares the encrypted random number returned by the mobile phone with the result of its own encryption process. If these numbers are the same, then the subscriber is authentic and is allowed to access the system.

This challenge – response authentication process is constructed so that the secret key is never transmitted between the mobile phone and the authentication center. An eavesdropper cannot use the transmitted random challenge and response to defraud the system, since a different random number is used each time the subscriber is authenticated.

The identifying information can contain information about the particular phone such as the number of calls made. This information is used in additional plausibility checks.

The cellular system also uses a cryptographic process to protect the voice data for privacy reasons. The secret key stored in the mobile phone is used to derive a session key from a random number. The random number may be the same one used in the authentication process. The data protection process is shown in Figure 14.12.

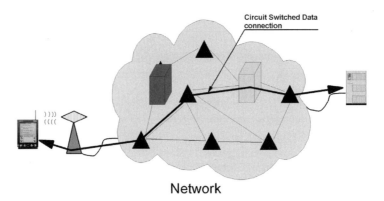

Figure 14.12:
Data Protection

Network

The session key changes each time a connection is established, so that even if an eavesdropper were able to crack the code and obtain the session key, it would not be possible to use that key to establish a new communications connection.

14.1.1.7
Connections

There are two major approaches for data transmission through a wireless network – circuit switched data and packet switched data.

Most cellular systems today use a circuit switched data approach. With this approach, an end point-to-end point channel for data transmission is reserved when a connection is established. Figure 14.13 illustrates a circuit switched data connection.

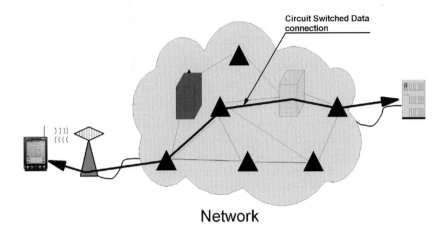

Circuit Switched Data connection

Network

Figure 14.13:
Circuit Switched
Data

The circuit switched data approach requires considerable overhead to establish a connection, but an established connection requires low overhead to maintain. The high overhead associated with establishing a connection may be acceptable for a voice connection that will be active for a longer length of time, but becomes unacceptable for data communication characterized by many short transmissions.

The routing path for the data is implied by the connection, so the individual data packets contain minimal addressing information. Since the circuit is reserved, a channel is always available when needed, simplifying the transmitter logic. In addition, the data packets arrive at their destination in a guaranteed sequence, so the receiving device requires no logic to handle out-of-order packets.

One disadvantage of the circuit switched data approach is that the channel remains reserved even no data is being transmitted (when nobody is talking, for example). This means that network capacity is not utilized fully.

With a packet switched data approach as illustrated in Figure 14.14, each data packet contains all of the addressing information necessary to reach its final destination. This is the same approach used by the Internet to achieve reliable and fault-tolerant data communication.

Each data packet takes its own route through the network. Since different paths through the network have different travel times, packets may arrive at the destination out of order. Under certain conditions, it can even occur that more than one copy of a packet or no packets at all arrive. This increases the complexity of the logic at the receiving end.

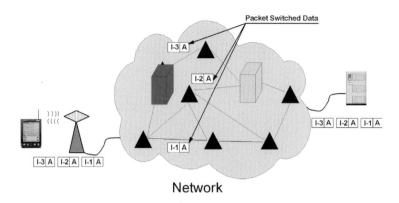

Network

With packet switched data, there is actually no connection as such. A device wishing to send a packet waits for an available slot and begins transmission. Although the overhead per packet is higher, a packet switched data approach can result in higher apparent responsiveness to the end user since there is no overhead involved with establishing a connection. Since no available transmission slots are wasted on reserved but unused connections, better use can be made of network resources.

As data-oriented applications such as web surfing become more prevalent, cellular systems will likely move more toward packet switched data services.

14.1.2
Major Digital Cellular Systems

There are four major digital cellular system types deployed worldwide.

- **PDC:** Personal Digital Cellular, previously known as Japanese Digital Cellular (JDC). This system is widely deployed in Japan.
- **TDMA:** Time Division Multiple Access, also known as IS-136. This system is deployed in the United States and South America.
- **CDMA:** Code Division Multiple Access, also known as IS-95. This system is widely deployed in the United States.
- **GSM:** Global System for Mobile Communications. This standard was developed in Europe by the Conference of Posts and Telecommunications (CEPT).

With an estimated 705 million subscribers worldwide as of June 2002, GSM is by far the most widely deployed and fastest growing system. CMDA follows with about 126 million subscribers, while TDMA and PDA have about 102 million and 59 million subscribers, respectively. The worldwide total number of digital cell phones is currently nearly 1 billion. This number is expected to double within the next 10 years.

This section will present some of the main features of the GSM system in order to provide an idea of the issues involved. Since many of the system-related topics are similar, only the salient features of the remaining systems will be presented.

14.1.2.1
GSM

In the early 1980's there were 9 analog cellular systems using 7 incompatible technologies deployed in Europe. The CEPT commission set up a working committee known as Groupe Spécial Mobile to develop a Europe-wide standard for mobile communication. In 1990 the first phase of the standardization work was completed, and the standard was renamed to Global System for Mobile Communications.

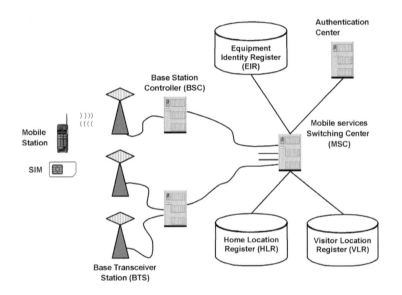

Figure 14.15: GSM System Components

This standard for digital cellular communication defines system components with well-defined interfaces in order to reduce the possibility of lock-in to a particular equipment manufacturer. These system components are illustrated in Figure 14.15.

The mobile station is composed of mobile equipment (often a phone) and a Subscriber Identity Module (SIM). The mobile equipment is programmed with International Mobile Equipment Identity (IMEI) information identifying the device type. All subscriber specific information such as International Subscriber Mobile Identity (ISMI), secret key for authentication and telephone book entries are stored in the SIM. This allows for subscriber mobility between mobile phones.

The mobile station communicates directly with a Base Transceiver Station (BTS) across the air interface. The base transceiver station provides the physical layer radio link to the mobile station. A Base Station Controller (BSC) can handle one or more base station transceivers. The BSC allocates the radio channels provided by the BTS and takes care of handovers between adjacent cells under its jurisdiction.

The Mobile Services Switching Center (MSC)[1] provides for mobility management. It allows mobile stations to access the system and acts as a gateway to the external PSTN. The MSC acts as a switch for calls between mobile phones and between mobile and fixed phones. The MSC also handles handovers between base station controllers.

The MSC uses information contained in the Equipment Identity Register (EIR) to determine if a mobile station is allowed to access the system and uses the Authentication Center (AuC) to validate the subscriber. This process was described in Section 14.1.1.5 Security.

The Home Location Register (HLR) is a subscriber database containing the subscriber identity and a subscriber profile as well as the current location of the mobile station. Each GSM system has a single home location register.

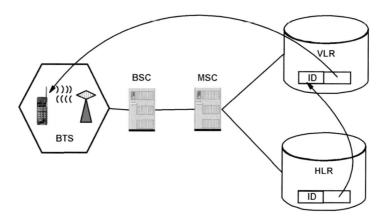

The Visitor Location Register (VLR) can be thought of as a type of working storage for administration of the active mobile stations within the jurisdiction of the corresponding MSC. The VLR contains a data record for each active mobile station.

When a mobile station registers itself with the system, its identification information is used to obtain data from the home location register. The MSC creates a record in the visitor location register for administration of the mobile station, and the home location register is updated to indicate the VLR responsible for

---

[1] The terminology used here is GSM specific. For this reason, it differs slightly from that used in Section 14.1.1 "Cellular Basics".

the mobile station. Figure 14.16 illustrates the situation where both the VLR and HLR belong to the same system.

If the subscriber is roaming out of bounds of his subscribed system, the locally available HLR will not contain the required subscriber information. In this case, the MSC must obtain the administrative data from a special MSC of the subscriber's system. This MSC is known as a gateway MSC.

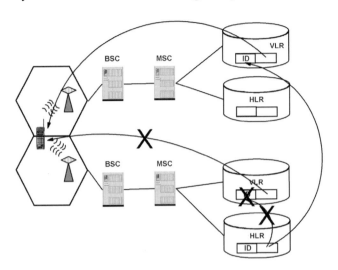

Figure 14.17: Roaming from MSC to MSC

Subscribers can also roam between cells controlled by different mobile service switching centers. In this case, the accepting MSC obtains the identification data from the original MSC and creates an entry in its VLR for administration of the mobile station. The original HLR entry is updated to indicate the new VLR. This situation is depicted in Figure 14.17.

Communication between mobile service switching centers, between the MSC and the base station controllers, and between the MSC and the PSTN uses Signaling System Number 7 (SS7), which is the same system used by the ISDN backbone system. This allows features such as fax, call forwarding, and caller identification to be offered.

GSM uses a combination of FDMA and TDMA to allow multiple access (see Section 14.1.1.2 "Multiple Access" for a discussion of these terms). The available frequency spectrum is first divided into carrier frequencies spaced 200 kHz apart. Each carrier frequency is then divided into slots that can be assigned to individual mobile stations or used to transmit control information.

Separate frequency ranges are used for transmission and reception. In the European 900 MHz band, the range 890 – 915 MHz is used for the uplink and the 935 – 960 MHz range for the downlink. The available 25MHz spectrum in each direction is divided into 124 carrier frequencies.

Figure 14.18:
The 900 MHz
GSM Band

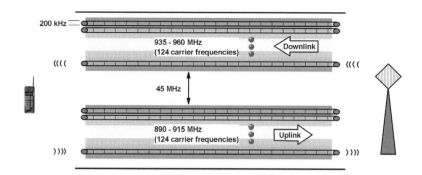

There are three main bands for GSM communication – 900 MHz, 1800 MHz, and 1900 MHz. GSM 900 is the most widely installed system globally. GSM 1800 is widely adopted in Europe and GSM 1900 is used predominantly in the Americas.

Voice Encoding

As someone once said, "the killer app for phones is voice". Since that is the case, the capability of the digital voice encoders determines the number of mobile stations that can be supported on each carrier frequency. A full-rate GSM voice encoder operates on a 20 ms speech sample to produce a 260 bit digital representation. This results in a data stream of about 13 kbps to represent the voice information. Half-rate voice encoders producing a data rate of about 7 kbps are also defined in the GSM specification.

Convolutional
encoding

Since radio transmission is subject to various types of interference, error checking and correction data must be added to the voice data stream to insure reliable communication. The convolutional encoding technique used for this purpose adds 196 bits to each 260 bit block for a total of 456 bits per 20 ms speech sample. This results a required data rate of 22.8 kbps.

Bursts

The raw data rate on each carrier frequency is about 270 kbps. The smallest unit of data transmission is known as a burst. Various types of bursts are defined, the most common being the normal burst. The normal burst is 156.25 bits in length and takes place in 15/26 (about 0.577) milliseconds. Each burst contains two 57-bit data blocks along with bits required for synchronization.

Frames

Eight such bursts make up a TDMA frame. Within a TDMA frame used for voice or data transmission, each burst is assigned to a different traffic channel. This results in eight full-rate traffic channels per carrier frequency.

26 TDMA frames are grouped together to form a 26-frame multiframe which has an internal structure. Frame 12 in each multiframe transmits control information (known as the Slow Associated Control Channel, or SACCH) and frame 25 is currently unused. This leaves 24 frames per 26-frame multiframe for voice or data transmission.

A 26-frame multiframe is sent every 120 milliseconds. Each traffic channel is allocated one burst per traffic frame, or 24 bursts per multiframe. This results in a data rate of 22.8 kbps, matching that required by voice transmission.

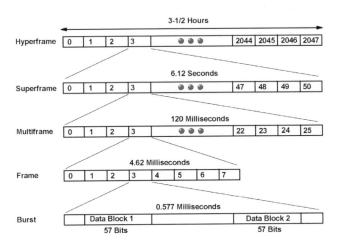

51 multiframes are packed into a superframe, and 2048 superframes make up a hyperframe. It takes about 3-1/2 hours to send a hyperframe. Within a hyperframe, each burst is specified uniquely by frame number and burst number within the frame. Figure 14.19 summarizes the relationships between frames and bursts.

The GSM standard defines a number of control channels that are used to synchronize the mobile station with the cellular network. Some of these control channels are associated with the traffic channels and others are transmitted on a dedicated broadcast frequency. Each base transceiver station has a dedicated broadcast frequency to carry control information. The burst and frame structure for the dedicated control channels differs from that used by the traffic channels.

Control channels

- **SACCH:** Slow Associated Control Channel. Used to regular transmission of control information, such as signal quality information between the mobile station and the BTS.
- **FACCH:** Fast Associated Control Channel. Used when a control message must be sent quickly. A data block normally used for voice data is used for transmission of control information. This results in loss of a voice data packet.
- **BCCH:** Broadcast Control Channel. Continually sends identifying and frequency allocation data from the BTS to the mobile station.
- **PAGCH:** Paging and Access Grant Channel. Notifies the mobile station of an incoming call. Since paging channel slots occur at regular intervals, the mobile station can sleep for the time interval between paging slots to conserve power.
- **RACH:** Random Access Channel. Used by the mobile station to request service from the system.
- **FCCH:** Frequency Correction Channel. Transmits a pure frequency burst from the BTS to the mobile station for synchronization purposes.

- **SCH:** Synchronization Channel. Transmits information about the current hyperframe from the BTS to the mobile station. The mobile station uses this information to synchronize its internal hyperframe counter with that of the cellular system.

14.1.2.2
IS-95 CDMA

The IS-95 CDMA system was designed to be compatible the existing AMPS (Advanced Mobile Phone Service) analog cellular system. Two new channels – the Digital Control Channel (DCC) and the Digital Traffic Channel (DTC) – were added to the existing Analog Control Channel and Analog Traffic Channel. This allowed dual-mode phones to be developed.

IS-95 Code Domain Multiple Access
As its name implies the CDMA system uses the Code Domain Multiple Access technique described in Section 14.1.1.2 "Multiple Access" to share the frequency bandwidth among many mobile stations. CDMA defines a radio channel with a bandwidth of 1.25 MHz to accommodate the digital transmission and allocates 64 unique Pseudo-random Noise (PN) codes for distribution of digital information across this frequency spectrum.

One 1.25 MHz radio channel is allocated for the downlink and a second for the uplink. Part of the downlink capacity is used to transmit synchronization and paging information to the mobile phone. Part of the uplink capacity is used to transmit access requests and other control information to the base station.

Channel Allocation
The entire 1.25 MHz radio channel has a data transmission capacity of 192 kbps. If this capacity were allocated evenly across 64 channels, it would result in a capacity per channel of 3 kbps, which would not be enough for voice data. Also, some of the channels are used for control information, further reducing the available channels for voice data.

Data Rates
The CDMA voice encoder produces digitized voice data at varying rates depending on speech activity. The achieved data rate ranges from 9600 to about 1200 bps. The transmission mechanism allows for this by varying the data rate available to each mobile station.

The data rate available for each mobile station may also be reduced to allow additional mobile stations to access the system. As the data rate is reduced, audio quality generally suffers, so data rate reduction can trade off audio quality for the ability to support a higher number of mobile stations. The ability to reduce the data rate is known as soft capacity limit, and allows 20 to 60 mobile stations to be supported on a single radio channel.

14.1.2.3
IS-136 TDMA

Like IS-95 CDMA, the IS-136 TDMA system commonly used in North America was also designed with compatibility to the existing AMPS analog system in mind. The TDMA specification also defines additional digital control and traffic channels. Many ideas from the GSM standard were incorporated into the IS-136 TDMA specification as it evolved from the earlier IS-54 specification.

However, TDMA uses a combination of Frequency Domain Multiple Access (FDMA) and Time Domain Multiple Access (TDMA) to share the frequency spectrum among multiple users. This is the same method used by the GSM system.

IS-136, IS-54 FDMA, TDMA

The TDMA system divides the available spectrum into radio channels each having a bandwidth of 30 kHz, which is the same bandwidth as an AMPS analog channel. Six time slots are allocated on each radio channel. Each time slot provides enough capacity for a single half-rate traffic channel, so that each 30 kHz radio channel can support up to six mobile stations.

Bandwidth

The base data rate for an entire 30 kHz radio channel is 48.6 kbps. After allowing for synchronization and control bits, each radio channel allows a full-rate channel data rate of about 13 kbps. Part of this remaining capacity is used for error detection and correction information, leaving about 8 kbps available for voice data transmission.

Data Rate

14.1.2.4
Japanese PDC

The Japanese Personal Digital Cellular (PDC) standard also uses a combination of FDMA and TDMA to share the frequency spectrum among multiple users. This system is quite similar to the IS-136 TDMA system.

Japanese PDC

The PDC system divides the available spectrum into radio channels each having a bandwidth of 25 kHz. Three time slots are allocated on each radio channel. Each time slot provides enough capacity for a single full-rate traffic channel, so that each 25 kHz radio channel can support up to three mobile stations.

Bandwidth

The base data rate for an entire 25 kHz radio channel is 42 kbps. After allowing for synchronization, control, and error detection bits, the available data rate per voice channel is about 8 kbps.

Data Rate

14.1.3
Advanced Cellular Radio Standards

Currently deployed second-generation digital cellular technology delivers data at rates of about 9600 bps. This is acceptable for voice data transmission and for data applications such as SMS that do not require large amounts of data. In order to offer the Internet web browsing, streaming audio, and streaming video applications of the near future, much higher data rates will be necessary.

The first-generation cellular technology was based on analog voice transmission. The second-generation technology was based on digital transmission, but was still heavily oriented towards voice data. Standardization work for third-generation cellular technology is nearly complete. Licenses have been awarded to service providers in many countries and deployment of the network infrastructure has begun.

3G Technology   The third generation, known as 3G, promises to connect up to 2 billion people worldwide by 2010 and offer data rates of up to 2 Mbit/second. Since 3G technology differs considerably from the current digital cellular technology, its deployment will require high investments on the part of the cellular system providers.

HSCSD, GPRS, and EDGE   As an interim step towards higher data rates, technology such as High-Speed Circuit Switched Data (HSCSD), General Packet Radio Service (GPRS), and Enhanced Data rates for GSM Evolution (EDGE) is currently being deployed. These are sometimes known as "2.5G" technologies. They use incremental advances in cellular technology to increase the capacity of the currently deployed network infrastructure. Figure 14.20 summarizes the evolution of wireless technology through the third generation.

Figure 14.20:
Digital Cellular
Technology
Evolution

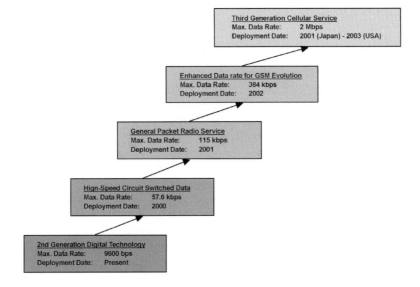

Packet switched routing   With higher data rate service, the cellular technology is also moving to packet switched data routing. Packet switched data routing offers a number of advantages. The available frequency spectrum can be used more efficiently since no transmission resources are allocated to a virtual circuit that may not be fully utilized. In addition, there is a greater immediacy since there is no connection setup overhead, as well as greater compatibility with the Internet protocol. Packet switched data routing will allow a "always on" user experience for data services similar to that provided by current web applications.

Packet switched by 2007   Industry representatives predict that by 2004 there will be more than 1 billion mobile phones deployed and that more than half the mobile wireless traffic will be packet switched data. Further, it is expected that all wireless traffic will be packet switched data by 2007 in the more advanced networks.

The current IPv4 Internet protocol addressing scheme uses 32-bit addresses to designate an Internet host. This limits the potential number of IP addresses to about 4 billion. The number of usable Internet addresses is considerably reduced through the internal structure of the Internet addresses.

Internet protocol addressing scheme

The next generation Internet protocol, IPv6, calls for 128-bit addressing, which will be enough to allow each device to have its own IP address. IPv6 will also define a quality of service scheme that will allow reliable delivery of real-time data such as voice. Wireless devices will likely migrate to the Internet addressing scheme as the Internet moves to IPv6.

IPv6

Most corporate networks use packet-based data routing in the form of the Internet protocol. Adoption of packet based service for mobile devices will allow seamless integration of such devices into existing corporate intranets.

14.1.3.1
High-Speed Circuit Switched Data (HSCSD)

HSCSD is the simplest step to higher data rates for GSM cellular network operators. Only minimal changes are required to an existing GSM network to enable HSCSD.

HSCSD increases the basic data rate of a single channel to 14.4 kbps. In addition, it allows bundling of up to four channels to enable data rates of up to 57.6 kbps for a single mobile station. The maximum data rate is close to the speed of a single ISDN channel. Figure 14.21 illustrates channel bundling.

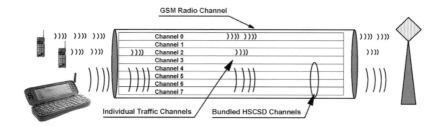

Figure 14.21: HSCSD Channel Bundling

The cellular operator can give voice data priority over HSCSD communication. When channels are available, the system will allocate additional channels to a high-bandwidth HSCSD device. When the system is busy, the system could allocate the HSCSD device a smaller number of channels.

For streaming audio (music) or video applications, HSCSD might be preferred over GPRS for reliability reasons. Once the HSCSD circuits are allocated, they are reserved for the mobile station until the connection is terminated. Due to this characteristic, fewer problems with lost or out-of-sequence data packets should occur.

Although moving to HSCSD is the easiest way to increase the capacity of an existing GSM network, many see it as a stopgap solution on the way to packet switched data.

14.1.3.2
General Packet Radio Service (GPRS)

The introduction of GPRS is the beginning of the packet switched data rollout for mobile devices. This technology will allow business users and private people to directly access e-mail, faxes, and Internet applications from mobile devices.

GPRS will support an "always-on" mode of connection. This characteristic of GPRS will provide a sense of immediacy to the user. It will not be necessary to dial a number and wait for a connection with a service provider to be established before accessing data applications. This characteristic will go far to further the acceptance of WAP.

Figure 14.22:
GPRS vs. Circuit
Switched
Connection

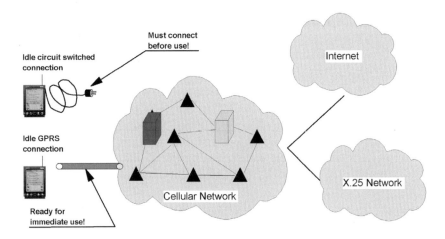

When using the Wireless Application Protocol (WAP) over a circuit switched network, the user must first dial into a service provider and then browse to the desired site to retrieve the information. This long dial-in time is often frustrating for the user. With a circuit-switched connection, the user is charged for the connect time. This encourages users to terminate the connection as soon as the desired information has been retrieved.

Packet networks   GPRS is designed for compatibility with X.25 and IP packet based networks, as shown in Figure 14.22. Due to the popularity of the Internet, connection with the IP-based networks will likely be emphasized initially. Due to this compatibility, a GPRS driver in the mobile device can appear as a network driver under a standard TCP/IP protocol stack. Typical Internet applications such as e-mail programs and web browsers can use GPRS transparently.

GPRS Support   Deployment of GPRS will require additional components in the cellular net-
Nodes   work. The Serving GPRS Support Node (SGSN) interfaces with the base station controller in the cellular network. It handles communication with GPRS devices. The Gateway GPRS Support Node (GGSN) routes traffic from the SGSN nodes in the system to the external X.25 and Internet packet networks. This is illustrated in Figure 14.23.

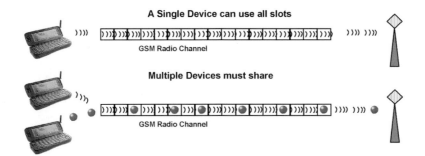

Figure 14.23:
GPRS Network
Components

Mobile services Switching Center (MSC)

To PSTN

Mobile Station

Base Station Controller (BSC)

Billing System

Serving GPRS Support Node (SGSN)

Base Transceiver Station (BTS)

Internet

Gateway GPRS Support Node (GGSN)

X.25 Network

Changes could also be required in the interfaces to the billing system. Since cellular networks have been connection-oriented before introduction of GPRS, billing has also been connection-oriented. With GPRS, it would be desirable to offer usage-based billing to reflect the connectionless aspect of GPRS. Such a billing system would charge the user according to amount of data transferred rather than for the length of time that the device is active.

GPRS uses the same GSM traffic channels as voice traffic, just as HSCSD does. Just as HSCSD, GPRS will also benefit from a system upgrade allowing 14.4 kbps data transmission per traffic channel. By using all eight traffic channels on a GSM radio channel for a single mobile station, GPRS can achieve a data rate of 115 kbps, which is faster than a standard fixed-line analog modem.

A Single Device can use all slots

GSM Radio Channel

Multiple Devices must share

GSM Radio Channel

Figure 14.24:
GPRS Devices
Sharing a Radio
Channel

At the beginning of GPRS deployment, each GPRS device will likely be limited to one or possibly two channels to avoid voice service deterioration. Later, when additional equipment has been installed to support more radio channels, service will be expanded to higher data rates.

GPRS is important to service providers as a stepping stone from 2G to 3G wireless technology because it introduces the important concepts of packet switched data and immediacy while reusing current network infrastructure.

14.1.3.3
Enhanced Data rates for GSM Evolution (EDGE)

EDGE changes the type of modulation used to encode digital data on the radio channel in order to achieve higher data rates. Current GSM systems use Gaussian-filtered Minimum Shift Keying (GMSK) to encode data at a raw rate of 270 kbps onto a 200 kHz radio channel. This results in a usable data rate of 13 kbps per traffic channel.

Figure 14.25: System Changes for EDGE

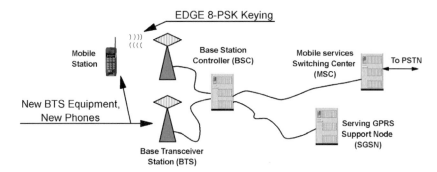

EDGE uses eight-phase-shift keying (8 PSK) to achieve a raw data rate of up to 473 kbps. After allowing for control and synchronization bits, this works out to a usable data rate of 384 kbps for the entire radio channel, or 48 kbps for each traffic channel. The change in keying accompanying EDGE will help prepare for deployment of third generation technology.

Since EDGE changes the fundamental way information is encoded on the radio waves, new base transceiver equipment will be required. In addition, new mobile stations will be required in order to take advantage of EDGE.

EDGE changes the air interface in the system. This change increases the raw data rate without changing the burst, frame, or channel structure of the information. The remaining system components will therefore require minimal changes. This implies that both GPRS and HSCSD will benefit from the introduction of EDGE. The maximum HSCSD data rate will be increased to 192 kbps, and the maximum GPRS rate will increase to 384 kbps.

The Universal Wireless Communications Corporation (the TDMA industry association) has developed a version of EDGE for use with IS-136 TDMA networks. This can be seen as a step towards convergence of worldwide cellular phone standards.

14.1.3.4
Third Generation Technology

ITU, WARC 92 The International Telecommunication Union (ITU) began standardization work for the third generation of wireless technology in the early 1990's. An initial study was presented to the World Administrative Radio Council in February 1992 (WARC-92). This study made an initial determination of the radio spectrum that would be required for third generation services. The ITU set a goal of completing the specification in the year 2000, so the standard was named International Mobile Telecommunications 2000, or IMT-2000.

The IMT-2000 workgroup defined a set of minimum requirements for a mobile radio standard. The goal was to create a single standard for future mobile communications allowing a truly global wireless infrastructure. The requirements call for global mobile phone coverage through use of satellite as well as terrestrial technology. The resulting infrastructure should provide users with high-speed digital mobile service from any point on the planet. This includes access from ships and airplanes.

IMT-2000

In the early 1980's, wireless service in Europe was provided by a number of incompatible systems. With GSM, a Europe-wide standard was created to promote compatibility among wireless systems. The worldwide wildfire adoption of GSM underlines the value of a single, global standard for telecommunication.

The mobile radio technology is especially interesting for developing countries. As of today, more than 80% of the world's rural areas have no coverage. The proposed satellite coverage will play an important role in expanding mobile service to these areas.

Through the single global standard and widespread deployment of the corresponding infrastructure, a user will be able to roam globally with a single handset. Standard service delivery mechanisms will insure that all applications will be available worldwide. The high data rates defined by IMT-2000 will enable applications such as streaming multimedia and real-time video conferencing.

If you wonder about the value of applications such as real-time video conferencing from the world's rural areas, just think of remote medical consulting, for example. A doctor on the road would be able to obtain expert advice visually regardless of location.

IMT-2000 characterizes target coverage in terms of environment, population density, speed of the mobile station with respect to the base station, and data rate. This is illustrated in Figure 14.26.

- **Pico-cell:** Calls for the highest target data rates of 2 Mbps. Mobile stations can be moving at pedestrian speeds at short range either indoor or outdoor.
- **Micro-cell:** Target maximum data rate of 512 kbps. Mobile stations can be moving at speeds up to 150 km/h.
- **Macro-cell:** Target maximum data rate of 384 kbps. Mobile stations can be moving at speeds up to 250 km/h.
- **Global:** Target data rate of 144 kbps.

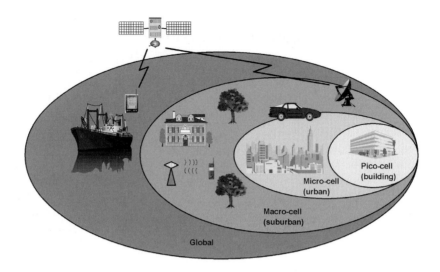

During the WARC-92 session, the IMT-2000 workgroup also proposed frequency bands for worldwide mobile communication. The proposed frequency bands allow for both terrestrial and satellite communication.

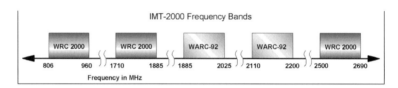

Unfortunately, some countries had already allocated spectrum in those bands to other purposes. Also, studies indicated that further airwave capacity would be needed to meet 3G radio demand. The World Radio Council convened in May 2000 to discuss this and other issues. WRC 2000 participants agreed to reserve three additional bands for third generation radio use.

The International Telecommunication Union makes recommendations about the use of radio spectrum. Country regulatory agencies have begun issuing 3G licenses. Some countries issue licenses based on technological and financial characteristics of the proposals while others sell the licenses to the highest bidder through an auction.

In early 2000, the UK Government staged an auction for 3G mobile phone licenses, expecting to net 3 billion pounds. The auction ended up bringing in 22.5 billion pounds for a total of five licenses.

The licenses must be awarded in 2000 to allow systems to be installed and tested in time for the projected start of service. The date for start of 3G service is country dependent. Service is anticipated to begin in 2001 in Japan, late 2001 – early 2002 in Europe, and 2003 in North America.

Submit Proposals

Evaluation

Consensus Building

Phase 1
Specifications

1997   1998   1999   2000   2001

Based on the requirements identified during WARC-92, the IMT-2000 workgroup issued a request for proposals. The IMT-2000 workgroup defined a schedule for specification completion as shown in Figure 14.28.

The boxes with the solid lines in Figure 14.28 indicate the original proposed schedule for creating the specifications. A total of 10 proposals for terrestrial radio and 5 for satellite based systems were submitted by June 30, 1998. These proposals have been evaluated and by the end of 1999 the IMT-2000 workgroup had made progress toward harmonization.

However, a number of issues prevented the harmonization and specification stages from being completed as planned. Difficulties in migration from existing systems, frequency spectrum availability in certain countries, and intellectual property rights issues hindered the harmonization efforts.

In March 2000, the IMT-2000 workgroup agreed to adopt a family of systems to provide 3G capability rather than continuing to work toward a single unified standard. The three proposals making up the family of systems are:

- **cdma2000:** Wide-band CMDA evolving from IS-95. cdma2000 was developed by the U.S. Telecommunications Industry Association (TIA).
- **UTRA:** UMTS Terrestrial Radio Access. UTRA was developed by the European Telecommunications Standards Institute (ETSI). UMTS stands for Universal Mobile Telecommunications System. It is being standardized by ETSI in the framework of the IMT-2000 proposals.
- **TD-SCDMA:** Time-Division Synchronous CMDA. TD-SCDMA was developed by the China Academy of Telecommunication Technology (CATT).

The family members are similar, and the IMT-2000 workgroup is now aiming to insure a high degree of commonality among the family members. This will enable interoperability and global roaming capability.

These systems all use CDMA technology for its superior spectrum efficiency and noise immunity. The systems differ in system parameter choices in order to optimize compatibility with and migration from existing systems. Compatibility with existing systems will allow development of multi-mode mobile stations that can work with either second or third generation technology. The cdma2000 proposal optimizes compatibility with the IS-95 CMDA standard. The UTRA and TD-SCMDA proposals emphasize compatibility with existing GSM and Japanese PDC systems.

Two of the developments that will aid deployment of 3G systems are smart antenna and software radio technology.

Figure 14.29:
A Smart Antenna

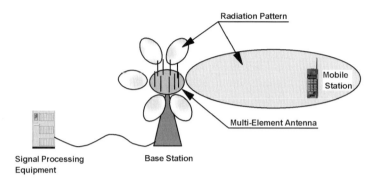

A smart antenna uses a multi-element antenna array coupled with signal processing equipment to steer the radio beam toward a target mobile station. This allows a wider range and lower power use than conventional antennas.

Software radio technology moves digital processing closer to the mobile station antenna. Traditional radios use analog circuitry for higher-frequency base-band processing. Digital processing is used for processing at relatively low frequencies. This affords software-controlled flexibility at the lower frequencies, but the high frequency characteristics are determined by the analog components.

Through use of very fast digital components, the digital signal processor can assume some of the high frequency processing. This allows very flexible determination of operating characteristics such as modulation and information encoding technique. Figure 14.30 shows the software radio concept.

Figure 14.30:
The Software
Radio Concept

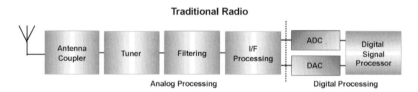

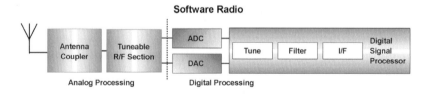

This technology could support development of handsets that can automatically adapt themselves to the radio standard supported by the base station within range. Software radios might also be able to adapt their transmission and reception characteristics to changing environmental conditions, making them less susceptible to noise and interference.

14.1.3.5
Wireless Application Services

The availability of high-speed digital wireless access will open the door for a new range of services. Service providers will allow access to Internet content through mobile devices. Corporate gateways will allow employees to access corporate data through portable wireless terminals.

A new breed of wireless information appliances will spring up. These devices will be small and portable; yet will support applications such as streaming multimedia through a wireless connection. There is already a wide range of such devices available. These range from simple mobile phones that support SMS messaging to hand-held computer devices with a small color display capable of graphic representation. Lightweight laptop machines with PC-sized screens will also be outfitted with wireless connectivity.

Wireless information appliances

With such a wide range of mobile devices, it would be tedious to define custom content for each device. Special software will be required to adapt the Internet and Intranet content to the capabilities of the access device. The process of converting information from an Internet format to a format suitable for a mobile device is known as content adaptation.

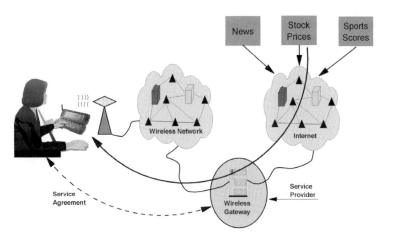

*Figure 14.31:
Gateway to the
Internet*

Certain content may require special viewer software on the mobile device. It would be convenient for the supporting software infrastructure to recognize the need for special software and provide the user with a means of installing it on his device. Particularly capable infrastructure software could automatically recognize the need for and install the needed additional software.

Already software suppliers are creating the necessary wireless gateway software for such applications. Figure 14.31 shows a wireless gateway forming a bridge to the Internet for a mobile user.

In this scenario, the user subscribes to a wireless information provider service, just as she would to an Internet service provider. A service agreement will take care of billing and will specify any special applications for the user. A portal functionality that would allow the user to configure the information display may also be included.

Figure 14.32:
Wireless Gateway

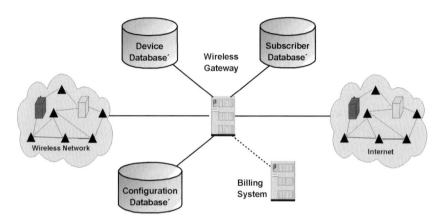

The wireless gateway must have access to information about the subscriber to set the proper access rights. The wireless gateway must also pass subscriber identification information to the billing system.

The gateway uses the device characteristics to adapt the content to a format suitable for the device. Device information may also be used to determine available applications or software updates needed. Some devices may be able to provide the gateway with an electronic serial number that may be used to help determine access rights.

Since 3G wireless technology will enable "always-on" access, they will be available for messages just as an Internet host. A wireless gateway could implement "push" functionality that lets the user subscribe to a particular type of information. When an event occurs, the user will be notified.

Figure 14.33 uses a soccer game as an example. When a team kicks a goal, the sports service sends the data to the gateway. The gateway prepares the content according to device type and sends the information on its way to the devices.

In a corporate setting, the wireless gateway would be positioned at the edge of the Intranet. The gateway provides mobile employees with access to corporate databases, email, and other information sources.

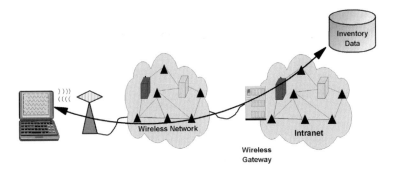

Goal!

Wireless Network

Internet

Wireless
Gateway

Service
Provider

Inventory
Data

Wireless Network

Intranet

Wireless
Gateway

14.2
Short Range Wireless Communication

This section deals with mechanisms for communications within a building or in a corporate campus setting. The distances covered range from under one meter up to a few hundred meters. The technology can be based on infrared transmission for very short-range communication or on radio technology.

14.2.1
DECT

DECT (Digital Enhanced Cordless Telecommunications) was designed by ETSI for use as a cordless connection standard for telephones and other office equipment. In the U.S., DECT is sometimes known as Personal Wireless Telecommunications (PWT).

DECT standard   The DECT standard is very complete, comprising the user interface for functions such as call forwarding and teleconferencing as well as the air interface. This allows interoperability among DECT handsets and base stations.

A DECT system is composed of a base station and at least one handset. The base station acts as a switch among the handsets as well as between the handsets and the PSTN.

The DECT standard defines enough capacity on the air interface to allow simultaneous conversations among internal handsets and between internal and external parties. The DECT concept supports other office devices such as FAX machines as well as telephone handsets.

Figure 14.35:
A DECT
Configuration

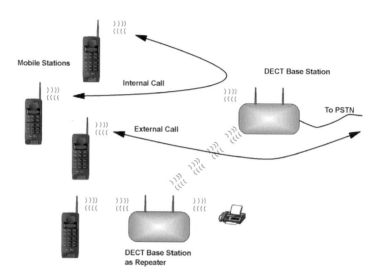

DECT as   DECT can also be used to implement a wireless LAN. Personal computers
wireless LAN can communicate with printers and other equipment through such a LAN. The basic data rate per DECT channel is 32 kbps. DECT supports higher data rate devices by allowing them to request use multiple channels.

A single base station may have a range of about 100 meters. Additional base stations may be added to the system as repeaters to increase the range and improve the coverage.

Hand-off between   When a handset is used in a multiple base station setting, it measures the sig-
base stations nal strength of each base station by comparing the error rates from each. It locks on to the base station with the strongest signal for communication. When the handset is moved within range of a different base station, a hand-off from base station to base station is performed. The hand-off is carried out seamlessly without loss of data or dropped calls.

Frequency ranges   In Europe, DECT uses 10 radio channels in the frequency band from 1880 – 1900 MHz. This frequency band has been allocated for DECT use in all European countries. In the United States, PWT generally uses 8 radio channels in the 1920 – 1930 MHz PCS (Personal Communications Services) band.

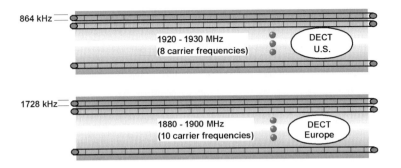

The U.S. and European systems also differ in modulation technique. DECT uses the same GMSK technique used by GSM. GMSK allows the transmission of one bit per phase change. PWT uses a more efficient technique that allows two bits to be transmitted per phase change. This allows PWT to allocate radio channels with a bandwidth of 864 kHz – only half the bandwidth of DECT while providing the same data capacity.

Modulation techniques

The DECT and PWT technology differs sharply from cellular radio technology when it comes to bi-directional communication. Cellular radio generally calls for the uplink and downlink to be implemented on separate radio channels. DECT implements the uplink and downlink on the same radio channel.

Each radio channel provides a raw data of 1.152 Mbps. DECT and PWT employ TDMA to divide the radio channel into time slots. Each time slot transmits 480 bits in 0.416 ms. Twelve slots are grouped together to form a frame, which is transmitted in 5 ms. Additional bits are used to transmit control and synchronization data for each frame. This allows support of up to 12 mobile stations on each radio channel.

Data rate

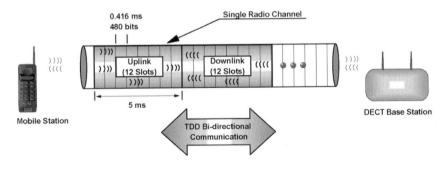

The system alternates the direction of data flow for each frame. First, the base station transmits a frame to the mobile stations, and then the mobile stations transmit to the base station. Each mobile station must transmit within its allocated slot. This method or alternating data direction on a single radio channel is known as Time Domain Duplex (TDD).

TDD & data rate   Since a frame is transmitted in 5 ms, each logical channel receives 100 slots per second. This works out to a raw data rate of 48 kbps, but some of the capacity is needed for error handling. The usable data rate is 32 kbps in each direction.

DECT Extensions   The ETSI standards group has completed work on extensions to DECT. These extensions define a wireless point-to-point protocol to enable Internet access through DECT. They also define a mechanism for providing ISDN services using DECT as a transport mechanism.

Additional work is being done to define a DECT Packet Radio Service (DPRS) and 2 Mbps service to support integrated data and voice applications. As an outgrowth of this work, ETSI submitted a proposal to the IMT-2000 workgroup for third generation mobile technology.

14.2.2
Bluetooth

Harald Bluetooth   Bluetooth is a technology for short-range wireless connections between devices. Bluetooth is an outgrowth of an effort began by Ericsson in 1994 with the goal of creating a low-cost wireless interconnection technology for mobile phones and their peripheral devices. In 1998, IBM, Toshiba, Nokia, and Intel joined Ericsson to found the Bluetooth Special Interest Group (SIG) to promote this technology. Bluetooth takes its name from a Danish king, Harald Bluetooth, who lived more than one thousand years ago.

Standard adoption   Bluetooth has proven to be one of the quickest standards to gain wide support. As of this writing, 1883 companies had pledged to support the standard with their products.

Despite its widespread support, adoption in actual products is proceeding more slowly than was originally expected. The slower than expected adoption has been caused primarily by high chip set prices – in 2001, prices for Bluetooth chips ranged from about $14 to $25. By the time of this writing one year later, prices had dropped considerably, with chips available for under $5. The lower cost should help speed adoption.

Design points   Major Bluetooth design points are low power, low cost, and the ability to support high-speed ad hoc networking. Companies will implement Bluetooth as a single-chip solution, each chip costing on the order of a few dollars after production has been ramped up.

Data rate   With a basic data rate of 1 Mbps, Bluetooth is suitable for interconnection of personal computers with printers, scanners, keyboards, and other devices. The communication channel is also suitable for real-time voice connections between a headset and a mobile phone.

Use scenario   One standard usage scenario is a mobile handset that automatically uses the lowest cost connection for communication. When the phone is within range of another Bluetooth-capable phone, it will automatically establish a connection directly with that phone. The public telephone network would not be used at all.

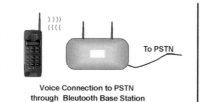

Bluetooth Voice
Connection

Voice Connection to PSTN
through Bleutooth Base Station

To PSTN

Voice Connection
through Wireless Network

*Figure 14.38:
Bluetooth Phone
Scenario*

When the mobile phone is outside the direct range of the target phone, but within range of a Bluetooth-capable base station connected to the PSTN, a connection will be established through the fixed PSTN connection. Finally, when the mobile phone is outside the range of a base station, it would establish the connection through the mobile phone network.

Another usage scenario is ad hoc connections between computer peripheral devices. When a user brings his Bluetooth laptop into an office environment, the Bluetooth module will automatically find the local mouse, keyboard, and printer. This will be much more convenient than physically attaching the PC to a port replicator or network and rebooting.

*Connecting
computer devices*

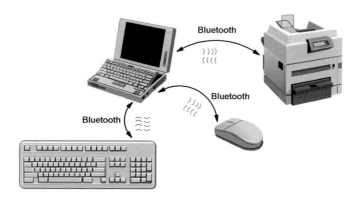

Bluetooth

Bluetooth

Bluetooth

*Figure 14.39:
Bluetooth
Scenario with PC*

14.2.2.1
Bluetooth Radio Technology

The Bluetooth radio interface is similar to DECT, which was discussed in the preceding section. The available frequency band is divided into radio channels designated by carrier frequency, and the available capacity on each radio channel is divided into slots. Capacity is shared by allocating the slots to individual devices.

Radio interface

In most parts of the world, the 2.4 GHz frequency band from 2.4000 to 2.4835 GHz is used for Bluetooth communication. This band is very noisy, since it is used by other devices such as garage door openers and baby intercom systems. The band is divided into 78 carrier frequencies spaced 1 MHz apart.

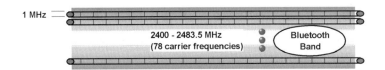

Each carrier frequency is divided into time slots of 625 microseconds length. The raw transmission rate is 1Mbps, so each slot transmits 625 bits. Some of these bits are used for synchronization, error correction, and control information, so the net data rate is reduced.

In order to achieve greater noise immunity and robustness, a frequency-hopping transceiver is used. Each consecutive data slot is usually transmitted on a different frequency. This makes the transceivers a bit more complex, since the pseudo-random frequency hop patterns must be synchronized to enable communication. The 625 microseconds slot length corresponds to a frequency-hopping rate of 1600 hops/s.

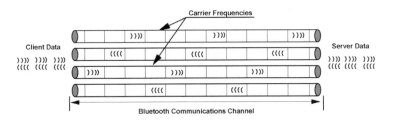

Bluetooth achieves two-way communication through time domain duplex (TDD) operation. A Bluetooth device receives data on the same radio channels it uses to transmit data. The transmission direction alternates with each packet.

Data is transmitted in packets that are generally one slot in length, but may be extended to span up to five slots. Multiple-slot packets are transmitted on the same hopping frequency. The packets have an internal structure consisting of an access code, a header, and payload data.

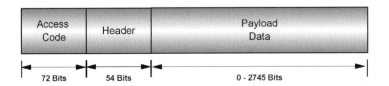

The packet header contains link control information. This includes a three-bit address for use within the ad hoc network, a four-bit packet type field, and other control information. The type field allows definition of up to 16 packet types. A number of packet types have been defined, among them control packets and data packets.

The packet type in the header defines the format of the payload data field. There are two main types of data payload – one used for synchronous voice data, and the other used for asynchronous data. Voice data always has a fixed length. Depending on packet type, the length of the synchronous voice data packet is either 80 or 240 bits. The asynchronous data packet has a variable length. These data fields are made up of a payload header, payload body, and a CRC code. The payload header can be made up of one or two bytes and contains a 5 or 9 bit length field as well as control information. The length field contains the number of bytes in the payload body. *Packet types*

The access code is used for synchronization and identification. In order to find other Bluetooth devices, a special access code with no header or payload data is sent out. This can either be a general inquiry access code (GIAC) to look for unspecified Bluetooth devices or a dedicated Inquiry access code (DIAC) to look for a specific type of Bluetooth device. After a connection has been established, a three-bit field in the access code identifies the communication connection. *Device inquiry*

Corresponding to the three-bit length of the identification field in the access code, up to eight Bluetooth devices can communicate among themselves. A group of communicating Bluetooth devices is known as a piconet as depicted in Figure 14.43.

Within a piconet, one device will act as the master and all others as slaves. The first device to initiate communication will assume the role of the master. The slave devices will synchronize their internal clocks and frequency-hopping sequences with those of the master. *Piconet*

A scatternet is formed by two or more overlapping piconets. Each piconet has its own master for synchronization. Piconets within a scatternet are not synchronized, but devices in one piconet may communicate with devices in another piconet. It is possible for the master in one piconet to be a slave in another piconet. *Scatternet*

Devices within the piconet set up links for communication depending on need. Two types of communications links have been defined.

The Synchronous Connection-Oriented (SCO) link represents a dedicated point-to-point connection between master and slave. The master device reserves slots at fixed times for communication. This guarantees resource availability for real-time applications such as voice transfer. A master device can support up to three SCO links with the same or different slaves. *Synchronous link*

The Asynchronous Connection-Less (ACL) link uses any slots not reserved for an SCO link. The master can communicate with a slave on a per-slot basis. A slave can only communicate with the master after the master has addressed it. *Asynchronous link*

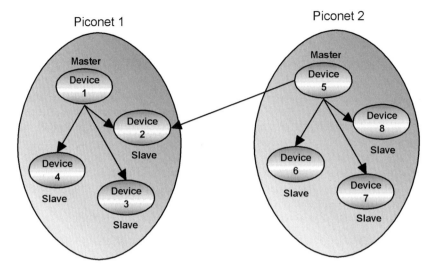

Channel capacity

Each of the three possible SCO channels can support up to 64 kbps voice transmission in each direction. When the entire capacity is used for the ACL link, either an asymmetric link supporting 723.2 kbps second in one direction and 57.6 kbps in the return direction or a symmetric transfer of 433.9 kbps in each direction can be supported. Bluetooth devices will also support a mixture of SCO and ACL channels.

14.2.2.2
The Bluetooth Protocol Stack

The Bluetooth specification goes much further than DECT to define the software layers used for communication on top of the radio link. The specification defines a number of Bluetooth-specific components, but uses existing protocols such as OBEX and TCP/IP as much as possible.

Protocol stack

The protocol stack is organized into layers as shown in Figure 14.44. The lower layer defines the Bluetooth specific components. The middle layer consists of industry standard protocols that were adapted for Bluetooth use. Use of existing protocols allows applications to be ported to Bluetooth more easily.

The top layer is the application layer. The arrows in Figure 14.44 illustrate that the underlying layers are visible to the application. In general, however, the applications will use the upper level protocols as much as possible.

The Bluetooth radio and baseband layers define the transmission characteristics and link level protocols that allow links to be established with other Bluetooth devices. These layers were described in Section 14.2.2.1 "Bluetooth Radio Technology".

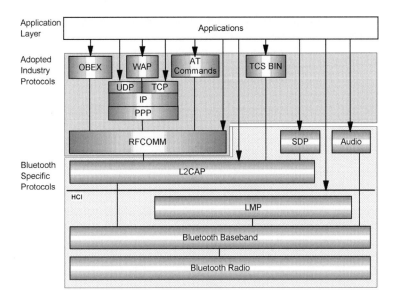

The Link Manager Protocol (LMP) defines messages that are exchanged between participating devices to setup and maintain links. These messages also control authentication and encryption. LMP messages are transported in the payload data field of a Bluetooth data packet. LMP is a data link layer protocol that communicates with its peer in the target device. LMP defines messages used by software in the master for polling the client even when the master has no data to transport.

Link Manager Protocol

Authentication is carried out using a challenge/response protocol between verifier and claimant. The verifier sends a random number (challenge) to the claimant. The claimant encrypts this number with its secret key and returns the resulting value to the verifier, which compares the result with the expected response. The verifier can be either master or slave.

Authentication

After the participating devices have carried out authentication, they may also agree on encryption. Participating devices must agree on common encryption parameters including encryption type and key. A master may require all slaves to use the same encryption parameters, or the parameters may be specific to one master – slave connection.

Encryption

The Logical Link Control and Adaptation Protocol (L2CAP) is used by upper layer protocols for data transport. It can be thought of as an adapter between the upper and lower layers. L2CAP messages are transported in the payload data field of Bluetooth data packets just as LMP messages are.

Logical Link Control & Adaptation Protocol

L2CAP is responsible for upper-level protocol data segmentation and reassembly as well as transport of quality of service information. L2CAP is a data link layer protocol providing both connection-oriented and connectionless services. L2CAP will accept blocks of data up to 64 kilobytes in length and will reliably transport them to its peer in the target device using the services provided by the baseband layer. It does this by creating a logical channel from one device to the other.

Bluetooth Audio Protocol

The Bluetooth audio protocol actually also belongs to the data link level. It is the only protocol to use the SCO link for synchronous communication. An application can obtain synchronous services simply by opening a voice data link.

Figure 14.45:
Bluetooth Data
Link Layer

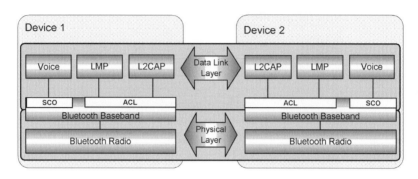

Service Discovery Protocol

The Service Discovery Protocol (SDP) is the final major Bluetooth protocol to be covered in Figure 14.44. This protocol is used to find new services as they become available and to deregister services that become unavailable.

The need for service discovery is widely recognized in the IT industry. A number of initiatives have been founded to address the issue – Jini™ and Salutation™, for example. The Bluetooth SDP provides a mechanism for using information from other services if it is available, but does not require it. The Bluetooth SDP is optimized for a fast changing environment.

Tracking services

A piconet is formed in an ad hoc manner as devices come into proximity of one another. When a new device enters the piconet, it brings services with it that can be used by the other devices. When a device leaves the piconet, its services become unavailable. It may also happen that a device already registered in a piconet makes new services available. This might be the case when a printer goes online, for example. A Bluetooth device must track all of these cases.

An application running on a Bluetooth device can request services according to class of service or through specification of service characteristics. It is also possible for an application to request a list of all available services. An SDP client application can issue requests to the local SDP server or to any SDP server in the piconet. The SDP servers in the piconet will return the relevant service records.

SDP Server

A Bluetooth device includes an SDP server application that keeps track of the services available on that device. A service record is kept for each service available on the device. The service record describes attributes of interest to an application. The Bluetooth specification describes a number of service attributes. Examples of service attributes include service class ID list, provider name, service name, icon URL, and service ID.

RFCOMM

The RF communication interface (RFCOMM) provides an RS-232 serial port emulation to the application or to higher-level protocols. RFCOMM provides a common programming interface for actual serially connected devices such as printers and modems as well as for communication through a Bluetooth radio link. RFCOMM supports up to 60 concurrent port connections.

Telephony Control

The Bluetooth Telephony Control protocol Specification – Binary (TCS BIN) defines the necessary call control signaling for establishing a voice connection between Bluetooth devices.

TCS BIN has three major components as shown in Figure 14.46. The call control component handles the establishment of a voice connection between two devices. Group management provides support for signaling among groups of devices in a piconet. The Connectionless TCS component provides signaling for applications that do not require a dedicated synchronous voice connection.

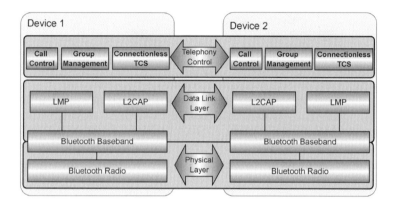

Figure 14.46: Telephony Control Protocol Specification Components

The Advanced Telephony (AT) commands also enable telephony control. These commands are generally used for modem control, but they also support FAX transfer.

Internet Protocol stack

The remaining protocol components in Figure 14.44 use the RFCOMM interface. An Internet protocol (IP) stack including UDP and TCP can be implemented using Point-to-Point Protocol (PP) as an interface to RFCOMM.

WAP Capability

The IP stack can be used to provide Wireless Access Protocol (WAP) capability on Bluetooth devices. WAP provides value-added services such as email delivery across wireless links. This can be useful when updating e-mail on a laptop, for example. The user would dial into his or her corporate network using a mobile phone. The laptop would communicate with the mobile phone using Bluetooth. The e-mail would be transmitted to the mobile phone and from the phone to the laptop using WAP as a transport protocol.

OBEX – Object Exchange Protocol

The final component from Figure 14.44 is the Object Exchange Protocol, or OBEX. The Infrared Data Association (IrDA) developed OBEX to exchange data objects over an infrared link. The Bluetooth specification includes the same OBEX interface used by infrared. OBEX provides a session layer service for applications such as synchronization and file transfer.

Figure 14.47 shows how OBEX can be implemented on Bluetooth. OBEX can either use the TCP/IP stack or go directly to the RFCOMM interface.

On a Bluetooth device, OBEX can be used for calendar or e-mail synchronization, generic file transfer, or object push. The latter would be used, for example, to push business card data from one Bluetooth device to another.

Figure 14.47:
OBEX
Implementation
on Bluetooth

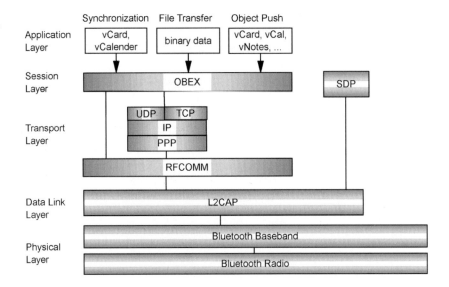

14.2.3
IrDA

IrDA
The Infrared Data Association (IrDA) has defined a number of standards governing infrared wireless communication. These include the IrDA-Data and IrDA-Control standards. IrDA technology is similar to Bluetooth in that it is useful for short-range communication. However, there are significant differences due to the characteristics of the transport medium.

Directional devices
Infrared light emitters such as light emitting diodes and laser diodes are directional devices. The emitted light propagates in a cone from the light source. This makes it easy to implement 'point-and-shoot' applications using IrDA. Imagine a room full of business people with IrDA devices. If two people want to exchange their vCard business card data, they just point their devices at each other and exchange the data.

Bluetooth radio transmitters are omnidirectional devices, so the same scenario is a bit different. If a number of people in the room all have Bluetooth devices, they will all find one another regardless of whether the devices are in use or are in someone's pocket. Either everybody would have to exchange business card data with everybody else, or some potentially complicated selection mechanism would be called for.

Beyond the directional characteristic, IrDA offers high data rates of 4 Mbps, with 16 Mbps technology in the offing. IrDA modules are simple and inexpensive. Over 170 million IrDA devices are expected to ship in 2000, rising to 240 million in 2001.

IrDA data rate

14.2.3.1
The IrDA Protocol Stack

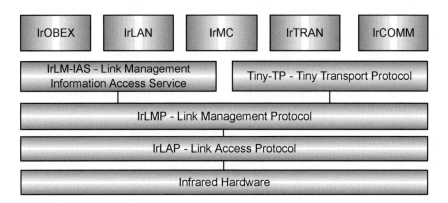

Figure 14.48:
The IrDA Protocol
Stack

Figure 14.48 shows the most important elements of the IrDA protocol stack.

Beginning with the physical layer, the infrared port allows half-duplex operation. The port cannot receive data parallel to transmission. IrDA achieves duplex operation by alternating the direction of the data link. First one device transmits then the partner device responds. Range of operation ranges from contact to about one meter. Low power implementations allow a range of up to 30 cm.

Half-duplex operation

The infrared hardware can operate in three modes:

- Asynchronous serial infrared with speeds varying from 9600 bps to 115.2 kbps.
- Synchronous serial infrared with a speed of 1.152 Mbps.
- Synchronous infrared using Pulse Position Modulation (PPM) to achieve data rates of up to 4 Mbps.

The **Infrared Link Access Protocol (IrLAP)** is a required component that provides connectionless and connection-oriented transport services to the upper layers. IrLAP provides for reliable communication between IrDA devices.

IrLAP

To begin communication, the software in one device will issue a IrLAP request. The initiator is known as the primary device. IrLAP generates a request frame and sends it to the receiver, which is known as the secondary device.

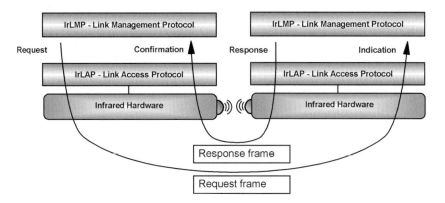

The IrLAP layer in the secondary device passes an indicator notification to the upper layer code. The upper layer software in the secondary device processes the indication and passes a response to its IrLAP layer. The receiving IrLAP layer returns a response frame to the primary IrLAP layer, which then returns a confirmation to its upper level code.

IrLMP   The **Infrared Link Management Protocol (IrLMP)** supports ad hoc connections with peer devices. IrLMP multiplexes data from multiple applications over the single data link. It also supports applications that request exclusive link access.

IrLMIAS   The **Link Management – Information Access Service (IrLM-IAS)** layer is responsible for discovery. It determines which services another device has to offer by retrieving data from its discovery database.

A number of high-level protocols provide services to applications.

- **Tiny-TP** is a transport layer providing connections over IrLMP that support data segmentation and reassembly. This service removes considerable complexity from applications.
- **IrComm** provides serial and parallel port emulation. Similar to the RFCOMM layer in Bluetooth, this support makes it easy to port applications written for such interfaces to IrDA.
- **IrLAN** is a protocol for implementing infrared LAN access. Since infrared hardware is capable of high data rates, an infrared connection to a LAN is practical.
- **IrTRAN** specifies transport of image data from digital cameras and similar devices.
- **IrMC** is a protocol for exchange of telephony and communication data.
- **IrBus** provides connection services for cordless peripherals such as keyboards, mice, and joysticks. It can also be used to implement communication between remote controls and television sets or VCRs.

14.2.3.2
IrOBEX

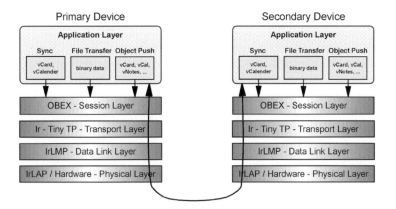

Figure 14.50:
IrDA OBEX Stack

The Infrared OBject EXchange protocol (IrOBEX) is likely the most impor-   *IrOBEX*
tant high-level IrDA protocol. As a general-purpose object transfer protocol, the
Bluetooth SIG has adopted it to promote interoperability between IrDA and
Bluetooth applications.

Figure 14.50 shows how OBEX is used in an IrDA setting. OBEX uses the
underlying layers to establish a session between the OBEX client and the
OBEX server. Object data is transmitted through the session between the primary and secondary devices.

OBEX is a binary protocol that allows data to be exchanged in a manner similar to HTTP. Object description data is transmitted among with the object data
itself. An OBEX transmission consists of headers describing the data followed
by the data itself.

OBEX defines a number of commonly used headers. In addition, any HTTP
header can be used, and new headers can be defined for special purposes.

Besides the object description language, OBEX defines messages for setting
up connectionless and connection-oriented sessions between an OBEX client
and server. These sessions support authentication, and extensions for session
encryption have been proposed.

14.3
Home Networks

Home networks allow appliances to communicate with one another, with a central controller, or with an external entity (a customer service site, for example).
Such home networks range in scope from simple systems that allow remote
control of lighting to complete home automation systems that integrate control
of heating and air conditioning, lighting, security, and entertainment systems
and additionally serve as a communication gateway to the outside world.

Many companies are working to add home networking capability to their products. To insure interoperability, there is a great deal of standards work going on in this area. This work, driven by groups of companies, is partly complementary and partly competitive. Some of the principal standards groups are:

- The CEBus Industry Council (CEBus) developed a power line carrier standard to transport messages between devices using existing electrical wiring. CEBus has also developed a Common Application Language (CAL) for home network products.
- The HomePlug Powerline Alliance is defining a standard for high-speed power line networking.
- The homeRF Working Group (homeRF) is working to establish an open standard for wireless digital communication between PCs and consumer electronic devices.
- The Home Phoneline Networking Alliance (homePNA) is developing specifications for interoperable home networking using existing telephone wiring.
- The Open Services Gateway Initiative (OSGi) is defining a gateway component for communication through the Internet.
- The LonMark organization promotes multi-vendor control networks based on LonWorks technology for industrial as well as residential networks.

Some of these organizations are defining standards for communication between devices at a physical and protocol level, while others are focused more towards system or gateway standards.

Connectivity options
These efforts are creating a wide range of connectivity options for communication among devices. Systems based on existing power line or telephone wiring use the existing infrastructure in the home. Wireless systems, based on radio or infrared technology, do not require physical links between the devices. Finally, the high-end solution calls for dedicated cabling – a realistic option only for new construction.

Radio frequency networking
Radio frequency networking uses high frequency radio signals to transport the data. These signals allow data rates of up to 1–2 Mbps, making radio frequency networking suitable for connecting personal computers to one another or to peripheral equipment. Sections 14.2.1 DECT and 14.2.2 Bluetooth cover the major technologies that will provide radio frequency networking in the home.

Power line networking
Power line networking uses the existing power grid in a home as a data transport medium. Various systems have been developed and are available on the market, but they all tend to be fairly slow in terms of data rate (< 10Kbps). This is due to the high amount of noise on the line and the high variability of the physical wire characteristics (aluminum or copper wire, number of outlets, devices connected, etc.).

Phone line networking builds on the prevalence of telephone jacks in the household. Many homes are wired with phone jacks in strategic locations such as bedrooms, the den, and recreation rooms. The phone line wiring can be used to transmit high-frequency signals from one room to another. Such signals enable data rates of 1 Mbps and above.

14.3.1.1
Power Line Networking

Standard household wiring carries high-power, low frequency (50/60 Hz) electrical energy. This energy is used to power the appliances, lights, TV sets, etc. in the home.

Since practically all appliances and devices are connected to the power distribution system in the home, it makes sense to try to use this existing wiring for communication.

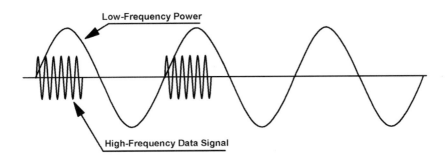

Low-Frequency Power

High-Frequency Data Signal

Figure 14.51: Data Signal Superimposed on Power Waveform

A common approach is to send the data as a high-frequency signal on top of the low-frequency power wave. This sounds like a very straightforward thing to do, but there are a number of practical problems. First, limited frequency spectrum is available for these types of applications. In addition, noise and attenuation on the power line can make reliable communication very difficult.

The Government regulates which frequencies can be used for power-line signaling. The European Cenelec regulations (EN-50065-1) provide four bands from about 10 kHz up to about 150 kHz for power line signaling.

The regulations reserve Band A for use by power companies and their licensees. Bands B, C, and D are available for consumer use. Regulations in the U.S. and Japan allow use of frequencies up to about 525 kHz, the beginning of the AM broadcast band.

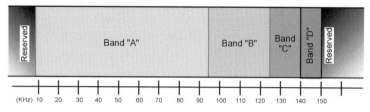

Noise, attenuation, and signal distortion on the power line cause problems for power line signaling applications. Devices such as vacuum cleaners, televisions, and microwave ovens leak high frequency noise onto the power line.

Figure 14.53:
Noisy Lines

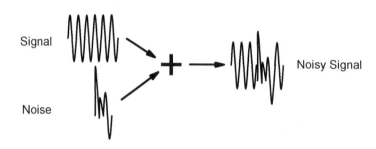

The inherent characteristics of the electrical line cause the signal to become weaker with distance. Some power strips have power line filters that short out exactly those frequencies that are of interest for power line signaling. Baby monitors and intercoms also emit carrier waves at those frequencies onto the power line.

Several methods of data transmission have been developed attempting to improve communication reliability. These alternatives are illustrated in Figure 14.54.

Narrow-band systems

The most direct way to go about it is through use of a single carrier frequency. A narrow-band system has the advantage of simplicity, but communication between devices can be very susceptible to noise.

Spread-spectrum

Spread-spectrum technology provides better noise immunity, but requires considerably more bandwidth to transmit the same amount of information. As Figure 14.52 shows, available bandwidth is quite limited, which limits the applicability of spread-spectrum technology.

Multiple carrier frequency

Another alternative is use of more than one narrow-band carrier frequency. When transmission on one frequency is jammed, the system can switch to a different one in order to avoid the problem.

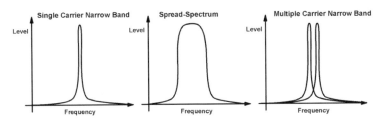

There are various systems on the market making use of these ideas. The X10 narrow-band system addresses primary control of lighting and small appliances in the home. The CEBus® Powerline Carrier system uses spread-spectrum technology and defines a more general-purpose method for sending information through home wiring. The Echelon Corporation manufactures dual-carrier power line transceivers for home and industrial use.

X10 is probably the most widely used system for home automation on the market today. Controllers and receivers are available from a number of manufacturers. They often take the form of plug-in units that can be installed without changing existing wiring.

X10

X10 controllers send commands such as "TURN ON", "TURN OFF", or "DIM" to the receivers. The commands contain addressing information organized as a 4-bit house code plus a 4-bit unit code to specify which receiver is to respond. This allows addressing of up to 256 devices.

The receivers are typically plug-in units that themselves provide outlets into which the device to be controlled is connected. The receiver switches the outlet according to the commands received from the controller.

Data is transmitted at a rate of about 300 bits per second using a single narrow-band carrier frequency of 120 kHz. While being sufficient for such applications as lighting control, the X10 protocol is too slow and its commands too limited for general data communications among devices.

The Echelon Corporation sells power line transceiver modules that use dual carrier frequency narrow-band technology. The carrier frequencies were chosen to minimize the possibility that both would be blocked due to the same line disturbance. If the primary carrier frequency becomes blocked, the device will automatically switch to the secondary frequency.

*Echelon
transceivers*

The two carriers were selected to be 132 kHz and 115 kHz. In addition to providing high noise immunity, this selection of carrier frequencies fits into the bands allowed by the European regulatory agency. Communication takes place at a rate of about 5.4 kbps.

The CEBus standard uses Spread-Spectrum Carrier™ technology to transmit information. The Electronic Industry Association has adopted the CEBus specification as the EIA-600 standard.

CEBus

The CEBus signal is swept through a range of frequencies in a period of 100 μS (the Unit Symbol Time, or UST) to generate a so-called "chirp". Symbols, such as "1", "0", and End-Of-Field, are made up of sequences of chirps. The data is transmitted as a series of such symbols.

In the U.S., the signal is spread over a frequency range of 100 kHz to 400 kHz, allowing a UST of 100 µS and resulting in an effective data rate of about 10 kbps. In Europe, the signal is spread over a frequency range of 20 kHz to 80 kHz. To retain the same signal processing advantage, the UST must be increased to 500 µS, resulting in an effective data rate of about 2 kbps.

Common Application Language

The CEBus standard goes beyond the physical layer specification for power line data signaling described so far. It also describes a packet transmission protocol, packet formats for peer-to-peer networking, and a Common Application Language (CAL) for application-to-application communication.

The data to be transmitted is partitioned into packets and the packets are transmitted over the power line. Since the CEBus specification defines a peer-to-peer protocol, any connected node may begin the communication. This being the case, it may happen that two nodes begin transmitting simultaneously, causing what is known as a collision. The packet data is garbled and must be retransmitted.

The CEBus standard uses a protocol known as Carrier Sense Multiple Access with Collision Detection and Collision Recovery (CSMA/CDCR) to handle packet transmission. This protocol is quite similar to that used with Ethernet systems.

When a node has data to be transmitted, it tries to detect a carrier on the line. If a carrier signal is detected, it will wait until the current transmission activity on the power line is finished.

Figure 14.55: Packet Collision

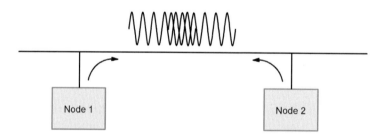

When there is no active carrier signal on the line, the node will begin to transmit a special packet preamble that is used for contention resolution. At the same time, it monitors the signal on the line to determine if another node also begins transmitting data at the same time. If another node begins transmitting data, then a collision is said to occur. Both nodes must immediately stop transmitting data and wait a random time interval before attempting to retransmit the packet. Since each node will wait a different length of time, it is likely that no collision will occur when the packet is retransmitted.

The following figure shows the CEBus packet frame as described for the power line medium. The packet starts with the preamble used for contention resolution and ends with a CRC code. The Link Protocol Data Unit (LPDU) contains a Network Protocol Data Unit (NPDU), which in turn contains an Application Protocol Data Unit (APDU).

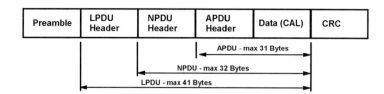

The LPDU header contains control, source address, and destination address information for the packet. The addresses are each four bytes in length, allowing addressing of over four billion nodes, and are divided into two portions – the System Address and the Media Access Control (MAC) address. The system address and MAC address are often referred to as the house code and unit code since many people are already familiar with those terms.

The NPDU header specifies how the packet is to be sent. It contains bit fields that provide privilege level, routing, and other information. The NPDU header also controls routing from one transport medium to another. A message could be routed from power line to a wireless medium, for example.

The APDU header contains information for the receiving application about the number of data bytes in the packet and desired response after processing. It is possible to request the CAL interpreter on the receiving side to respond with a receipt acknowledgement, for example.

According to the CEBus specification, the data field must contain Common Application Language tokens. These tokens represent CAL objects within the nodes, actions that can be performed on the objects, literal data, or programming language elements such as IF, REPEAT, BEGIN, and END.

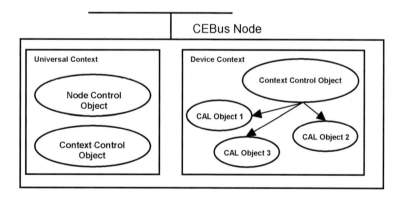

Figure 14.57:
CEBus Node

Figure 14.57 shows the internal logical structure of a CEBus node. Each CEBus node contains two or more contexts, each of which contain two or more CAL objects. The CEBus standard defines context and object identifiers, which are used in the APDU data field to gain information about the node and to change the state of the node.

The Universal Context contains a node control object providing administrative information such as manufacturer name, serial number, and addressing information. It must always be present.

The device contexts each contain a Context Control Object, which lists additional CAL objects for device control. The node type determines which additional CAL objects must be present. A simple light switch might have a Lighting Control context with one Analog Control object, for example. A more complex device, such as a stereo receiver, might have multiple contexts, each with multiple control objects.

Figure 14.58 shows the general structure of a CAL object. CEBus defines methods and instance variables for each type of object. Instance variables may be read-only or read-write. The methods define operations to be performed on the instance variables. The device is controlled by changing the state of the CAL object instance variables.

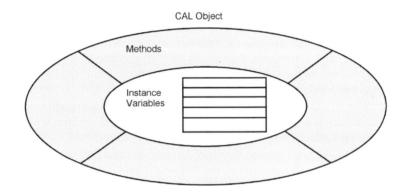

Figure 14.58:
CEBus CAL
Object

14.3.1.2
Phone Line Networking

Phone line networking focuses on providing local area network functionality in a home setting through use of existing telephone wiring. It should be possible to share resources such as tape units or printers in a home setting just as within the corporation. Sharing a common Internet connection among multiple household PCs through use of a gateway is another major requirement.

Technology for phone line networking must be compatible with installed telephones, modems, answering machines, and FAX machines on the shared telephone line infrastructure. They may not interfere with normal operation of the telephone network nor allow operation of a telephone or modem to interfere with operation of the phone line network.

Figure 14.59 shows a possible home phone line network. Every house has a unique network topology depending on the telephone jack layout and the installed devices. The resulting variation in electrical characteristics poses a challenge for phone line network operation. Varying line lengths and characteristics of the installed devices cause differing line impedance and noise levels. In addition, these operating conditions may change whenever a telephone is taken off the hook.

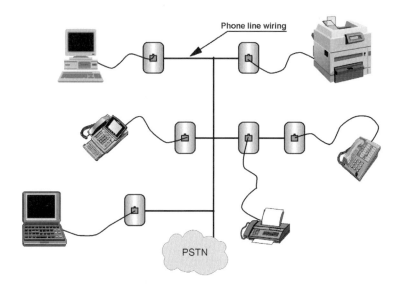

Figure 14.59: Home Phone Line Network

As for power line networking, phone line network devices transmit data using higher frequencies than those needed by traditional telephone system devices. Voice transmission in a standard telephone system occupies the range from 20 Hz to 3.4 kHz. Advanced telephone services such as DSL occupy the range from 25 kHz to 1.1 MHz. Phone line networking occupies the frequency range from 5.5 to 9.5 MHz.

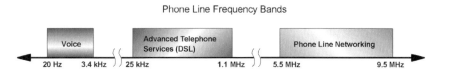

Figure 14.60: Frequency Bands

Regulations governing use of phone line infrastructure vary from country to country. The frequency and phone line use characteristics described in this section pertain specifically to the United States.

Use of high frequencies on a standard phone line became possible through use of digital signal processor hardware. The base transceiver hardware uses algorithms that can adapt to changing line conditions to maintain a high data rate. The base data rate of typical systems is about 1 Mbps, with 10 Mbps systems under development.

The phone line network operates by transmitting data packets between devices. Since several devices share a single medium, the network adapters must deal with packet collision, just as with power line networks. As with power line networks, Ethernet technology was adopted to allow multiple access to the common communications medium.

14.4
Further Readings

Cellular Communications

Harte, Lawrence et al.: Cellular and PCS: The Big Picture

This book provides a technical overview of current wireless technology and Personal Communications Systems (PCS).

Schneiderman, Ron: Future Talk: The Changing Wireless Game

This book provides a high-level overview of cellular and PCS systems.

International Telecommunication Union Main Site

http://www.itu.int

This site provides a wealth of information about telecommunication standards in general.

ITU IMT-2000 Site

http://www.itu.int/imt/index.html

The International Mobile Telecommunications – 2000 web site provides information including the draft standards for third generation wireless technology.

European Telecommunications Standards Institute (ETSI) Main Site

http://www.etsi.org

This is another good site for obtaining descriptions of standards. Many standards are available for free download.

GSM World

http://www.gsmworld.com/index1.html

Provides information about the GSM system.

The Official Bluetooth Web Site

http://www.bluetooth.com

First stop for information about Bluetooth technology.

The Infrared Data Association Web Site

http://www.irda.org

A very good source for information about IrDA technology.

Home Phoneline Network Alliance Web Site

http://www.homepna.org

Describes technology associated with phone line networking.

CEBus Industry Council Web Site

http://www.cebus.org

Provides information about CEBus, Home Plug and Play, and the CEBus Common Application Language.

Intellon® Technical Articles Web Page

http://www.intellon.com/products/ssc/techarticles.html

Provides in-depth articles about the CEBus standard and power line networking.

LonMark® Interoperability Association Web Site

http://www.lonmark.org

Provides information about home and control networking solutions based on Echelon® LonWorks technology.

15 Service Discovery

The number of intelligent devices in offices or at homes is increasing rapidly. All these devices are able to provide specific services to you or to other devices. For example: A printer could offer a printing service to all other devices, not only to a PC, but also to a set-top box or service gateway. Another example might be the TV set, which is able to display information provided by other appliances within home or office.

In the old days it would have required a system administrator to configure and maintain these devices as well as to make them all interact with each other. However, nobody can afford or even wants to spend a lot of time configuring these devices.

There is a great need for a mechanism that allows these devices to dynamically interact with each other and to offer services to other devices, as well as to enable other devices to query for a specific type of service that it might require at a certain point in time.

Recently, a number of software architectures emerged which try to solve this problem. They are all heading for simple, seamless, and scaleable device interaction and propagate "Service Discovery".

Three of the most promising ones are Jini, Universal Plug and Play (UPnP), and Salutation. They have been proven to be stable since the year 2000 and were adopted by the industry.

Service Discovery in general means that a subset of the following capabilities is supported by a device:

- The capability to make other devices aware of its own existence and presence in the network.
- The possibility to make other devices in the network aware of the services offered by the device and to describe these services to them.
- The ability to search for a service in the network.
- Zero-administration.
- Interaction with other devices in the network to fulfill a function.

The following three standard focus primarily on the making services available between smaller or embedded devices. Web services enable service on an application layer between servers. More about web services can be found in Chapter 13.

15.1
Universal Plug and Play

Microsoft invented Universal Plug and Play (UPnP) as a sort of extension to Microsoft Plug and Play. Then Microsoft founded an industry group, the UPnP Forum, which is now maintaining the standard with Microsoft leading the group.

UPnP is based on TCP/IP as the communication protocol. It uses XML to describe the services and capabilities a device is offering.

Each device must have an IP address, which is assigned by a Dynamic Host Configuration Protocol (DHCP) server; therefore, each UPnP client must also have a DHCP client. In a network with no DHCP server, the client uses Auto IP to obtain an address. Auto IP is a mechanism to retrieve an unused address from a range of reserved addresses. As soon as the device has obtained an IP address, the UPnP protocol can start.

Figure 15.1:
UPnP
Architecture

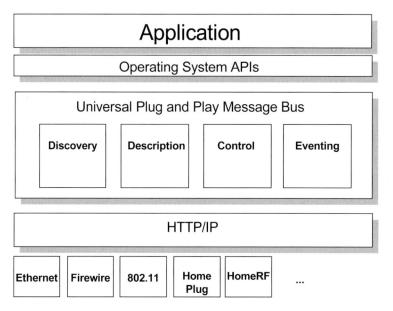

The UPnP protocol consists out of the following five steps:

- Discovery
- Description
- Control
- Eventing
- Presentation

Discovery is the first step of the UPnP protocol sequence. A control point is a central registry that stores the information about devices and services in the network neighborhood. A network could have several of these control points. As soon as a new device enters a network, it sends a broadcast or multicast message to the other devices in the network to advertise its embedded devices and services. In case a control point is interested in this sort of device, it adds the device to its list.

Discovery

At the time a control point enters a network, it searches the network for interesting devices by sending a message to devices in the network. Each device that matches the search criteria has to respond to the control point. This mechanism is defined by the Simple Service Discovery Protocol (SSDP).

The next and second step is Description. During the Discovery step, the device sends an URL to the control point. Now, the control point uses this URL to retrieve the device description document from the device. The UPnP description includes vendor-specific manufacturer information, like name and model, serial number, name of the manufacturer, and so on. All data is related to the device itself, the description of each service is retrieved in the next step. In addition to this, it also contains the URLs required for the following three steps: Control, Eventing and Presentation. The control point is not able to search for specific devices based on the device description in this step.

Description

The Description is an XML document provided by the device vendor and based on the UPnP Device Template.

Control is the next step. The control point retrieves the descriptions of these services that the device offers. The description for a service includes a list of actions (the UPnP term for feature), which the service offers, as well as the arguments for each action. The state of the service is represented by a set of variables. The service documents are XML documents that are provided by the device vendor and are based on the UPnP Service Template.

Control

After one or more of the services' state variables changed, the device is publishing these changes using the Eventing mechanism. A control point can register itself for these events by sending a subscription message to the device. The device returns the duration of the subscription and it is up to the control point to renew the subscription, if necessary. There is also the possibility that some state variables are not evented. In this case, the control point would need to poll the state of the variables from the service. This might be useful in case a variable changes too frequently.

Eventing

The control point could load a specific URL into a browser to present the user some device information or allow the user to control the device. The control point retrieved this URL from the device as part of the device description.

Presentation

UPnP does not cover the area of service invocation. The devices and the services they offer are described using XML documents, but it is still up to the devices to find a way to interact with each other and to use each other's services.

The Bluetooth Interest Group has defined an Extended Service Discovery Profile which enables the interoperability of Bluetooth devices and UPnP services.

15.2
Jini

Jini is the service discovery system based on Java. It is Sun's technology to build a network of distributed devices, which could use each others services.

A Jini system consists of the following parts:

- components that make up the infrastructure,
- a programming model, and
- services.

Infrastructure The Jini infrastructure is based on distributed devices, each of them with a Java Virtual Machine and the capability of using RMI (Remote Method Invocation) to call objects which reside on other devices. Proxies can be used to integrate devices into a Jini system, in case these devices do not have their own Java VM. The proxy acts on behalf of this device and forwards the commands to it.

A discovery and join protocol is used to allow a device to enter a Jini system and to make the other devices aware of the services that this new device is offering.

A lookup service is the central database, which stores the information about all devices and services in a Jini system. The lookup service stores this information as Java objects. These objects can be downloaded by a device, which wants to make use of the service this object belongs to.

Programming
Model The lookup service leases entries in its database to a service only for a specific period. It is the responsibility of the service to renew the lease or the entry will be removed by the lookup service from its database. This mechanism helps the lookup service to keep its entries accurate. The lookup service broadcasts events to these devices, which have registered interest in receiving these notifications. Events are for example broadcasted if a service joins or leaves a lookup service. These event and notification interfaces are based on the event model used by JavaBeans components.

Jini also provides transaction interfaces that could be used to ensure that out of a set of commands only all or none of them are executed successfully. The transaction protocol consists of two steps: In the voting phase, each object involved in the transaction is asked if it completed the commands successfully. If this is the case, then in the next step the coordinator sends a commit request to all involved objects.

A service provides some functionality to other objects in a Jini system. An object can make of use of such services through interfaces, which reflect the type of service a service object offers.

Services

Discovery, join, and lookup are the core protocols of a Jini system. At the time a new device enters a Jini system, it first uses the discovery protocol to find a lookup service with which to register. The device broadcasts a message on the local network to ask every lookup service to notify the service from its existence. As soon as a lookup service is located, the device uses the join protocol to register with the lookup service. Using this protocol, the service sends a service object to the lookup service. This service object contains Java language interfaces, which are used in the next step.

Discovery, Join

A client who wants to make use of a certain service can use the lookup protocol to search for a service by its type and some descriptive attributes. As soon as a service is selected, the corresponding service object is downloaded into the client.

Lookup

Now, everything is ready and the client can use the service. The client uses the service object to directly communicate with the service provider using RMI.

RMI

The possibility to query for devices based on attribute/value pairs during lookup differentiates Jini from UPnP and helps to get better results from a lookup. The second big difference is the ability to upload not just descriptions to the lookup service, but also Java classes, which the user of the service downloads and applies to access the service.

15.3
Salutation

Salutation is an industry organization, which defined the Salutation architecture to enable devices to discover and use services provided by other devices in the network.

The Salutation Manager is the core component of the Salutation architecture. Each device registers with a Salutation Manager, which could be local on the same device or remote in the network. The Salutation Manager handles discovery of services as well as the communication of the client with the service.

The Salutation Manager itself is network protocol independent. The transport dependent parts are encapsulates in a so-called Transport Manager. A Salutation Manager uses the Salutation Manager Transport Interface to communicate with a Transport Manager.

A registry is part of each Salutation Manager, which contains information about the locally connected services. In addition to this, the registry could also contain information about remote services that are connected to other Salutation Managers. This allows storing information about important remote services locally to ensure faster service discovery.

Registry

Service Discovery is done by comparing the required service types with the characteristics of the locally registered devices and with the characteristics of specific or all services registered with remote Salutation Managers.

At the time a client wants to use a service, the Salutation Manager establishes a Service Session between the client and the service. The client and the service are using messages defined by the personality protocols to exchange data. The management of the session is usually handled by the client and by the service directly, but it is also possible for the Salutation Manager to perform this task in case the client or the service is not able to do so.

Compared to Jini and UPnP, Salutation seams to be the more flexible and viable architecture also due to its independence of the underlying transport layer and the programming language used.

None of the three architectures described is currently leading the market and the future will show which will win the race and conquer the market.

15.4
Further Readings

Jini

http://www.jini.org

The Jini Community is a discussion forum around Jini. It provides a high-level overview on Jini as well as a developer forum to exchange source code, questions, problems and ideas. A news section informs about future Jini Community events.

Directory of Jini Resources

http://www.litefaden.com/sv/jd

This site is a good list of different Jini resources on the Internet. The resources are categorized into articles, reviews, FAQs, books, tutorials and examples, documentation and so on. In addition to this, the site also offers a Jini chatroom.

Salutation Consortium

http://www.salutation.org

The Salutation Consortium is a non-profit industry group that develops and promotes the Salutation Architecture. The Salutation Architecture is free of any royalties and license fees.

The Salutation homepage provides the complete specification, a list of product supporting Salutation, and technical tips.

The membership in the Salutation Consortium is open to interested parties. The membership benefits are listed in detail and an on-line registration form provides a quick way to get involved in Salutation.

Sun's Java Developer Resources

http://www.sun.com/developers

This is the right place to start developing for Jini. On these pages, Sun provides all the required tools to develop for the Java platform, including Jini.

Sun's Jini Connection Technology

http://www.sun.com/jini

Sun has developed Jini and provides on its Jini homepage a technical overview on Jini, a FAQ, and a few demos. The demos show some scenarios how Jini could help in the home, office, or with home audio and video.

Universal Plug and Play Forum

http://www.upnp.org

Microsoft has founded the Universal Plug and Play Forum to get input and feedback from the industry on UPnP. The UPnP Forum provides specifications, white papers, presentations, and may be in future some sample code to the public. To get all benefits and to be able to provide input to the future development of UPnP one has to become a member and sign a membership agreement.

Part IV
Back-End Server Infrastructure

Back-end systems are definitely strategic components of Pervasive Computing setup, since they provide the access to valuable data and content. Tasks and requirements for these systems are changing with Pervasive Computing. Four important issues can be identified:

Tasks and requirements

- Interoperability:
  Upcoming new devices are manifold and have very different capabilities. Various operating systems harden interoperability. The applied communication protocols differ. There is a strong emphasize on mobile networks not based on the TCP/IP standard, which is widely used in today's personal computing.
- Manageability:
  Devices are temporarily disconnected and access servers from changing locations. They run various applications, which need to be deployed and maintained.
- Scalability:
  There is a vast number of registered and concurrent users in pervasive networks. Server systems need to keep pace with the tremendous increase of network connected devices and appliances.
- Security:
  Since IT has left secure closed shop computing islands and has built up a worldwide Internet based network infrastructure, security concerns are more important than ever.

To meet these requirements the existing hardware and software needs enhancements. SAP, Visual Age, DB2, and MQ Series are products, which have been enabled for the new arising tasks. But also entirely new components need to be developed. The Nokia WAP Server, the IBM WebSphere Everyplace Suite, and the SecureWay Wireless Gateway are such products designed for Pervasive Computing.

Typically, the back-end infrastructure comprises three tiers (Figures IV.1 and IV.2):

Tiers of the back-end infrastructure

- The connectivity gateway is the interface between a server and the devices. A gateway is an adapter mapping various device specific networks protocols, such as WAP, to the common TCP/IP based communication protocols applied on the server side.

- Access servers or portal servers access information stored on the content servers on behalf of a client. The server receives HTTP requests, manages communication and application sessions, executes business logic, and interacts with the appropriate back-end systems. This tier is often split up into a Web server and a separate Web application server.
- Finally, the back-end content servers provide data stored in databases or Enterprise Resource Planning Systems, as well as other kinds of web content.

Figure IV.1: Overview Back-End Server Infrastructure

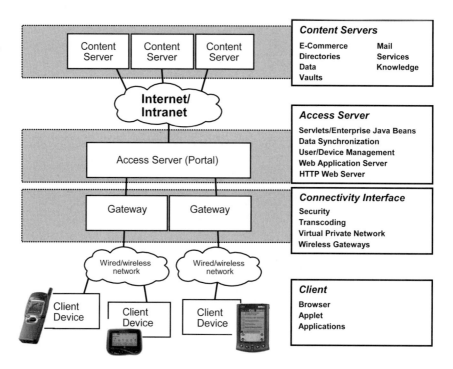

Applet versus servlet

Generally spoken, two different models of providing applications and content to end-user devices can be applied: the applet and the servlet model.

The client-server based applet model executes the application on the client device itself. The server sends the entire application and the corresponding content to the client, which processes it.

Typically this model is applied with devices and applications, which can operate intermittently connected: the user works offline and periodically synchronizes his local data with the remote server databases. Buffered data entries and changes are processed each time the device connects to the network. Examples for this way to work is writing an email, or creating new orders.

Since applications run locally without communication delays, that model is efficient especially when voluminous application code is downloaded only once and stored persistent on the device. A significant disadvantage is the effort needed to deploy and administer the application on all client devices.

The browser based servlet model processes the entire application on the server side. The client just queries data entries and displays the results using a general purpose browser. Since the entire application logic and code is kept on the server there is no need to distribute and maintain the applications on every single device. Certainly, the online connection is always required to run the application, although there are some mechanisms for caching.

Typically, this model is applied when either the deployment of the application to all clients is not feasible or inconvenient for the user, for example when he only uses that application occasionally.

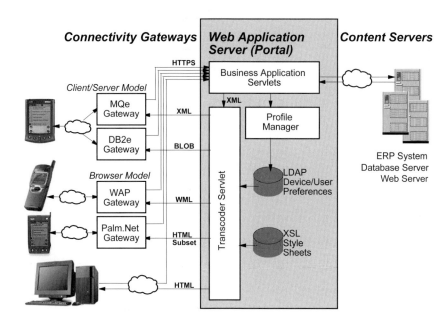

Figure IV.2:
Typical Server
Back-End
Infrastructure as
Applied by IBM

16 Gateways

Pervasive Computing is dominated by a variety of non-IP network protocols, which are either proprietary or specific to a particular industry. As soon as devices want to access the Internet or other computer networks a protocol conversion is required. This task is done by a *connectivity gateway*. Gateways are the point of entry into a server network. They isolate the individual communication channels and their specific requirements from the rest of the back-end systems. Gateways, like a WAP Gateway or a Palm Webclipping Proxy Server convert TCP/IP to protocols optimized for wireless data transmission.

Besides protocol conversion, content can also be adjusted for specific classes of devices. Using appropriate *transcoding gateways*, a web application server can deliver content in various formats to different types of clients. The Proxi-Ware Server is one example for this type of gateway, which will be covered later in this chapter.

Gateways can either be part of the back-end system or they can belong to an user's personal setup. The first possibility could be a gateway which is installed by a service provider offering wireless access to his customers.

The second possibility covers workstations or other systems taking the role of a gateway between devices and the Internet. Local gateways can act as a focal point for personal application management and integration of manifold subordinated devices. One example is the classical PC to which a handheld companion connects. Another example is a *residential gateway* acting as a local interface between appliances connected to an in-home network and the external network, like the Internet. The residential gateway can dispatch external services entering a home network to the connected devices. Or it can route wireless appliances within an office to a high bandwidth Internet line.

16.1
Connectivity Gateway

This section covers three examples of connectivity gateways: the Palm Webclipping Proxy Server, a WAP gateway, and the SecureWay Wireless Gateway.

16.2
Palm Webclipping Proxy Server

The developers of the Palm VII handheld computer decided not to use regular web browsing for wireless Internet access: Initially the Internet was not designed for small devices. To achieve a better performance they introduced web clipping. Instead of using a generic Internet browser, web clipping needs a corresponding standalone client application for *each* particular web site. This application implements the required browsing capability and has to be downloaded and stored locally on the device first. It holds the hard coded request forms for each query to the corresponding web page. Web site developers need to develop and deploy such a query component complementary to the regular HTML files. Web clipping uses HTML version 3.2 with some extensions and limitations. For example, there are special tags to identify a device. Applets, Java Script, frames, and cookies are not supported. Only simple queries and no complex hyperlinks to other web clipping sites are possible. The HTML files are converted into a binary format, which can be rendered on the Palm VII device. That technology minimizes storage requirements and the amount of data transferred. The benefits are lower transaction costs and longer battery life.

Palm.Net Data Center
All Palm VII communication goes through a proprietary connectivity gateway, the Palm Webclipping Proxy Server. That proxy server is part of the Palm.Net data center, which provides wireless Internet access to Palm VII users (Figure 16.1).

The communication between a Palm device and the gateway is based on the UDP protocol, which allows one packet to be sent as a request to the server and multiple packets to be replied. The data is compressed and encrypted to be optimized for wireless transmission with low bandwidth and latency. When a Palm VII sends an UDP query to Palm.Net, the gateway translates that query to a standard Internet request and forwards it to the corresponding web server. The server returns a response, which can be displayed on the Palm VII browser.

Frequency agility, network authentication, and the Certicom elliptic curve cryptographic algorithm are used to establish a secure connection between the Palm VII and the Palm.Net Data Center. The connection from the Palm.Net data center to an Internet server can be protected with SSL using a 40 or 128 bit key. The Palm.Net Data Center itself is protected by firewalls.

Mobitex Wireless Network
Bearer for the Palm VII wireless communication is Bell South's Mobitex Network. It only operates within the US and uses the 900 MHz frequency. Every Palm VII device has a built-in radio frequency transceiver with an unique Mobitex access number. A direct line to Palm.Net Data center is configured.

One obstacle of the applied communication technology is that the radio frequency bearer allows only about 8 kilobytes per second for data transmission. This limits the amount of information, which can be exchanged in a reasonable response time. About 50 bytes for the query and 500 bytes for the response are the recommended sizing.

Although the Palm VII device only accesses the Mobitex network from Bell South, the architecture is basically open for devices based on other network technologies and providers.

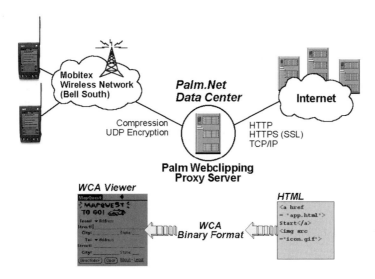

Figure 16.1:
The Palm.Net
Data Center

16.3
WAP Gateway

In order to connect WAP phones to the Internet the standard HTTP Internet protocol used by web servers must be converted into WAP's Wireless Session Protocol (WSP). As described in Chapter 11, this task is done by a WAP gateway. Additionally the content, which must be provide in the Wireless Markup Language (WML) format is translated into a binary content representation (WBXML) to optimize transmission.

Besides content and protocol transformation, the server initiated client notification needs to be supported, which was introduced with WAP 1.2. This "push"-mechanism is a bit circumstantial, since normal HTTP servers can only respond to requests received from clients. The initiator of a push acts as a HTTP client of the WAP Gateway, which takes the role of a Web server, referred to as Push Proxy Gateway (PPG). The PPG forwards the push notification to the WAP client over WSP. Each push message consists of two components: the content to be pushed to the WAP client and a control entity, which holds information, like the message expiration date.

Since WAP is an open standard and independent from a specific telecommunication network any gateway should work with any WAP compliant phone.

Among the available commercial gateway products is *Nokia's WAP Server*. Connections from WAP clients to the gateway can base on the bearers Short

Nokia WAP Server

Message Text (SMS) and Circuit Switched Data (CSD). It is possible to implement other protocols than HTTP on the server side, for instance, when legacy systems need to be supported. An application programming interface (API) enables a service provider to develop particular stand-alone applications, like servlets, directly on top of the gateway. Additional back-end server systems can be omitted this way.

Figure 16.2:
WAP Gateway

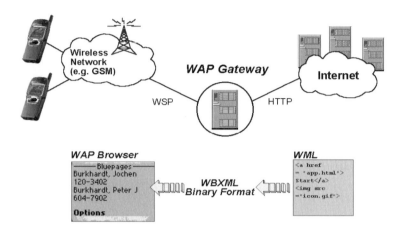

16.4
Wireless Gateway

Gateway   The *IBM SecureWay Wireless Gateway* tunnels the IP protocol through a diverse set of wireless and wireline networks. Applications on client systems can communicate with enterprise networks based on TCP/IP, although this protocol is actually not supported by the communication bearer. Any application developed for the standard TCP/IP interface, will run on top of SecureWay Wireless without modifications.

Client   The *SecureWay Wireless Client* is the counterpart of the gateway and must be installed on the user's client system. A user interface allows to initiate the communication to the gateway. The mobile user can select the network type, which meets best his current needs, without making any changes to the installed applications.

In order to setup the network transparent to the application, different network characteristics are encapsulated by SecureWay Wireless. On the client side, all network specific details are hidden beneath the common TCP/IP interface layer. SecureWay Wireless works like an additional network device driver. It intercepts the TCP/IP communication and routes the data traffic over the chosen non-IP networks. Inside the gateway, Mobile Network Interfaces (MNI) expose the access to a particular network type.

The list of supported network protocols include:

- Public and private Packet networks (CDPD, DataTAC, GPRS, and Mobitex).
- Analog and digital cellular networks (GSM, CDMA, TDMA, and AMPS).
- Satellite networks (Norcom).
- Wireline (Cable modem, DSL, PSTN, ISDN, and Dial-up ISP).
- LAN connections (Ethernet, Token Ring, and wireless LAN).

Besides the simple IP-based communication between client and gateway, SecureWay Wireless optimizes the network traffic using data compression mechanisms. Increasing efficiency is extremely important for low speed wireless networks. The benefits of less data traffic are better response times and often lower network fees. Additionally, an idle connection can be disconnected automatically and will resume as soon as communication continues. The user will not notice that interrupt (short-hold mode). *Efficient*

For reliable data privacy, all data transmitted through public networks from the client device to the gateway can be encrypted with RSA. Two way client/gateway authentication mechanisms prevent unwanted access. *Secure*

Besides gateway and client, IBM provides *the SecureWay Wireless Gatekeeper.* This tool enables system administrators to configure multiple wireless and dial-up network, and define users, resources, and access rights. LDAP is used to store user profiles and network management data.

WebExpress is an enhancement for HTTP communication through SecureWay Wireless. WebExpress features HTTP header reduction, as well as HTML data stream compression and content caching.

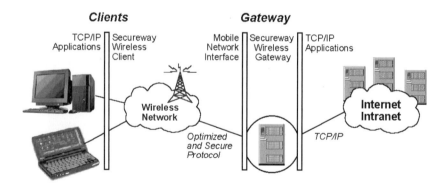

Figure 16.3: SecureWay Wireless Gateway

16.5
Transcoding

Transcoding is a content adaptation technology, which tailors information for a specific device by transforming its format and representation. Multimedia data as it can be found on Web pages, is filtered, converted, and reformatted, until it matches the capabilities of the displaying device.

Why transcoding?
The purpose is obvious: When providers deliver content to various pervasive clients, they need to accommodate device specific constraints such as limited memory, slow data transmission, and small screens. Transcoding automatically translates content into different representations for each class of receiving client systems. This simplifies authoring, deployment, and maintenance dramatically! An arbitrary multimedia web page can be provided to Internet TVs, handhelds, and WAP phones without change. It is even possible to adapt content from legacy systems into a standardized Internet representation, or transcode Web pages to the proprietary format of a specific client device.

Besides adapting content according to particular device capabilities, another aspect of transcoding gains importance: information is personalized for a particular user, depending on his preferences and organizational policies. For instance, content of interest can be transcoded into individual user interface styles or a preferred language.

Transcoding Proxy
Transcoding is typically done by a gateway or proxy, which intercepts the data exchange between the content server providing multimedia data and the client rendering the received information (Figure *16*.4). The transcoding proxy analyses the content and identifies the included media objects. In order to determine graphical navigation buttons, images, and related textual content, the semantics of each object is examined. According to defined preferences, appropriate methods for content manipulation are invoked. Preferences can be device capabilities, user profiles, communication bandwidth, and others. They indicate what kind of information has to be transcoded in which way.

Modality and fidelity
This content adaptation comprises two dimensions:

- *Modality*:
  Multimedia includes video clips, images, text and audio data in manifold formats. Since not all of these modalities are suitable for all devices, transcoding translates data according to the targeted devices needs.
- *Fidelity*:
  Reducing the fidelity of information is another approach to reach even small devices with multimedia information. Compressing and filtering are techniques applied in this context.

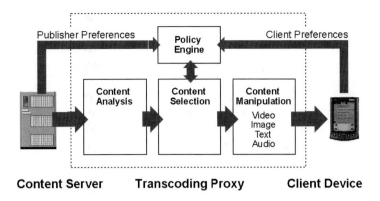

Content Server   **Transcoding Proxy**   **Client Device**

The following list shows several common methods for content manipulation:

Content manipulation methods

- Conversion of images to hyperlinks, allowing the user to retrieve them separately only when explicitly needed.
- Compression and reduction of scale and color level of images.
- Substitution and removal of images.
- Conversion from one image format to another.
- Translation of XML documents according to a given style sheet.
- Video skimming.
- Video-to-image conversion.
- Text-to-speech conversion and speech recognition.
- Text summarization from detailed text to headlines.
- Text reformatting from tables to simple lists.
- Language translation.

16.6
InfoPyramid Framework

An implementation of transcoding technology is the IBM InfoPyramid Framework. This framework allows to manage media objects with different modalities and fidelities. Media objects can be either stored in different representations or be created on-the-fly in order to meet the capabilities of a requesting device.

Media objects correspond to cells in an *"InfoPyramid"* (Figure *16.5*). Within the pyramid, the fidelity of information representation decreases from bottom to top. The lower cells refer to detailed information, such as high-resolution images or comprehensive texts. The upper cells correspond to strongly compacted information, such as simplified images or summarized text. When moving horizontally within the pyramid, the modality changes. For instance text-to-speech conversion changes a text based representation to audio.

The framework includes methods to manipulate media objects. These are referred to as *transcoders* and are distinguished between *translators* and *summarizers*. Translators convert modality, while summarizers change fidelity. The framework is extensible and allows to add new transcoders using a provided developer toolkit. It is also possible to define additional modalities, such as different languages of text.

The framework has been implemented as a stand-alone proxy server as well as a servlet. The proxy server intercepts HTTP requests and responses between the user and any selected Web server. The servlet version runs directly on a particular Web application server.

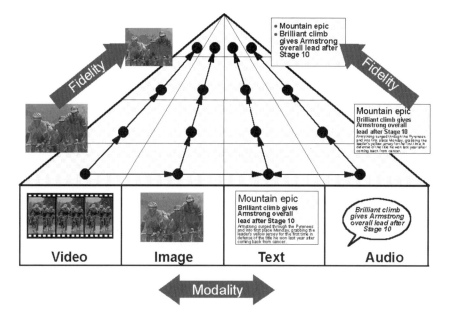

16.7
ProxiNet Transcoding Gateway

ProxiNet's transcoding technology is based on the *ProxiWeb* browser for client devices and the *ProxiWare* gateway. Any selected web site is retrieved, transformed, and delivered in a very compact representation to the client's ProxiWeb browser. ProxiWeb is an ultra-thin browser for Palm OS and comprises only about 120 kilobytes. When the device is connected to an Internet provider, the browser gives immediate Web access. Downloaded sites are kept within a cache and can be viewed offline at a later time again.

16.8
Residential Gateway

The residential gateway is a computer that acts as a central bridge between the outside world and the networked devices in the home. Figure *16.6* shows a residential gateway with several connectivity options in a home. There are two major aspects to the residential gateway.

On one hand, the residential gateway integrates "traditional" communication media – telephone and cable TV – into the home audio-visual system. Through a caller ID function such a system could, for example, announce a caller's identity through the home audio system. The resident could then decide whether or not to get out of his Jacuzzi to pick up the phone.

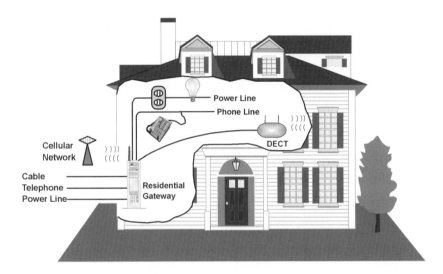

Figure 16.6:
Home Networks

On the other hand, work is carried out to define the gateway as a bridge between the home and the Internet. Valid services should be able to access the home systems. For example, the refrigerator repair center should be able to access the appliance in the home for diagnostic purposes. However, unauthorized access must be blocked – it must not be possible for unauthorized persons to determine information about the residents.

16.9
Further Readings

IBM Transcoding Solution

http://www.research.ibm.com/networked_data_systems/
transcoding

This site includes documents and publications related to transcoding. An online demonstration transcodes arbitrary web pages, which can be selected by entering their URL.

IBM Networked Home

http://www.ibm.com/pvc/nethome

This web page includes a scenario of a networked home environment.

Information Delivery Infrastructure in the Mobile, Wireless Age

http://www.proxinet.com/technology

This white paper covers the ProxiNet transcoding approach as well as a generic analysis of different technologies to deliver content to small devices [FOX99].

Nokia WAP Gateway

http://www.nokia.com/corporate/wap/wap_server1_1.html

The technical features of the Nokia WAP gateway can be accessed through this link.

Palm VII – Wireless Internet Access

http://www.palm.com/pr/palmvii/7whitepaper.pdf

This white paper explains Palm's approach to provide wireless network access through the Palm.Net gateway [PALMVII].

Panja Home Systems

http://www.panja.com

This site features a photo tour through a networked home installation.

Point Clark

http://www.pointclark.net

A commercial provider for residential gateway software and services.

ProxiNet

http://www.proxinet.com

The place where users can subscribe to the ProxiNet transcoding services for Palm devices.

Residential Area Highway Project

http://hoegaarden.ec.ele.tue.nl

This page includes interesting links to standardization activities, documents, and projects related to in-house networks and residential gateways.

SecureWay Wireless

http://www.ibm.com/software/network/mobile

Information about IBM's SecureWay Wireless product.

17 Application Servers

In recent years, the World Wide Web has evolved from a network of static information to a dynamic application deployment and management system.

17.1
Architecture and Components

Today's e-business solutions are usually designed in a multi-tier architecture. In this design paradigm, the Web Application Server is the key component that connects a Web browser front-end with back-end applications (Figure 17.1). Web applications connect to databases, Enterprise Resource Planning Systems, or other external data repositories to get access to the requested content.

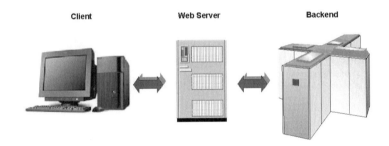

Client   Web Server   Backend

Figure 17.1:
Multi-tier
Architecture
of e-business
Solutions

The success of this architecture is based on the fact that the Web browser is a low-cost, universal, and easy-to-configure frontend client.

Today, several technologies are used to develop applications for Web servers:

- Static HTML Requests: The basic procedure of serving HTML requests is the most common and the simplest task a Web Server can perform. The Client sends an HTTP request to the Web Server, the Web Server retrieves the data from the server's file system and sends it to the client.
- Common Gateway Interface (CGI): CGI was an early solution for the problem of creating interactive web-based applications. These applications can call CGI programs through the Web Server. They assemble HTML web pages dynamically, which are returned to the client.
- Enterprise Java Beans (EJBs): A very powerful part of a Web Application Server is EJB support. EJBs provide an infrastructure of services and func-

tionality to access databases and other back-end data repositories. They help the programmer to focus on the application without having to concern about the server specific infrastructure.

- Servlets: Similar to CGI programs servlets are Java programs which dynamically produce HTML pages for given client requests.

More details on EJBs and Servlets can be found in the following sections.

17.1.1
Java Servlets

Servlets can be defined as server side Java classes which are platform- and protocol-independent and run in a Java enabled Web server to provide dynamic HTML content to clients. Initially, servlets were supported in the Java Web Server from JavaSoft.

Figure 17.2:
The Basic Servlet
Process Flow

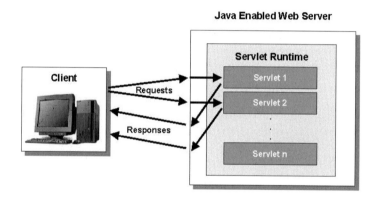

The basic process flow in a Servlet takes the following steps:

1. The requests are sent by the client to the server.

2. The server sends the request information to the servlet.

3. The servlet generates the response, which is passed to the server.

4. The server sends the response back to the client.

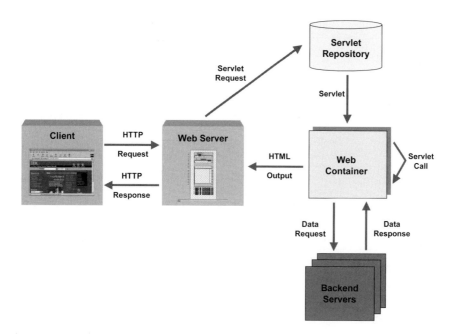

The server has to initialize, start, and destroy the service methods and servlet instances. When servlets are requested for the first time, they can be loaded dynamically. Another possibility is to configure the Web server in a way that defined servlets are loaded during initialization time of the Web server.

There are different possibilities for a Web browser to access servlets:

- A common way are HTML forms where user data is sent to the servlet using POST or GET methods.
- An other possibility is the use of the SERVLET tag. The HTML SERVLET tag is supported by some Web servers. In this case the SERVLET tag in the HTML page is replaced by the output of the servlet's service.
- A third way of accessing servlets is through hypertext links. A link invokes the service or doGet method of the servlet.

Servlets use some special packages defined in the Java servlet API. The API consists of two packages:

- javax.servlet.http: This package contains classes and interfaces for the support of HTTP servlets.
- javax.servlet: This package contains servlet classes and interfaces for protocols supporting the request/response paradigm like SMTP and FTP.

The communication between the Web server and the servlet takes place via the servlet interface, which defines the methods described in the following table.

Method	Description
init	When a servlet is loaded for the fist time, this method is called. The method has to be implemented and called only if setup tasks have to be performed once and to be completed before requests are handled.
service	This method is called each time a client makes a request and a response has to be generated. A ServletResponse and a ServletRequest object is passed to this method.
destroy	This method has to be implemented to perform cleanup tasks. It is called whenever the Web server unloads the servlet.
getServletConfig	In order to return the ServletContext or initialization parameters a ServletConfig instance is needed. The getServletConfig method returns this instance.
getServletInfo	This optional method returns defined information about the servlet.
doPost	When a servlet method is called, it determines whether the call is a POST or a GET request. If it is a POST request, the doPost method is called.
doGet	When a servlet method is called, it determines whether the call is a POST or a GET request. If it is a GET request, the doGet method is called.

Here is a simple example of a servlet handling HTTP doGet requests:

```
public class SmallServlet extends HttpServlet
{
 /*
 * Servlet Example for handling a HTTP doGet request
 */
 public void doGet (HttpServletRequest req,
                    HttpServletResponse resp)
 throws ServletException, IOException
 {
    String      heading = "Welcome to the demo Servlet";
    PrintWriter  output;

    // initialization of response field
```

```
    resp.setContentType("text/html");


    // write the response data
    output = resp.getWriter();


    output.println("<HTML><HEAD><TITLE>");
    output.println(heading);
    output.println("</TITLE></HEAD><BODY>");
    output.println("<H1>" + heading + "</H1>");
    output.println("<P>This is the text Servlet's content");
    output.println("</BODY></HTML>");
    output.close();
  }
}
```

The SmallServlet class extends the HttpServlet class, which implements the Servlet interface.

The doGet method in the HttpServlet class is overwritten by the servlet. The request is represented by an HttpServletRequest object, the response is represented by an HttpServletResponse object. The reply is sent using the Writer object obtained from the HttpServletResponse object because text data is returned to the client.

Java Server Pages (JSPs) are a way to easily create servlets for programmers who are familiar with HTML. Although JSPs are quite different from servlets during development, they are actually precompiled to servlets at runtime. JSPs are stored in the document hierarchy of the Web server. When the JSP is invoked for the first time, it is compiled to a servlet by the JSP compiler. After this action the JSP is treated like a servlet for the rest of its life cycle. The JSP is automatically recompiled whenever the Web Application Server detects changes.

Java Server Pages

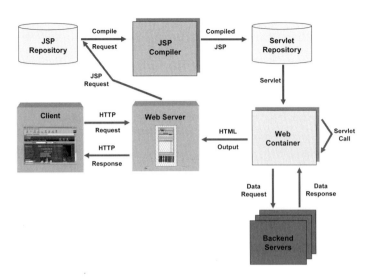

17.1.2
Enterprise Java Beans

In March 1998, Sun Microsystems announced the Enterprise Java Beans 1.0 specification. Many of the companies developing products in the domains of databases, transaction monitors, and CORBA, announced to implement the specification.

The Enterprise Java Beans (EJB) specification defines the EJB classes and the interfaces between the EJB technology-enabled server and the component. The specification shows a methodology for the separation of business logic and the technology of handling persistency, transactions, and other middleware-related services. Thus, one goal of EJBs is to simplify the process of accessing enterprise data via the Web without having to code all of the middleware.

Now, programmers can write large building blocks that can be reused, updated, and combined in order to implement new programs. Applications can be created concentrating on the business part which is packaged in different EJBs. Another advantage of this component-based programming model is the fact, that the programming, test, and maintenance of solutions built of EJBs and their distribution across distributed servers, becomes less time consuming.

The EJB specification defines a container model, a definition for each of the services that this container provides to an EJB, and the management capabilities of the container.

EJB Components   EJB classes are Java classes representing business-logic components. EJBs run in an EJB container, which can contain one or more EJB classes. These classes are identified by their home interface, which allows clients to create, find, and remove instances of an EJB. The home interface can be given a name for identification by using the deployment descriptor during the EJB development time or later on by using deployment tools.

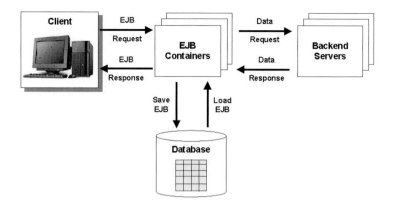

There are two types of EJBs which have been specified:

- Session Beans
- Entity Beans

Session beans usually show a lifecycle behavior like the one of their clients. The session bean is a process running on a server. Clients can request services from the session bean. This means that every client requesting a service will create its own instance of the bean. Session beans contain conversational states that are not persistent and they have to communicate their management mode to the container. Management modes of session beans are separated in two types: stateful and stateless.

Session Beans

The session bean is called stateless if it can be destroyed in case of a memory shortage. No client related information will be stored in that case. Stateful session beans keep the client related information.

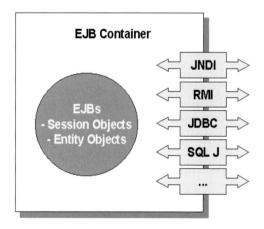

*Figure 17.6:
Enterprise Java
Beans (EJB)
Components*

Entity Beans   Entity beans are used for handling persistent data. The fields of entity beans are mapped to a data source. In the most common case, the data source could be a row or a table of a database, but it is also possible that the data source is another form of data storage or representation.

Entity beans can be separated into two types according to the management of the data persistence: Container-managed persistence and Bean-managed persistence.

- With Container-managed persistence, it is hidden to the user which data source provides the persistence. This means that for these EJBs their container takes over the calls to the data source. The programmer only has to define the data fields which have to be persistent. Container-managed persistence is on one hand easy to use for developers, on the other hand the difficult management parts are "transferred" into the container. This means, that the container provider also has to provide a set of tools, e.g. for mapping EJB fields to the data source.
- Bean-managed persistence means, that the management of calls to the data source has to be implemented within the bean itself. It is obvious that this kind of EJB is not suitable for large applications because it is connected very tightly to the underlying architecture.

EJBs must only be accessed via a proxy provided by the EJB container to be handled properly by the server. The container then can control the persistence, security, caching, and connection management. The EJB Home and EJB Object interfaces, respectively the instances of these interfaces, facilitate these tasks. The server performs management tasks under the cover by mapping calls to these interfaces, which then call the EJB itself that controls the transactions and the database access.

Figure 17.7:
Enterprise Java
Beans (EJB)
Architecture

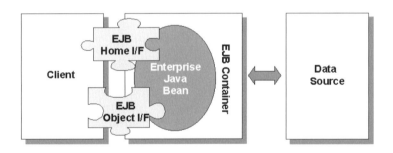

The EJBHome   Instances of the EJBHome interface allow clients to find EJBs. Session
Interface   beans as well as entity beans have methods in the EJB Home interface which are able to create an instance of an EJB Object in a EJB Container. For each EJB class one EJBHome interface instance exists in an EJB Container.

Instances of the EJBObject interface control the access to the methods provided by an EJB component. The EJBObject class has the role of a proxy which handles the communication between the EJB and the client.

The EJBObject Interface

17.2
IBM WebSphere Application Server

IBM WebSphere comprises a set of products with the goal to help developers deploying Web applications (Figure 17.8).

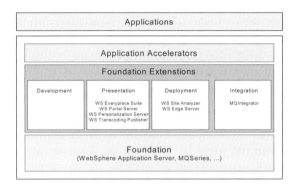

Figure 17.8: IBM WebShere Platform

In this chapter the focus is on the WebSphere Application Server (WAS) itself. The WebSphere Application Server provides tools to facilitate the management and deployment of Web applications. It provides Java 2 Platform, Enterprise Edition (J2EE) compliance and built in support for the key Web Services open standards. As this server is Java based, the components of the applications are usually Java servlets, Enterprise Java Beans or Java Server Pages. These applications can be installed and managed by a Java-capable Web browser.

The main components of the IBM WebSphere Application Server are the application server, the Web container, and the EJB container (Figure 17.9).

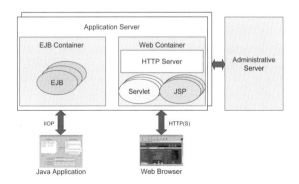

Figure 17.9: Major Components in WAS

These architectural components and the way they map into the J2EE component architecture will be explained in more detail.

The application server returns customized responses to a client's request. The code of an application including Servlets, JSPs, EJBs and supporting classes run in the application server. According to the J2EE component architecture, Servlets and JSPs run in a Web container, EJBs run in an EJB Container. It is possible to define multiple application servers, each running on its own Java Virtual Machine. An administrative server also runs on its own Java Virtual Machine.

The Web container processes Servlets, JSP files and other type of server-side includes. Each Web container contains a single session manager. When handling Servlets, the Web container creates a request object, invokes the Servlet's service method and the servlet's destroy method when appropriate. After the Servlet is unloaded, the JVM performs the garbage collection.

The Web container also supplies the PageListServlet to call a Java ServerPage (JSP) by name. This Servlet uses configuration information to map a JSP name to a Uniform Resource Identifier (URI), and the URI specifies a JSP file in the Web module.

The WebSphere Administrative Console can be used to edit the configurations of Web containers.

The EJB container provides all the runtime services needed to deploy and manage Enterprise Java Beans (EJBs). This container runs in a server process that handles requests for session and entity beans and provides low-level services, including threading and transaction support.

The latest version of IBM's WebSphere Application Server (Version 5) supports dynamic application integration via messaging services like the Java Messaging Service (JMS), container managed messaging, and the support of the Java Connector Architecture.

Another focus of the new WAS version are the Web Services support and the extension of XML-based deployment and administration.

WebSphere Everyplace Suite WebSphere Everyplace Suite (WES) is a presentation extension of the WebSphere Application Server. WES integrates a set of IBM products into one consistent product specially designed for a pervasive environment. It comprises the following modules:

- Connectivity: integrated connectivity gateways allow to communicate with wireless and wired networks including GSM, WAP, CDMA, TDMA, and TCP/IP. MQe is a very robust messaging technology, which can transfer data even to intermittently connected devices in a reliable way.
- Content handling: Transcoding plug-ins can translate directly one data format to another, in order to adapt content for specific device capabilities. The IBM Transcoding Publisher is part of WES. A synchronization engine is provided to keep content on devices and data servers consistent.

- Security: end-to-end security is achieved with authentication and access control mechanisms. There is a device independent login mechanism for users. WES supports Virtual Private Network technology (VPN). A virtual private network connects the resources of one network with another network by tunneling through the Internet or a public network.
- Optimization: load balancing and caching are applied to achieve scalability and face a rapidly growing community of pervasive devices.
- Management Services: device and subscriber management, billing, accounting, customer care, and provisioning help to control connectable devices. Tivoli Subscription Manager is the underlying technology, which is applied for these purposes.
- Base Services: underlying functionality comprises the WebSphere HTTP Server, directory services, and a common installation procedure for all components. A web console allows integrated management of the server setup.

WES applies open standards, wherever possible in order to achieve a flexible, extensible product, which is highly interoperable: For example the HTTP server part is based on top of the open source Apache server. SyncML, WAP, LDAP are other standardized components.

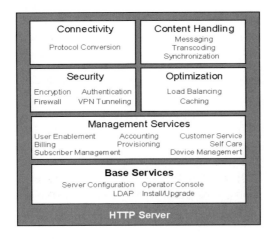

Figure 17.10: WES Components

17.3
Oracle9i Application Server

Like most of the Applications servers that are introduced in this book, the Oracle 9i Application Server Release 2 (Oracle 9iAS) is standard based and provides a Java2 Enterprise Edition environment. Oracle 9iAS provides an environment that supports the development and deployment of Web services, Enterprise Portals, Wireless and Business Intelligence Applications (Figure 17.11).

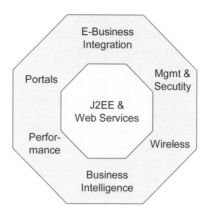

Oracle 9iAS uses an Apache-based HTTP Server to serve static and dynamic content. In addition to the features provided by the standard Apache Server, the Oracle HTTP Server provides services like single signon, public key security, CGI performance improvements and rewriting of Web requests. Through a Plug-in, the Oracle9iAS also supports other Web listeners like Netscape or Microsoft's Internet Information Server.

For the development of J2EE applications the Oracle Application Server provides containers for the implementation of Servlets, Java Server Pages and Enterprise Java Beans.

Web Services enable loosely couples, reusable software components that semantically encapsulate discrete functionality and are distributed and programmatically accessible over standard Internet protocols. Oracle9iAS provides simple and Complex Web Services, where simple Web Services provide basic "request/response" functionality and support the three primary Standards – WSDL, UDDI and SOAP. Complex Web Services can be characterized as multi-party, long-running transactions that involve sophisticated security as well as business-to-business collaboration. Complex Web Services comprise primarily ebXML and RosettaNet standards.

Oracle9iAS
Wireless
With the Oracle9iAS it is possible to license an option called Oracle9iAS Wireless. This option provides tools for the deployment of Web content to mobile devices. It includes mobile services such as PIM and e-mail, location based and messaging services.

17.4
BEA WebLogic Platform

The BEA WebLogic Platform 7.0 includes BEA WebLogic Server, BEA WebLogic Portal and BEA WebLogic Integration (Figure 17.12).

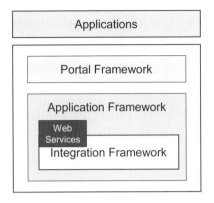

Like the Oracle9i AS and the IBM WebSphere Server, the BEA WebLogic Platform supports Web Services. The Web Services support in the BEA Platform includes the Portal Framework for the presentation and consumption of Web Services, the Integration framework for the orchestration of Web Services around business processes, a runtime framework for deployments and a development framework to build, deploy and test Web Services. Like IBM and Oracle, BEA contributes to and uses standards in the Web Services area like UDDI, SOAP, WSDL and WSRP.

The work generated during development with the BEA Platform also results in standard-based J2EE business logic.

17.5
Sun ONE Web Server

The Sun Open Network Environment (ONE) Web Server 6.0 formerly known as iPlanet Web Server contains the following software modules: Content Engines, Server Extensions, Runtime Environments and Application Services (Figure 17.13).

The Sun ONE Web Server contains three content engines: The HTTP engine, the Content Management engine, and the Search engine.

These content engines make up the Web Publishing layer of the Sun ONE Web Server. The HTTP engine represents the core of the Sun ONE Web Server. From a functional perspective, the rest of the Web Server resides on top if this engine for performance and integration functionality. The Content Management engine supports the creation and storage of HTML pages, Java Server Pages, and other files. The Search engine enables users to search contents and attributes of documents on the server.

Content Engines

Sun ONE Web Server extensions are used to replace or extend functions of the server to suit special customer needs. In the core Web Server architecture, the Common Gateway Interface (CGI), the Netscape Server Application Programming Interface (NSAPI), Java Servlets and Java Server Pages are used as extensions.

Server
Extensions

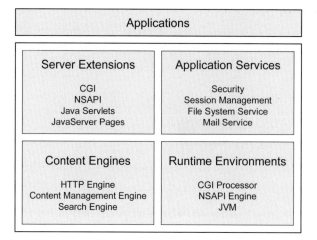

The runtime environments included in the Sun ONE Web Server are directly related to the Server extensions. Thus, there is a CGI processor, a NSAPI engine and a Java Virtual Machine to support the Server extensions.

Finally there are a set of application services, like security, session management, file system and mail services, which are included to the Web Server architecture.

17.6
Further Readings

Apache

http://apache.org

The renown open source web server is presented at this website.

Application Framework for e-business

http://www.ibm.com/software/developer/library/security/index.htm

A paper describing how security requirements can be addressed in a web server environment.

Application Server Zone

http://www.appserver-zone.com

A list of available web application servers with links to the corresponding vendors.

BEA Systems

http://www.beasys.com

Information about the BEA WebLogic server.

Servlet Fundamentals

http://developer.java.sun.com/developer/onlineTraining/Servlets/Fundamentals/index.html

A short course explaining the fundamentals of servlets.

Enterprise Java Beans

http://java.sun.com/products/ejb

Sun's Enterprise Java Beans website containing developer information, documentation, and more.

JavaServer Pages

http://java.sun.com/products/jsp

Sun's website on JavaServer Pages containing developer information, documentation, and more.

Microsoft

http://www.microsoft.com/servers

The Microsoft site for Windows based server systems.

Web Developers Journal

http://www.webdevelopersjournal.com/hubs/prophub.html

Articles about technical web development issues, like JavaScript, CGI, etc.

WebSphere

http://www.ibm.com/software/webservers/appserv

Downloads and Information about IBMs WebSphere Application Server.

WebSphere Everyplace Suite

http://www.ibm.com/pvc

IBM's homepage for Pervasive Computing activities.

18 Internet Portals

Portal technology is playing an increasing role in computing. Service providers are rolling out portals to allow users to create customized web sites that display exactly the information of interest. Corporations are rolling out portals to provide employees and business partners customizable access to corporate information.

Portals aggregate information from diverse sources to make it available through a web interface. The information sources can be backend systems within the company; news feeds from external providers, or email and calendar information from corporate communication systems. The portal makes information from these sources available to the user. In general, the user will be able to place the information on web pages according to his or her needs. For example, one portal page may be dedicated to communication and may contain email, calendar, and instant messaging information. A second web page might provide access to a corporate billing system, etc.

Portals aggregate information

Many corporations have mission critical applications running on legacy systems that are too large and complex to be retrofitted with modern point-and-click user interfaces. Portal technology can be used to provide a web interface for such applications. Programs running within the portal obtain access to the information through the usual interfaces for that application and format the information for web display. If the application requires user input, the web page can contain entry fields to solicit data from the user. The portal programs capture the user input and format it appropriately for the backend application.

Web interfaces for legacy applications

Under the older two-tier client-server computing model, the client devices connected directly to the backend system. The backend system had to prepare data for display on each supported client device. Since corporations typically have many applications running on many different systems, a great deal of effort is needed to enable the applications to support multiple devices. If there are N applications to be made available on M different devices, then specific device support modules must be written N x M times.

A portal is a middleware component that runs on the application server tier between the client tier and the data storage tier according to the three-tier computing model. The portal can obtain data directly from the data storage tier or can access the data indirectly though additional application servers. The applications themselves have to be adapted at the most one time in order to provide information to the portal.

Portal as middleware

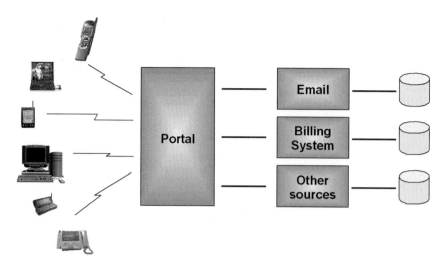

As illustrated in Figure 18.1, the portal acts as a mediator between the information sources an the client devices, it is well situated to adapt the content to the specific characteristics of the client device. When the portal is accessed, it can recognize the type of client device sending the request. Characteristics such as preferred mark-up language, screen geometry, color capability, and input capability can be retrieved from the portal configuration data for the client device. This information can be used by the portal programs to optimally format the data for display.

18.1
Portal Functional Overview

The portal itself consists of a standard framework and application-specific programs that run within the portal runtime environment. The framework provides services that are needed by all applications. These services include user authentication and security functions, services for configuration and administration of portal pages, and connectors for obtaining information from backend systems.

The application-specific programs running within the portal are known as portlets (see Figure 18.2). The portlets use the data access connectors and other services to obtain data from the backend applications they support. They use portal facilities to format the data optimally for the client device. The framework allows many portlets to be placed on a single page. When the framework renders the page, it obtains the information to be displayed from each portlet in turn and integrates it onto the page.

Figure 18.2:
Portlets

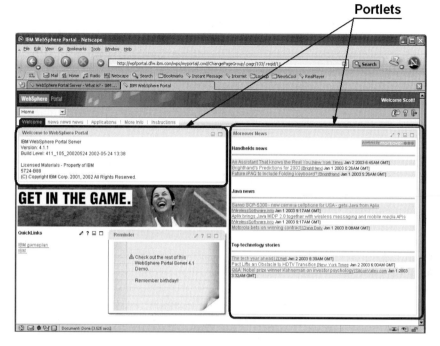

Depending on the portal implementation, it may be possible or even necessary for portlets to produce output optimized for specific devices This is especially true when more than one mark-up language is supported. When being rendered, the portlets will be able to obtain information about the target device or the desired mark-up language from the framework.

Portlets

Users can add, delete, and move portlets around on the portal pages. The user-specific configuration data for the pages is stored locally to the portal. Each portlet can also have user-specific settings that are also stored locally. For example, a POP3 email portlet might have a setting for the destination mail server and user ID. These settings would be different for each user.

Portlet configuration

For portals that support multiple devices, it is important that pages are rendered appropriately for the device. Rather than displaying portlets side-by-side, the portlets may be displayed sequentially on client devices with limited display capability such as WAP or i-mode phones, for example. Such portals may allow separate pages to be set up for devices with limited display capability, since not all portlets may support all devices.

Page rendering

Each user is required to logon to the portal so that the user-specific data can be retrieved. During the logon process, the user is authenticated. This is usually done through user ID and password, but can also be carried out through use of a smart card or other security token. Once the user is authenticated, a session between the portal and the client device is set up.

Authentication

With most portals, not all users have the same rights to create, delete, and configure resources. Regular users may be able to change their private pages, but may have view rights only for common pages. Some users may be able to view pages containing confidential information while others cannot. When the user attempts to access a resource, the portal can check with an authorization system to determine whether the access should be allowed.

Portal administration
With users to be administered, access rights to be set up, and pages to be configured, the portal is a complex system. The portal must provide facilities for system administrators to perform these tasks. The portal may even allow for a hierarchy of administrators, with low-level administrators being allowed to make certain changes but not others, while higher-level administrators are allowed make any necessary changes.

Web services
Web services allow portlets to be shared between portals. The actual programming logic of the portlet can reside on one portal while the user while the formatted portlet output is made available to users through use of a portlet proxy on a second portal.

For example, the human resources department might create a portal allowing employee access to personal records. The administrators of an employee portal in a manufacturing plant might want to make that same portlet available to plant employees without allowing access to the complete human resources portal. This can be done by installing a portlet proxy for the desired HR portlet on the plant employee portal.

An effort is underway to standardize web services for portals. The Web Services for Remote Portlets effort is described in Chapter 15, Web Services.

The following sections will describe portal technology and its uses in more detail. Before delving into the topic more deeply, it should be noted that the term "portal" is new to computing and can mean different things to different people. For some, the term simply refers to a web document management system. For others it refers to a product suite that can be used to create customizable web sites – a large amount of programming effort is required to customize the products for use.

For the purposes of this book, the term "portal" will refer to a middleware product that provides a runtime environment for portlets as well as an engine that aggregates information from many portlets onto customizable pages as described in the preceding paragraphs.

18.2
Types of Portals

Portals can be divided into two broad categories – service provider portals and enterprise portals.

Service provider portals
Service provider portals offer information to subscribers. They are characterized by large numbers – even millions – of users accessing information of general interest such as stock quotes, news articles, sports scores, cooking recipes,

etc. The information offered by such portals typically does not originate from within the service provider company itself, but is obtained through the Internet from external sources.

Service provider portals must also be highly available. Significant downtime could cause subscribers to switch to another information provider. The portal installation must also be highly resilient in terms of peak load. During popular sporting events or in times of crisis the load on the system can climb to several times the average load.

Self-registration is another feature of may service provider portals. Many service providers to register with the web site in order to receive an email account or premium services. Since Internet users would not tolerate long delays associated with postal or telephone registration procedures, it must be possible for users to automatically register with the service provider portal. This is generally done through filling out an HTML form on a web page.

Self-registration

Enterprise portals are characterized by smaller numbers of users accessing information that originates from backend systems within the corporation. Information from external sources may also be provided, but the main business need is for corporate information access. The portal may make the information available to employees through the corporate intranet or through the public Internet. Figure 18.3 illustrates the information source differences between service provider and enterprise portals.

Enterprise portals

Many companies install special portals for specific purposes. A human resources portal may personal and salary information available to employees. An engineering portal might pull all of the design and development information for a large project into a central location. A manufacturing plant portal coupled with workflow technology might help coordinate production processes by providing workers at assembly stations with timely information about the work to be done. A supplier portal might make internal stock information available to suppliers so that resupply can automatically take place when stock runs low.

Naturally, it is also important for enterprise portals to be highly available. However, for many types of enterprise portal, it is possible to schedule downtime for maintenance purposes – for example, on weekends or over holidays.

Self-registration capability is also less important for enterprise portals than for service provider portals. Corporations generally have authentication and directory systems, often LDAP-based, in place for employees and business partners. Arbitrary people cannot just sign up for a typical enterprise portal. Instead, users are added by administrators as a result of business decisions. Enterprise portals are often set up to use the standard corporate authentication system.

The following sections describe common categories of enterprise portals.

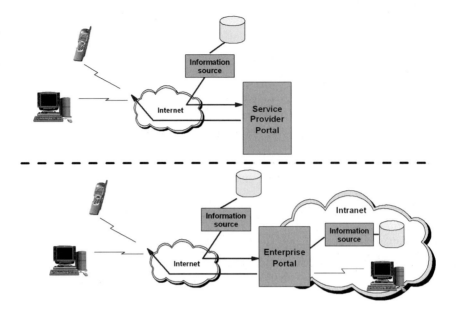

18.2.1
B2E Portals

Business to Employee (B2E) portals are built by corporations for use by their employees (see Figure 18.4). These portals are internally oriented and are accessed through the corporate intranet. B2E portals are integrated into the corporate IT infrastructure and use corporate directory and authorization stores to determine which employees are allowed to access information. Some of the uses for B2E portals were already mentioned above.

Employee information portal

An employee information portal can act as a focal point for news about the company, internal telephone number and address information, descriptions of processes and procedures, and management to employee communications. The information comes mostly from internal sources, although news from external sources can also be offered. Employees can customize portal pages by adding, removing, and editing portlets so that job relevant information is displayed.

Although the B2E portal is accessed through the intranet, it can be used by employees who are not on-site, because many companies have gateways that allow employees to access the intranet either through the Internet or through a dial-up connection. Through use of corresponding gateways, the data can be sent via the cellular data net to i-mode phones, WAP phones or PDA devices.

Portal information for mobile employees

The B2E portal can be used to make corporate information available to mobile employees. For example, a logistics company might create a portal for its delivery personnel. Portlets would format delivery and appointment information appropriately for display by on-board systems in the delivery vans or by hand-held computers carried by the delivery personnel. When a package is de-livered, special hardware could scan the identifying bar code or read the smart label affixed to the package and send it to the portal so that the company can maintain very accurate parcel tracking.

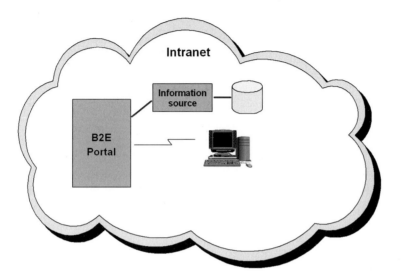

Figure 18.4: B2E portal

Security is very important, so the portal must insure that only employees ac-cess the portal, and that only employees with a "need-to-know" can view sensi-tive data. Corporate LDAP (Light-weight directory Access Protocol) directories often organize employees into user groups that help the portal control access to information. Portal administrators can restrict access to the information through user groups. For example, a corporate inventory system portlet may allow all employees to view the equipment assigned to them, but allow only managers to change the equipment assignments.

Security

18.2.2
B2B Portals

Business to Business (B2B) portals connect portal users in one organization with internal information from another company. The portal is connected to the Internet through a firewall. The firewall blocks unauthorized attempts to access IT resources within the organization protected by the firewall. This is illustrated in Figure 18.5.

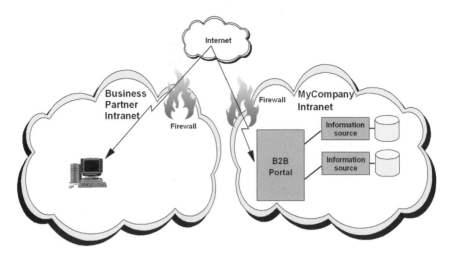

Portlets installed on the portal allow users within the business partner organization to access selected corporate information. The users are authenticated by the portal infrastructure and can be assigned resource access rights according to their roles.

Administrative roles

It is possible to define administrative roles in a hierarchical fashion so that the high-level administrators within the company can define the set of all portlets visible by the business partner users. These administrators can also define the basic page layout containing the banner, company logo, colors and fonts, etc.

Delegated administration

The high-level administrators can delegate administration rights to lower-level administrators within the business partner organization. The administrators within the business partner organization can restrict the rights of users to view and configure portlets or portal pages.

Take the case of a parts supplier, for example. The parts supplier would create portlets that display parts specifications, offer online help and discussion with the parts designers, and make online ordering possible.

The business partner may be a development and manufacturing organization. Within the business partner organization, the engineers may be given rights to view the parts specifications and obtain online help from the supplier organization's designers. However, only the procurers would be allowed access to the ordering portlets.

The development and manufacturing organization may also have a portal to allow the parts supplier access to internal information. When a product is in production, the supplier could check the quantities of parts in storage at the manufacturing plant and could arrange a new shipment of parts as soon as they run low.

18.2.3
B2C Portals

The Business to Consumer (B2C) portal is illustrated in Figure 18.6. The company sets up a portal that allows consumers to view product information through the Internet.

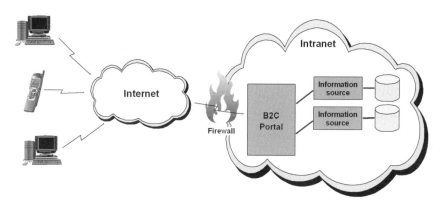

Figure 18.6:
B2C Portal

A typical B2C portal example would be an online merchant that makes product information and ordering capability available through the net. The merchant might provide portlets for specific product categories. The customer would select portlets to monitor categories of interest.

The merchant could also use collaboration functionality provided by the portal to create places for discussion of products or topics of interest. The customer could sign up for topics of interest and immediately see when others of like interest are online. The customer could use instant messaging to exchange ideas with those online.

18.3
Portal Infrastructure

Previous sections discussed some of the ways portals can be used by an organization. This section will provide a few details about the technology used to create a portal.

Although there are portal products based on other technology, industry standard portals are based on the Java™ 2 Platform, Enterprise Edition (J2EE). This and the following sections will discuss industry standard portals based on J2EE and will use the IBM WebSphere® Portal architecture to illustrate some of the fundamental portal features.

*Industry
standard portals*

18.3.1
J2EE

J2EE is a standard platform for creating multi-tier architectures for web-based applications. It is designed for server-side programming and allows for separation of the presentation logic from the business logic. See Figure 18.7.

Figure 18.7:
J2EE Containers

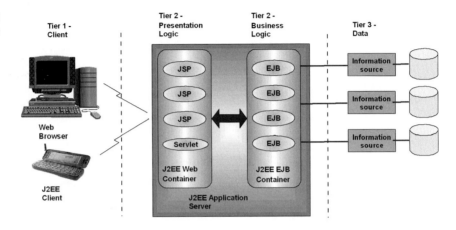

J2EE extends the Java 2 Platform, Standard Edition by adding support for Enterprise Java Beans™ (EJBs), JavaServer Pages™ (JSPs), Extensible Markup Language (XML), Java Servlets, and other technologies. J2EE applications consist of reusable components whose runtime environments are known as containers. JavaServer Pages, Enterprise Java Beans, and Java Servlets are components that run in J2EE containers.

Containers   Containers invoke the components when needed and provide services that help them carry out their tasks. Services provided by the containers include security, transaction processing, Java Naming and Directory™ (JNDI) lookup, and resource pooling. Security services allow components to be configured so that they can only be accessed by authorized users. Container transaction management allows multiple component method calls to be treated as a single, atomic transaction. JNDI lookup services allow components to access the corporate directory. Many of the services are configurable per component at component deployment time through information contain in the enterprise application's deployment descriptor. The components can use many other APIs included in the J2EE platform in addition to the services provided by the container.

Servlets   Just as applets are small Java applications that run on a JVM within a web browser, servlets are small Java programs designed to run on a server JVM. Servlets were developed as an alternative to CGI scripting for generating dynamic web content. They run in the web container of an application server and are generally associated with the presentation logic portion of an application. Servlets can invoke other servlets, JavaServer Pages, or Enterprise Java Beans.

Servlets are written by extending the appropriate Java base classes and implementing the required methods. They generate output streams as a response to input requests. Servlets are often used to dynamically create web pages in response to HTTP requests.

JavaServer Pages, like servlets, are used to create dynamic web content. Essentially, JSP technology embeds executable Java code within static HTML pages. When interpreted, the static output is copied to the output stream. When Java code is encountered, it is executed and the resulting dynamic output is added to the stream. A JSP can invoke other JSPs and can incorporate their output in its response.

JavaServer pages

Java code can also be indirectly referenced in JSP pages. To do so, JSP tags similar to HTML tags are used. Tag libraries associate JSP tags with the corresponding Java code. When the JSP processor encounters a tag, it retrieves and executes the code from the tag library. By creating tag libraries, commonly used functions can be written once and reused on many JSP pages. Use of JSP tags and tag libraries also isolates the JSP pages from changes in the underlying Java code.

The WebSphere Portal provides tag libraries to support common functions. For example, the following tag can be used in JSP pages that must be translated into multiple languages

Tag library

```
<h1><wps:text key="heading.hello" bundle="mystrings"/></h1>
```

At run time, the text string in the appropriate language is retrieved from the resource bundle and is used as input to the enclosing HTML header 1 tag.

Enterprise Java Beans are made up of Java code that conforms to certain programming standards to enable persistence and componentization. EJBs are often used to encapsulate the application business logic and database access. The use of EJBs to implement business logic helps enforce a separation of business rules from the presentation code.

Enterprise Java beans

Application servers implement the J2EE specification to provide open platforms for middleware application development. The WebSphere Portal consists of JSP, servlet, and EJB components that run on the IBM J2EE offering – the WebSphere Application Server.

Application Servers

18.3.2
Portal Structure

Figure 18.8 shows the high-level conceptual structure of WebSphere Portal. Additional information is available from the IBM Portal Zone web site (see section 18.6 Further Readings).

The WebSphere Portal consists of a number of functional blocks. The portal uses the member subsystem and security features of the WebSphere application server for user authentication. When a user logs in, the member subsystem looks up the user ID in the corporate directory, checks the password to authenticate the user, and retrieves information about the user such as group membership from the directory. If the installation uses a directory product that does not provide LDAP interfaces, custom code can be written to allow the portal to obtain the necessary information.

Portlet container   The portlet container invokes the portlets during page rendering. Since portlets play such a central role in the portal, the portlet container will be covered in more detail in the next section.

Figure 18.8:
Portal structure

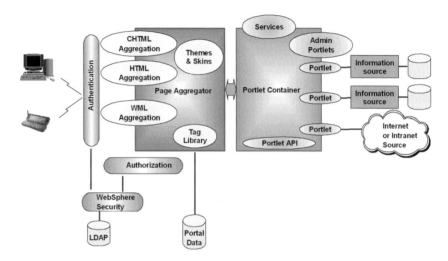

Page aggregation   The page aggregation code controls the page rendering process. After authentication, it uses WebSphere security functions as well as information contained in the portal database to verify that the user is authorized to perform the requested operation. Instead of the portal database, an external security system such as IBM Tivoli Access manager can be used for access control list storage. The portal enforces access control for portal resources including pages, page groups, and portlets.

The request sent by the client contains client device identification information that allows the page aggregation code to decide which type of markup should be generated. Markup generation is controlled by pluggable aggregation modules. The WebSphere Portal provides aggregation modules for HTML, WML, and CHTML markup. As we shall see in a later section, aggregation modules for additional markup languages can be added by other products.

Page layout   The page aggregation code retrieves the page layout definition from the por-
definitions   tal database. The layout information tells the aggregation code which components and portlets are to be placed on the page. The page aggregation code goes

through the layout definition, retrieving the designated banner, navigation, and visual control artifacts and adds them to the output stream.

Many of the artifacts consist of JSP pages that executed to obtain the appropriate markup for the output stream. These JSP page typically include tags from the portal tag library to carry out their tasks.

When the layout definition calls for a portlet to be placed on the page, the page aggregation code invokes the portlet through facilities provided by the portlet container and integrates the resulting markup into output stream.

During the page rendering process, the page aggregation code retrieves the theme and skin definitions and applies these settings to the output markup. The theme is used to define colors, fonts, and other visual elements for the entire page. Themes consist of cascading style sheets, JSPs and images. Skins are visual adornments such as control icons, borders, and icons that are placed around individual portlets.

Skins and themes

18.3.3
Portlets

Portlets appear to users as rectangular areas with a control bar on top and bounded by borders that on the web page. A single web page can have many portlets. Portlets share the page with logos, menu bars and navigation facilities. For the programmer, portlets are reusable Java components that create markup language output for display of application, web content, and other data.

18.3.3.1
Portlet Display Modes

Since portlets are complete applications, they can have persistent state and configuration data that is stored through services provided by the portlet container. The portal allows for several portlet operating modes. The user activates these modes by clicking the appropriate icon on the portlet menu bar.

Display modes

- **View mode** – This is the default mode for a portlet. When invoked, the portlet provides its operational content.
- **Help mode** – When this mode is activated, the portlet displays information that will help the user work with the portlet.
- **Edit mode** – This mode allows the user to change portlet settings. With an Internet mail portlet, for example, the user may have to enter the mail server name and his user ID.
- **Configure mode** – This mode allows he administrator to change portlet settings that are common for all users. For example, a portlet in an enterprise portal that accesses Internet data may require a proxy server name.

In any of the above modes, the portlet window can be minimized, maximized, and restored using the menu bar icons. When minimized, only the portlet menu bar is displayed.

18.3.3.2
Out-of-the-Box Portlets

A number of portlets are installed with the WebSphere portal product. Java developers can also write their own portlets. Through the WebSphere Developer's Studio, tools are provided for rapid portlet development. The IBM Portal Zone web site offers a catalog containing a large number of portlets written by IBM and IBM business partners that companies can download and use. The following list shows a few of the portlets that are available from IBM.

Microsoft Exchange Portlets	A family of portlets that display Exchange mail, calendar, to do list, and contact list information.
Banner Ad	Displays a banner ad image
Web Clipping	Displays selected portions of a web page
Sametime	Allows portal users to launch the Lotus Sametime client in a separate window
SQL	Displays the results of an SQL query against the specified database.
POP3	Used for Internet email access.

Administrative portlets

The product also provides portlets for a wide variety of administrative tasks. Administrators use access rights management portlets to control resource access on a user level, portlet management portlets to install, remove, and update portlets, and page customization portlets to create new page layouts, just to name a few.

Web desktop

Developers can write portlets for a very wide variety of application scenarios. Using Citrix technology, portlets can even allow Microsoft Windows® applications running on the server to be used by web clients through the portal. This opens the door for the portal to become a true web desktop.

18.3.3.3
Portlet Container

Portlets are Java applications that run within the portal context. They are extensions of servlets so that they can use all J2EE facilities available to servlets. For example, they can use the JDBC interface to retrieve data from a database or the Java Connector Architecture (JCA) to obtain data from an Enterprise Resource Planning (ERP) or Customer Relationship Management (CRM) systems.

The portlet container (shown in Figure 18.9) provides an execution infrastructure for portlets just as the web container does for servlets.

Portlets rely on the container for deployment, instantiation, initialization, and destruction. Portlets are deployed with the help of XML deployment descriptors, just as servlets are. The deployment descriptor describes the portlet and the resources it uses.

Portlet life cycle

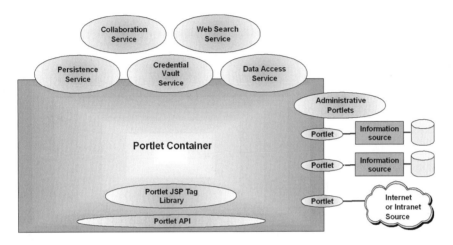

Figure 18.9: The portlet container

Portlets must be instantiated and initialized before they can be used. The portlet container takes care of this work automatically. The container also handles portlet instance destruction in order to minimize performance impact associated with frequent creation and destruction of objects.

The container also provides a mechanism for discoverable services. The product provides discoverable services for portlet data access, persistent TCP/IP connections, and credential vault access. Through the data access service, portlets can save and retrieve configuration and other data. Through use of persistent TCP/IP connection service, an Internet connection can be kept alive even when the portlet is not currently being displayed. The credential vault service allows sensitive information such as user IDs and passwords to be stored in a secure manner. Developers can write and install additional services to add capability such as collaboration or web search.

Discoverable services

When the user clicks buttons or icons, or enters data into portlet fields on a portal page, the request that the client sends to the portal contains information about the actions to be performed. The portlet container makes sure that the actions are routed to the proper portlet and provides a mechanism through which the affected portlet can retrieve associated data. For example, when the user selects a list element, the affected portlet can retrieve the selected element through the data provided with the action.

Portlet actions

The container provides a mechanism for communication between portlets. A portlet can send a message to another portlet on the same page. This is useful for the implementation of families of portlets that work together. For example,

Inter-portlet communication

an inbox portlet could display a list or newly arrived mail. When the user selects a particular mail for reading, the inbox portlet could send a message to a mail view portlet, causing it to display the selected mail.

18.3.3.4
Portlet API

Portlet API
Standardization

The portlet API provides standard interfaces to access the functionality provided by the portlet container. IBM is working with other companies to standardize the portlet API through the Java Community Process (JSR 168).

Many of the larges portal vendors – Sun, BEA, Epicentric, and Plumtree to name a few – support the standardization effort. When the standard is complete, content and enterprise application providers will be able to write portlets for their products that run on the portal platforms offered by all major portal vendors.

18.4
Extensions for Mobile devices

The portal aggregates information from many sources and that makes it an ideal middleware component for adapting the aggregated content to mobile devices.

The WebSphere Portal was built with mobile devices in mind. In addition to HTML markup, WML and CHTML markup languages are supported out of the box. The aggregation system is extensible, allowing support for new markup languages to be added. Developers can add portlets that support the new markup languages. The filter capability of the portal aggregation system can be used to adapt output for new devices.

Mobile
computing
support

The IBM WebSphere Everyplace Access product takes advantage of the portal to provide comprehensive support for mobile computing devices. It adds mobile device enabled portlets, transcoding, synchronization, device management, and notification capability as well as mobile database support (see Figure 18.10). Administrative portlets are provided for some of the Everyplace Access server components. Everyplace Access also provides client software for synchronization and device management.

Many portlets included with WebSphere Portal are enhanced through Everyplace Access to optimize information display for mobile devices. The information format and visual adornments were optimized for the more limited displays of mobile devices. Enhanced portlets include POP3 and IMAP Internet mail, Microsoft Exchange, Lotus Domino, banner ad, and Moreover news applications.

Transcoding technology is added as extension of the core portal system. *Transcoding*
Transcoding allows portlets that were written for display on a PC to be used
with mobile devices. It does this by translating the portlet output according to
defined rules. For example, transcoding technology can translate HTML portlet
output into WML for display on a WAP phone.

Offline browsing capability allows users to view portal content with their *Offline browsing*
mobile devices even when a connection with the server is not possible. Synchronization technology is used to copy selected portal content to the mobile
device while the device is connected to the server for later offline viewing. The
offline capability includes support for forms as well as static web content.

Everyplace Synchronization Server allows content on mobile devices to be *Synchronization*
synchronized with corresponding content on the server. For example, the user
may insert calendar entries on his handheld, while at the same time, his secretary changes other calendar entries on the server. Through the synchronization
process, changes made on the handheld are propagated to the server, and visa-

versa. The synchronization server is based on industry-standard SyncML technology, insuring interoperability with synchronization products from other vendors.

Intelligent notification

The Intelligent Notification Services allow users to be automatically notified when certain defined events take place. Users can subscribe to information services and define how and when the information is to be delivered. For example, the user can be notified when news articles containing predefined keywords arrive.

DB2 Everyplace

The DB2 Everyplace Access component allows for synchronization of a database on the mobile device to be synchronized with a server database. An application for a traveling salesman might use a database for storing price, product, and ordering information on his handheld computer. When the handheld is online, the ordering information entered by the salesman is transmitted to the server and any new or changed price and product information is downloaded to the handheld.

Device manager

The Device Manager component helps manage software on mobile devices. It can be used to distribute new software packages, update existing software, or delete software that is no longer needed. The device manager uses a database to keep track of software levels on user devices throughout an organization.

Client programs

Not all of the data transmitted to mobile devices through Everyplace Access is in browsable markup language format. The information transmitted is not meant for immediate display. When a database is synchronized, for example, the information transmitted from the server to the client is meant for later use or viewing. The mobile device must run special client software to handle these types of data exchange. The TrueSync Plus and Everyplace Client software run on mobile devices to fulfill this need.

TrueSync Plus

TrueSync Plus is a multi-tier synchronization agent. In a typical enterprise environment, the user must synchronize his or her PC PIM applications with the server, and then synchronize his handheld with the PC. If information is to flow from the client to the server as well as from the server to the client, one of the synchronization steps must be repeated. The user must synchronize handheld to PC, PC to server, and then PC to handheld again in order to assure the same state in all locations. TrueSync solves this tedious synchronization problem in a single step.

WebSphere Everyplace Client

The Everyplace Client software contains client applications for Personal Information Management (PIM) applications, DB2 everyplace database synchronization, offline browsing, and device management functions. The PIM applications include support for the Lotus Domino and Microsoft Exchange products.

18.5
Extensions for Voice Access

Voice support in the portal

Adding support for mobile devices with text and image display functions is a fairly obvious extension of the portal. What may be less obvious is the extension of portal technology to provide voice support. As it turns out, this can be done similarly to the mobile device extensions by adding corresponding aggregation support to the portal along with enhanced portlets and supporting server products.

Figure 18.11: IBM Voice Application Access

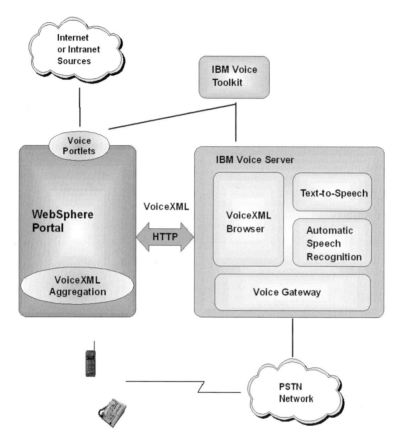

A voice-enhanced portal allows application access through standard telephones, cell phones, and car phones. A customer representative could have email read to him while he is in the car on the road to his next appointment. A mobile worker could call the portal from home to listen to his calendar entries rather than fire up the PC to download and view his current calendar. Moreover, for many workers with visual disabilities, application access through voice is an essential technology.

The IBM Voice Application Access product opens the door for these types of applications. It adds voice support to the portal through VoiceXML aggregation and voice server technology as shown in Figure 18.11. *IBM Voice Application Access*

Voice Application access consists of Voice-enabled portlets, a VoiceXML aggregation module for the portal, the voice toolkit, and the voice server. The VoiceXML aggregation module allows portlets and navigation text to be rendered appropriately for the voice server. The voice toolkit supports voice portlet and voice application development through a VoiceXML editor, a speech recognition grammar editor and a pronunciation editor. The grammar editor defines the allowable vocabulary and word ordering for the voice input to be recognized.

Voice server
The voice server renders the VoiceXML into speech output and recognizes spoken text for conversion into portal requests. The voice server contains several functional blocks.

The Text-To-Speech (TTS) engine converts text blocks from the voice browser to speech. The Automatic Speech Recognition (ASR) engine works the other way around, converting spoken words into text.

VoiceXML browser
Just as a visual web browser renders output for the human eye, the voice browser renders the VoiceXML output from the portal into speech. When the user speaks commands to navigate through the portal page, the voice browser uses recognized text from the ASR engine to create VoiceXML request for the portal.

The gateway portion of the server provides software interfaces to control hardware voice-over-IP (VoIP) gateway machines attached to the public telephone network.

18.6
Further Readings

IBM Corporation Main Site

http://www.ibm.com

This site provides a wealth of information about many computing topics.

IBM WebSphere Portal Site

http://www.ibm.com/websphere/portal

This site provides technical information, white papers, and case studies about the IBM WebSphere Portal.

IBM WebSphere Portal Zone

http://www7b.software.ibm.com/wsdd/zones/portal/

Technical resources for administrators and developers of WebSphere Portal. The site includes a catalog of downloadable portlets as well as links to support sites.

IBM WebSphere Voice Zone

http://www7b.software.ibm.com/wsdd/zones/voice/

Provides information about IBM voice technology.

IBM WebSphere Everyplace Access Site

http://www-3.ibm.com/software/pervasive/products/mobile_apps/ws_everyplace_access.shtml

Provides complete information about the IBM WebSphere Everyplace Access product.

Sun J2EE Site

http://java.sun.com/j2ee/overview.html

This site provides information about the J2EE programming platform including technology and documentation downloads. A J2EE tutorial is also available here.

Java Community Process Site

http://www.jcp.org/en/home/index

The JCP site offers information about the standardization process as well as descriptions of standards currently in progress. Of particular interest is JSR 168, the Java Standardization Request for the portlet API.

19 Device Management

The need for device management is obvious: When the PC was introduced in enterprises two decades ago, the management costs of these decentralized enterprise clients soon overtook their hardware costs. The vast of even more decentralized pervasive devices will strengthen that issue. The number of clients a management system must be able to administer is much larger than it used to be in a PC centric world. Although pervasive devices are much less complex to configure than a PC their diversity makes device management a complex issue: Management and support is driven by specific operating systems, interfaces, peculiarities, and capabilities of each individual device. Heterogeneous devices, applications, and users need to be administered from remote in an invisible and self-evident way. Especially mobile devices must operate reliable and simple without having any system administrator available.

19.1
Tasks of Device Management Systems

Device management systems keep track of issued devices, as well as manage applications and services delivered to these devices. They need to enroll subscribers to particular services, roll out new applications, supply latest maintenance updates, and monitor usage for billing and customer care purposes.

For a device management system, which is dedicated to Pervasive Computing, the following typical tasks can be identified:

- First of all, there is the pure device management itself: The entire life of the device from deployment and installation to deactivation is covered by this category. The arising tasks are quite similar to traditional PC system management. They include keeping track of the current configuration and processing maintenance requests. For mobile devices, location tracking can be an additional requirement.
- IT network infrastructure configuration and management is a task, which does not change fundamentally with Pervasive Computing, but is getting more complex. The number of different networks operating at the same time increases.
- Application management includes deployment, maintenance, and updating of software. Especially beyond the boundaries of enterprises, this task is challenging. Users might install applications from different providers. Each

provider needs to manage its applications independently from all others. Issues arising from mobility are, that there is often no continuous connection between the management system and the managed device. This impedes software updates by the application provider. For instance, an electronic purse on a smart card might be used for months for paying various retailers, before connecting again to the management system of the issuing bank.

- Subscriber enrollment, management, and billing are typical tasks service providers need to perform. Service delivery requires a more active engagement than just reacting on support requests and keeping track of the user's applications. Gathering information about the people using specific products is a valuable information, when offering individual services or targeting campaigns and promotions for a particular audience.

- Accounting and authentication information needs to be managed independently from particular devices. For instance a mail account might be accessed via phone, handheld, or set-top box.

All these tasks need to be done individually for each single user, application, and device.

19.2
Tivoli Device Support Infrastructure

To leverage these tasks Tivoli has developed the *Device Support Infrastructure* (DSI). DSI extends Tivoli's existing enterprise management framework and provides a common infrastructure for managing any form of device. The framework is based on CORBA and allows enterprise-wide, heterogeneous computing.

In the traditional Tivoli Framework, the managed computer systems run Tivoli Management Agents (TMA), which interact with the management application through a Tivoli Management Gateway. The TMA provides full functionality down to the endpoints. He is responsible for management operations, such as monitoring local operations or receiving software distributions from a Tivoli Profile Manager. The TMA executable requires about 1 to 2 MB of storage.

Device Agent   When extending the scope to PDAs, phones, and set-top boxes, the TMA is no longer suitable as the endpoint of system management. Although the TMA is already a thin client, this component is still too complex to run on small devices with limited processing power, small memory footprints and manifold operating systems. This is why DSI introduces a new component: A *Device Agent* ("Actuator") is a spin-off of a TMA and is installed on every administered device. An enhanced TMA acts as gateway and connects the Actuator to a *Device Manager*, which is the core component residing on the Tivoli Management Server. Once a device is connected to the TMA, the Actuator exchanges the management info through the TMA with the Device Manager. A management console allows the system administrator to interact directly with the monitored devices (Figure 19.1).

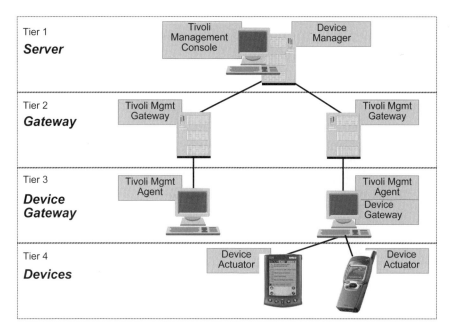

Figure 19.1:
Tivoli Device
Support
Infrastructure

Since managed devices can be represented with the traditional Tivoli management system, no changes to existing installations are required. Nevertheless the management of devices differs in some aspects:

- A single device cannot be modeled as an individual Tivoli object, due to scalability reasons. Instead the available device information is stored and maintained by the Device Manager within a repository.
- Devices of the same type can be bundled as Device Groups. These groups are the smallest entity, which can be managed as a regular Tivoli object. For instance, groups can subscribe to a profile manager or apply a security policy. The Device Group object leverages the management of a huge number of devices – both from a system scalability and an administration usability point of view.
- The vast amount of clients, require a new mechanism for identification. A unique device ID comprises the TMA label and a local ID. That ID is assigned and maintained by the Device Manager.

Tivoli targets handheld computers, cable modems, and set-top boxes, as well as POS devices, ATMs, and self-service terminals. The Tivoli Manager for Retail has been the first product implementing DSI. It targets POS devices and fea-

Early products

tures the centralized management and monitoring of all store controllers. A second product is the *Tivoli Device Manager for Palm Computing Platform*. This version supports installing and removing applications, maintaining inventory information, and performing configuration management functions. The Device Management System ensures that all Palm OS based devices within one organization can be supplied with the same software version and set of corresponding data [KUM00].

19.3
User Profiles and Directory Services

Device Management Systems need access to user, device, and network profiles. These profiles include information about network addresses and characteristics, user passwords and accounts, owned devices, their respective capabilities, resources, and current locations. More sophisticated profiles can even hold personal information, like individual preferences, billing plans, bookmarks, or security credentials.

Lightweight Directory Access Protocol
This vast amount of data need to be structured and stored, in order to be able to find and retrieve the information as fast as possible. Directory services such as the standardized Lightweight Directory Access Protocol (LDAP) are the basis for storing and accessing profiles. LDAP combines a query protocol with a flexible data schema to represent the profile information in a tree based structure. The LDAP scheme allows to specify objects, such as persons, groups, services, computer names, organizations, and organizational units. Each object needs an unique Distinguished Name (DN), followed by mandatory and optional attributes. A binary protocol specifies how to access and use the objects in a directory in a standardized and platform independent way. LDAP allows to retrieve directory information without needing to understand how and where the entries are stored. For instance, an arbitrary device can load its configuration from a standard directory. LDAP can be used by multiple applications across a variety of systems and networks in parallel. This helps to avoid redundant storage and maintenance of directory information within an enterprise.

An example

```
dn: dc=de, dc=ibm, dc=com
objectclass: organization
o: IBM Deutschland Entwicklung
l: Boeblingen
postalCode: 71032
streetAddress: Schoenaicher Strasse 220
telephonenumber: 07031 16 0
```

The following example illustrates the usage of LDAP. First a root organizational unit is defined. DC stands for Domain Component, o refers to the organization name, l is the location name.

```
dn: ou=pvc, dc=de, dc=ibm, dc=com
objectclass: organizationalUnit
ou: pvc
description: Pervasive Computing
```

Next, a sample organization unit is described. OU defines the name of the unit.

Finally, one sample user information is introduced. CN and SN refer to the users name. UID is the user name.

```
dn: uid=tstober, dc=de, dc=ibm, dc=com
objectclass: person
ou: pvc
cn: Thomas Stober
sn: Stober
mail: pvcbook@web.de
```

LDAP is available as an open source implementation from OpenLDAP Foundation. This eases interoperability between products and encourages synergy between different vendors. The APIs are exposed both in C and Java. The LDAP directory can be distributed on multiple servers to be scalable. Another possibility for LDAP support is the Java Naming and Directory Interface (JNDI), which is delivered with the Java 2 Enterprise Edition.

OpenLDAP Foundation

Storing user information with LDAP instead of using a usual relational database has two essential advantages:

First of all, LDAP is independent from a particular operating system or device. While relational databases have only standardized interfaces such as ODBC or JDBC, LDAP specifies the entire protocol, which is built directly on top of TCP/IP.

Second, LDAP implementations are primarily optimized for performing of few dedicated task, while relational databases focus on offering generic and flexible ways to store, access, and modify data. This approach helps to speed-up the directory service and tune performance as well as availability.

19.4
SyncML Device Management

As described above, Device Management is a clear need in a world where mobile and wireless devices are getting more and more capable of taking over tasks which a few years ago only were possible on PCs. Hand in hand with the possibility to download applications to a device goes the need to update and configure it.

Not every mobile phone user is a computer specialist and wants to deal with the software on a device, as well as wireless service providers have a clear need for seamlessly working devices, with no need to any user invention, even if something needs to be reconfigured, may be due to a change in the ISP's infrastructure.

At the time the industry was looking for a way to solve these problems early in the year 2000, all the major mobile device manufactures as well as the major server software providers already gathered in the SyncML Initiative and were in the process of releasing a set of specifications for a interoperable and efficient way to synchronize data on mobile devices with data on servers. As Device Management is a sort of "special" one-way synchronization, where the server is setting and updating data on the client as well as requesting data from the client (but not updating the server database with that data), it was a natural move to start the effort of developing an interoperable device management framework based on the SyncML Synchronization specifications as part of the SyncML Initiative.

19.4.1
SyncML Device Management Architecture

The SyncML Device Management Architecture is based on the existing SyncML Data Synchronization Specification. It uses the DTD defined by the SyncML Representation Protocol and defines an own Device Management Usage Specification with slightly different meanings of the tags.

Using the same DTD for Data Synchronization and Device Management enables the reuse of the same base components, like XML document generators and parsers to be used for both purposes.

On top of the Representation Protocol, the Device Management Protocol defines the protocol flow during a device management session. One notable difference between the Device Management Protocol and the Data Synchronization Protocol is that during a device management session only data on the client can be updated, as well as the server can read data from the client. But the client can't update data on the server, like during synchronization.

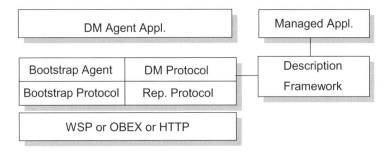

The Description Framework allows the definition of a hierarchical structure of management objects, the entities which hold the data managed by the DM Server.

Before a device can be managed by a DM Server, the DM client on that device must be configured in a way to be able to interact with that DM Server. This initial configuration of the DM client is called bootstrapping. This can happen in two ways:

- WAP provisioning
- SyncML-message-format based

WAP provisioning is based pushing a WAP message to the client using SMS. This method was defined by the WAP Forum as part of the WAP specifications. This is the preferred method to provision mobile phones, as it is a very compact message and fits in one SMS.

SyncML-message-format based provisioning is based on a SyncML-DTD based message and therefore could be decoded with the already existing SyncML parser. Because it is based on XML, the message usually does not fit in one SMS and therefore should be used for devices which are connected to for example a TCP/IP based network which allows addressing clients.

The Bootstrap Agent receives the provisioning message, decodes it using the Bootstrap protocol and finally configures the DM client. Now everything is ready to start a regular device management session.

The notification phase is needed in case the server starts the device management session. It is doing so by notifying the client using SMS or WAP Push that the client should start a device management session with the server. In this phase it is especially important that the client can authenticate the server to ensure that no attack can be started this way.

The setup phase is initiating the device management session between the client and the server. The message from the client to the server identifies the client to the server, contains details about the device (manufacturer, model, serial number / IMEI,...) and informs the server if the client is initiating the session, or if the client is responding to the server's notification.

Figure 19.3:
Device
Management
Session Flow

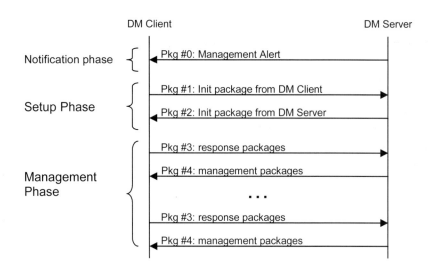

During the management phase the actual management of the device takes place. The server can set, update, and delete data (like configuration settings, software,...) on the client as well as query data from the client to make decisions about what needs to changed or updated.

More in depth details about SyncML Device Management can be found in the SyncML Device Management Specifications available at the SyncML Initiative's website, as well as in the book SyncML – Synchronizing and Managing your Mobile Data [SML02].

19.5
Further Readings

Pervasive Management - Expanding the Reach of IT Management to Pervasive Devices

http://www.tivoli.com/products/solutions/pervasive

This white paper describes how Tivoli extends its management framework to pervasive devices [TIV00].

End-to-End Management with Tivoli - Managing PDAs

http://www.redbooks.ibm.com

A redbook about using device management for Palm OS based devices [KUM00].

Directory Interoperability Forum

http://www.directoryforum.org

A web site with resources related to directory services.

SyncML Initiative

http://www.syncml.org

The SyncML Initiative defines the SyncML Device Management Specification, which are available for download.

20 Synchronization

20.1
What Synchronization Is All About

Is it not always possible to maintain a persistent connection between a mobile device and the wired network of enterprise servers. In order to execute mobile applications while not connected, at least a subset of the data must be copied to a local data store on the device. When offline, the mobile application can read and manipulate this local copy of the data. As a result, the user gains mobility by achieving independence from the network while still having access to his relevant information. But a side effect is, that various copies of the same data are now widely distributed among different mobile devices and various backend servers. For example, contact information is stored on your phone as well as on the central directory database of your company, maybe in a Lotus Domino Public Names Address Book database. Calendar information and Emails are kept on your PDA in addition to your mail database on the enterprise Domino or Exchange server. Mobile users need to maintain a local copy of the corporate data they are processing as well.

Working offline

The challenge is to keep all copies of the same piece of information consistent and up-to-date, regardless where it is stored. To achieve this, the device must connect to the network intermittently and compare its own version of the data with the other copies stored on other systems. Any changes made locally on the device are sent to the network servers and any new server side updates are retrieved. Ensuring that two sets of data are identical, even if different changes are made in parallel to both sets is called *data synchronization*.

20.2
Usage Models of Synchronization

There are three basic usage models of synchronization, which can be distinguished: local, remote, and passthrough synchronization (Figure 20.1):

Local Synchronization

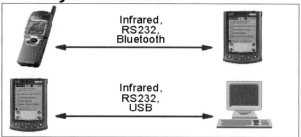

Remote Synchronization

Pass-Through Synchronization

Local
Synchronization

Local synchronization reconciles corresponding data on two directly connected system. The connection is done through serial port or wireless using Bluetooth or infrared. A network is not involved. An example for this usage model is the very common sync between a PDA and a desktop computer. Numerous products such as ActiveSync, HotSync or IntelliSync offer solutions to synchronize the native PIM applications on the device with Notes or Outlook on a personal computer.

This usage model also includes the synchronization between two devices, e.g. when exchange data between a phone and a PDA.

Remote
Synchronization

Remote synchronization links data of local systems with a server database through a wired or wireless network. Available products such as IBM WebSphere Everyplace Access synchronize PIM applications on mobile devices with enterprise software like Domino or Exchange. Other products like Starfish TrueSync power server based PIM applications of service providers such as Yahoo or Excite. TrueSync allows the subscribed user to access the server and synchronize their calendar and contact information to their mobile device.

Not all devices are capable of accessing a remote server through a network themselves. In those cases the device can establish a local connection to a third system, which is connected to the network and the server. That third system acts as a passthrough gateway. Virtually, the device can now execute a remote synchronization by routing the data flow though the passthrough system to the server. Typically a PDA without network capabilities would connect using a cradle and a serial port to a desktop computer. The desktop would forward the communication via LAN to the server. In a different configuration, the PDA would connect to a mobile phone with infrared or Bluetooth. The phone will then establish a connection to the remote server through a wireless network.

Pass-Through Synchronization

20.3
The Challenges of Synchronizing Data

You may think that keeping data in-sync on multiple systems can not be too difficult. If two systems are of the same sort, if they use the same application software, and if you use only one system at one time, this can be fairly easy. In this simple case, you can just copy the data files from one machine to the other.

The first potential problem arises, when the synchronized systems have hardware or software differences. Here are just a few examples:

Different software and hardware

- The data in a native PIM application of a device is represented in a completely different data structure than in a server application. A mapping of corresponding information is required to match the content between the two systems properly.
- Devices usually have strong limitations in functionality. The content of each data entry needs to be adjusted and simplified specific for the capabilities of the target device. A mobile phone for example, might only have the possibility to store a shortened form of the name and the phone number of a person, but is not able to save any address information.
- Because of memory restriction often only a subset of the server side entries can be stored on the device. Filtering functionality is used to select the kind of entries the user wants to load into the device.
- On pervasive devices, entries are typically identified by a single byte local identifier. On the other hand, on your desktop computer each record might be identified by a 16-byte (or larger) globally unique identifier. During synchronization the relationship between the identifiers on the mobile device and the corresponding ones on the desktop must be maintained.

An additional set of difficulties needs to be solved, when changes are made to the same data on different systems at the same time. To avoid loosing the changes on either system, a tool is needed, which detects and reconciles the differences. This tool is called *synchronizer* or *synchronization engine*. The synchronizer identifies which individual data entries have been added, changed or

Merging data changes

deleted on each system and propagates the newer version to the other system. The synchronizer is usually executed on the server side rather than on a device in order to take advantage of the more powerful processing capabilities.

Conflict resolution
It can happen that one entry has been modified in different instances of a same data at the same time, e.g. by two different users working concurrently. The synchronizer is responsible to resolve such conflicts and ensure that only one consistent level of data remains on all synchronized systems. There are several choices in such cases:

Rule bases resolution
One version of the data set can be defined as the dominant one in case of any conflict. The dominant version is kept and overwrites the other version, which will be lost. Typically there are three alternative options:

- A rule can be defined that the server side changes should always be kept. This options is referenced as "server wins"
- Of course a corresponding opposite rule would define, that client changes should be prioritized ("client wins").
- The third option specifies that the most recent change has to be preserved ("time-based conflict resolution").

Usually the user can configure his preferred behavior in a profile of the synchronization server.

Conflict documents
A different approach is to duplicate a record whenever a conflict occurs. This ensures that no changes are lost, but it adds additional records to a database, which eventually need to be reconciled manually by a user.

Tracking Changes
A synchronization server must be able to detect which entries have been changed in a database since the last successful synchronization. Most of the synchronized applications on desktop computers or network servers generally can keep a detailed track which records have changed. One approach is to document all modification made to the database in a *change log*. A different way is to store the time of its last modification with each record of a database. Yet again, devices offer not that amount of sophistication. Usually they only set a simple "*dirty*" flag on each record, which has been modified since the last synchronization.

The time of the last successful synchronization is referenced as synchronization anchor. The synchronization server needs to remember this time in order to be able to search for any changes since that time. The complexity of maintaining the sync anchors varies, depending on the system topology (figure 20.2).

One-to-one
The most simple case is, when one server synchronizes data with one dedicated client. A typical scenario of this one-to-one synchronization is a Lotus Notes database on a desktop computer, which interacts with the PIM applications on a PDA. The sync anchor is only a single value and can be easily stored and updated after each successful sync.

One-to-Many
It gets more complicated, when multiple clients synchronize with one server. This applies for instance when one user wants to use the data of his Lotus Notes database on more than one device, maybe on his phone and his PDA as well.

Another scenario is a service provider which offers a synchronization server for many client accounts. This one-to-many or star synchronization requires the server to store a different sync anchor for each individual device, since every device will synchronize with the server at different times. Whenever a device accesses the server, the synchronizer must know when that particular device has synchronized the last time. The device must get all updated entries since that time.

The many-to-many synchronization is also referred as peer-to-peer synchronization. It is rarely used, because of the involved complexity. There is no longer a central server, which can provide the synchronize and maintain the sync anchors. Now, each device needs to store sync anchors itself and remember which data has already be synchronized with which other device.

Many-to-many

One-To-One Synchronization

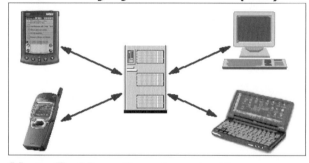

Figure 20.2: Synchronization topologies

One-To-Many Synchronization (Star)

Many-To-Many Synchronization

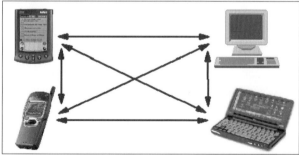

The problems outlined in the above section are just a few of the problems that arise while trying to keep two databases in-sync. We hope we were able to demonstrate to you that data synchronization is not as simple as it might look like.

20.4
Industry Data Synchronization Standards

There arc many proprietary synchronization products on the market today, which only allow synchronizing between selected devices and applications. For example, between an address book on a Web portal and the contacts on a Palm handheld computer, but not with the phone book in a mobile phone or the personal address book on a desktop computer. This section provides an overview of a number of current industry efforts to specify data synchronization standards. The next section will then give a short exemplary overview on some products, which are in the market today.

20.4.1
Infrared Mobile Communications (IrMC)

The Infrared Data Association (IrDA) mobile communications committee defined the "Specification for Infrared Mobile Communications" [IRMC99], to provide information exchange over infrared. IrMC defines five levels of information exchange:

- Level 1 (Minimum Level)
- Level 2 (Access Level)
- Level 3 (Index Level)
- Level 4 (Sync Level)
- Level 5 (SyncML Level)

Level 1 to Level 4 of the IrMC specification defines the exchange of a limited number of different objects, such as business cards, calendar data, messages, and email data over personal area networks with connection-oriented or connectionless links. They do not support the synchronization of other forms of data such as relational databases or table data. Since the IrMC specification was initially designed for local data synchronization, the methods proposed are not optimized for data synchronization over wide area networks, such as synchronizing the phone book on a mobile phone with one of the corporate public address book databases over the Internet.

Level 5 was added in 2001 and defines SyncML as the synchronization technology used in IrMC. The SyncML Initiative is closely working with IrMC in order to produce a Synchronization Profile that even more strongly recommends SyncML as the preferred synchronization technology.

20.4.2
Wireless Application Protocol (WAP)

In June 1997, Ericsson, Motorola, Nokia, and Unwired Planet (now known as Openwave) founded the Wireless Application Protocol (WAP) Forum as an industry group for the purpose of extending the existing Internet standards for the use with wireless communication. By Spring 2002, the WAP Forum counted more than 500 member companies from all parts of the industry, including network operators, device manufacturers, service providers, and software vendors.

The WAP Specification Version 1.1 was released in the summer of 1999, and the first WAP devices and services were available as early as the fourth quarter of 1999. Version 2.0 of the WAP Specifications was approved and released to the public in February 2001.

Since Version 2.0, WAP requires all WAP Servers and WAP Client Devices supporting data synchronization to support data synchronization over WAP via SyncML. WAP also requires clients and servers to pass the SyncML Conformance testing as a prerequisite for passing the WAP Conformance testing, in case synchronization is supported.

20.4.3
Third Generation Partnership Program (3GPP)

The Third Generation Partnership Program (3GPP) was established in December 1998 as an organization of all partners interested in the evolution of mobile systems to the third generation based on the GSM technology. GSM is looked at as the second generation, GPRS and EDGE as the 2.5 generation, and UMTS, UTRA, W-CDMA, and FOMA as the third generation of mobile communication technology.

SyncML and 3GPP are closely working together and SyncML technology is mandated as the method of choice for data synchronization since Release 4 of the 3GPP specifications.

20.4.4
SyncML

The SynML initiative was launched in February 2000 by a group of mobile computing industry leaders who had a common vision for a universal data synchronization format and protocol. It was founded by Ericsson, IBM, Lotus, Motorola, Nokia, Palm, Inc., Psion, and Starfish Software. In the first four months of the initiative, more than 275 organizations have endorsed their support for the initiative. The SyncML initiative released preliminary specifications and code for a reference toolkit in May 2000. Version 1.0 of the specifications and open source for the reference toolkit was released to the public in December of 2000. The current version is 1.1.1 and was made available in October 2002.

SyncML includes a universal data synchronization format that is defined by an Extensible Markup Language (XML) document type definition (DTD). This format is exchanged as SyncML messages between network devices. A SyncML message is just an XML document. SyncML is independent from the underlying transport layer and can be used in wireless as well as wired environments.

In addition, SyncML defines a synchronization protocol. This protocol specifies how SyncML conformant messages are exchanged to synchronize databases on different network devices. The synchronization protocol supports both one-way, as well as two-way data synchronization. The SyncML specifications also define HTTP, OBEX, and WSP transport bindings, which describe the minimum set of features that a SyncML compliant transport implementation must support.

With SyncML as a common synchronization protocol, all parties involved in building and using a data synchronization solution, like end users, device manufacturers, solution providers, and application developers will benefit. A synchronization solution just needs to know how to generate and understand the SyncML protocol to synchronize with any other device and application that supports SyncML data synchronization. The times where an end user had to install several synchronization products on a device, just because each of them supported a limited set of data object types is hopefully over now.

20.4.4.1
SyncML Data Representation

A SyncML based synchronization operation is based on the exchange of a set of well-defined SyncML messages. Each SyncML message is represented by one XML document. An application can use the SyncML MIME type to identify a SyncML message. One SyncML package can consist of one or more SyncML messages. This package contains a complete synchronization between two devices.

20.4.4.2
SyncML Framework

SyncML also defines a conceptual framework for identifying the system components involved in data synchronization. The SyncML Reference Toolkit is the open source code for an implementation of this SyncML Framework.

The SyncML Framework consists of the SyncML objects, a conceptual SyncML Adapter, and the SyncML Interface.

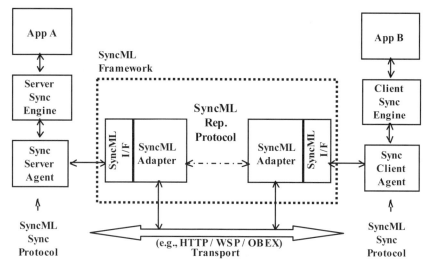

The SyncML formats are registered MIME (Multipurpose Internet Mail Extensions, http://www.ietf.org/rfc/rfc2045.txt) media types, specifically defined for use in data synchronization.

SyncML MIME Type

There is a clear-text XML MIME media type, as well as another media type for representing a tokenized, binary representation of XML, called WBXML. MIME media types are well defined for transfer over a wide range of transports.

The SyncML Adapter is a conceptual entity within each of the SyncML compliant devices. The SyncML Adapter is responsible for maintaining a transport connection with the other network device. In addition, the SyncML Adapter is responsible for marshalling synchronization commands and data into the SyncML format.

SyncML Adapter

The SyncML Interface is an API programmable mechanism for communicating with the SyncML Adapter, which is implemented by the SyncML Reference Toolkit. It will likely be a widely accepted tool for implementing SyncML support in network devices.

SyncML Interface

The Sync Agent is responsible for handling the SyncML-based data synchronization. The Sync Agent provides the SyncML support to generic data synchronization engines.

Sync Agent

The Sync Agent uses the SyncML Interface to access the services of the SyncML Adapter, like marshalling data to be send into the SyncML format or for un-marshalling received data.

20.5
Synchronization Solutions

20.5.1
PIM Synchronization

PIM information is the most important focus of sync technology. Virtually every mobile device is already preloaded with some PIM application and offers synchronization solutions to the common data stores, such as Domino and Exchange. There is a wide range of products available on the market, which synchronize the device data either with the user's desktop computer or with his account on network server. Well-known product names among many others are PumaTech IntelliSync, Lotus EasySync, ESI XTNDConnectPC, Starfish TrueSync, and IBM Websphere Everyplace Access (WEA).

To illustrate the implementation of a synchronization server for PIM information, the architecture and major components of the IBM WEA Synchronization Server are described in this section in more detail (figure 20.4).

Synchronization Adapter
When devices connect to the synchronization server, the synchronization adapter acts as the entry point to the enterprise network. The adapter is responsible for the protocol management. Two different protocol layers exist:

- A basic communication protocol is used to as the carrier for any interaction between server and device. In the case of WEA, the applied communication protocol is HTTP and the synchronization adapter is implemented as a servlet running on a Websphere Application Webserver.
- The second layer on top of that basic network protocol is the standard SyncML Synchronization Protocol, which handles the message exchange and data flow when synchronizing information between device and server. To manage this protocol, the synchronization adapter needs to keep track of current sessions with clients and it must maintain the time of last synchronization (sync anchors) for every client. That information is stored in a DB2 database on the server.

The exchanged XML or WBXML messages are parsed and generated by the synchronization adapter according to the SyncML format. The adapter also converts the synchronized data objects, which are embedded in the SyncML message: an internal representation of the server is mapped to standardized formats used by SyncML. Common data formats are vCard, vCalendar, vToDo, vJournal, and RFC 822 emails.

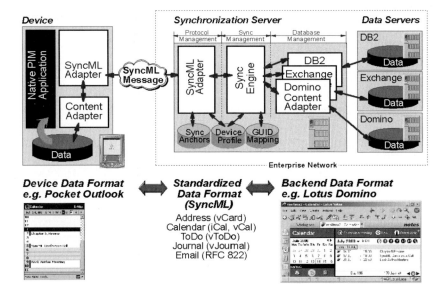

Device   Synchronization Server   Data Servers

Device Data Format
e.g. Pocket Outlook

**Standardized
Data Format
(SyncML)**

Address (vCard)
Calendar (iCal, vCal)
ToDo (vToDo)
Journal (vJournal)
Email (RFC 822)

**Backend Data Format
e.g. Lotus Domino**

*Figure 20.4:
IBM WEA
Synchronization
Server*

The synchronization engine is the core component of the server and manages any differences in the data sets on client and server. Input for processing are the current client and server version of a document. These are reconciled by the synchronization engine, based on a configured policy. The resulting new version will be stored afterwards both on the client as well as on the server. As part of this, the synchronization engine maps the document ID used on the server to a corresponding local ID on the client. These mappings are stored persistently in a DB2 database.

*Synchronization
Engine*

By using standardized data formats, SyncML limits at least the number of different data formats exchanged between the synchronization server and devices. But when the synchronization server is accessing backend data stores in a heterogeneous network environment, the complexity is increasing. Each backend system technology requires an own specific way of interaction. For instance, a relational database can be accessed through JDBC calls, whereas a Domino server exposes its own API. For this purpose, the WEA Sync Server allows to integrate a number of different content adapters. Each content adapter implements the specific way of retrieving and writing information to a particular remote data store. A content adapter maps the proprietary data format of the backend system to an internal representation used by the sync server. The adapter also provides functionality which is very specific to a particular backend system, such as authentication and access control. As a benefit of encapsulating these tasks in a content adapter, the other components of the sync server can be implemented in very generic way and have no dependency on a particular backend system.

*Database
Adapter*

By providing different content adapters, WEA can support synchronization of PIM information between SyncML devices and a variety of different backend systems:

The DB2 Content Adapter gives access to relational databases. These are typically used by service providers who would store large amounts of data for a large number of subscribed accounts.

Adapters for Domino and Exchange allow to synchronize with enterprise messaging systems used in a corporate environment. These provide more powerful workflow applications with more sophisticated calendar and scheduling functionality.

Cache   To optimize performance the WEA Sync Server uses a DB2 database to cache the data from the various backend system locally. The cache information can be accessed very fast, since no remote network transactions to the backend are necessary during synchronization. The cache is updated regularly to reflect any changes on the backend

Device   Each user account requires specific configuration which is reflected in a de-
Profiles vice profile. The profile defines the device type, language, and timezone. It is also possible to set the preferred conflict resolution policy, the user credentials to be used for authentication, as well as the location of the synchronized backend databases. Important is the option to define filter criteria to reduce the amount of data being transferred to the mobile device. Filtering can be done by time, e.g. to select just the most recently changed documents. Other filter selections are by document size or by specific values of certain fields. The device profile can be accessed and modified by the end user with a browser through a web interface.

Security   Three security mechanisms protect the data on the backend system during synchronization:

- The basic communication protocol between device and the web server can be secured using SSL and HTTP authentication schemas. SSL includes encryption to prevent the eavesdropping of the data which is exchanged.
- The SyncML protocol provides an additional layer of user authentication by verifying the user credentials.
- The backend system performs an additional authentication and verifies that the sync user is a user of that backend system as well. Addition, the backend system enforces any defined access restriction to particular databases or documents.

Device   A SyncML server can be accessed by any SyncML compliant device. The
Components necessary device components are very similar to the server components, which are already described above: The SyncML adapter leverages communication protocols which are available on the device and manages the message flow as specified by the synchronization protocol. A simplified XML/WBXML parser and encoder is required to process the exchanged SyncML messages.

The database adapter converts the standardized data objects of a SyncML message to the entries in the device specific PIM data store. After updating the device data store during synchronization, the PIM application will access up-to-date information.

A synchronization engine is not necessary, since the conflict resolution is performed on the server. The synchronization component on the device usually exposes some kind of user interface to configure the connection to the server, including the credentials for authentication.

20.5.2
Synchronizing Non-PIM Databases

The most common data to be synchronized to mobile devices is PIM information. But there are other business applications as well, which require to synchronize non-standard and customized data schemas too. For instance, solutions for the mobile sales force need to use and manage offline customer profiles, order status, and more. Medical applications distribute patient records or prescriptions on various portable devices. Synchronizing that kind of data is more challenging, because each application uses its own unique data schema. One option to deal with such application specific data is to implement an own proprietary synchronization component tailored to the application needs. A different, more generic approach is the Lotus *Domino Everyplace Enterprise Server* (DEE), which extends the usage of the Lotus product *Domino* to mobile devices.

The workflow system Domino is based on a proprietary database technology. *Domino*
Each application consists of a set of databases on a Domino server, which contain the actual user data as well as corresponding user interface elements like forms and views. The user accesses the data through a Notes client software on his desktop computer. Both, the data structure and those user interface elements have been created by an application developer with the help of a development tool called Domino Designer.

DEE, the mobile version of Lotus Domino, includes a special version of the *Mobile*
Designer tool. The mobile Designer supports user interface elements targeted *Application*
for mobile devices as well as a subset of the scripting language offered by the *Designer*
regular desktop Designer. Once a mobile application with all its design elements is defined, the tool compiles the application information into a XML representation.

The content of the server side database is synchronized into a local DB2e database on the mobile device of a user. This synchronization includes two steps: *Synchronization*

- First the compiled XML version of the application information with the designed user interface elements is loaded onto the device.
- Afterwards the actual user data from the server database is exchanged with the device. For that purpose, the data is converted from the Domino format into a XML representation.

Instead of saving sync anchors indicating the time of the last successful synchronization, DEE maintains a version document listing a history of changes of each data document (change log). During each synchronization the version document is examined to determine which modified data documents need to be updated.

Mobile Notes   After synchronization, all user data and application information is stored in XML format in a local DB2e database. *Mobile Notes* is the client component, which renders the user data on the device screen as defined by the provided user interface elements (figure 20.5). With Mobile Notes the end user can view and edit his data on the device. Any changes are saved in the local XML data and returned back to the server during the next synchronization.

Administration   Similar to the Websphere Everyplace Synchronization Server described in the previous section, DEE uses an administration database, which contains user profile information. The user profile stores the Device ID, filter criteria and a list of mobile applications to be synchronized to the user's device. There is also a deployment utility, which adds new applications to a group of user profiles.

The overall architecture of DEE is shown in figure 20.6.

Figure 20.5: Rendering of an application by Mobile Notes

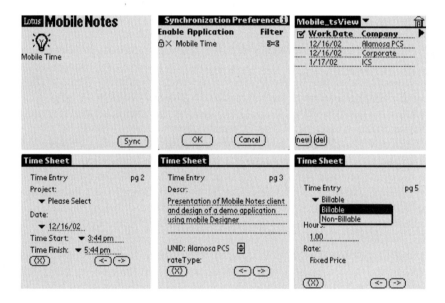

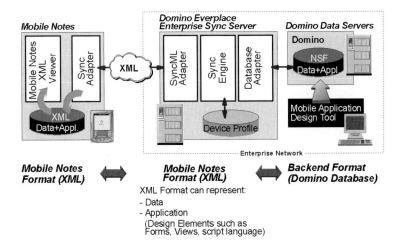

Figure 20.6:
Lotus Domino
Everyplace
Enterprise Server
and the Mobile
Notes Client

20.6
Further Readings

IBM WEA Sync Server

http://www.ibm.com/

Further information and news related to the WEA Synchronization server.

Infrared Data Association

http://www.irda.org

The Infrared Data Association is defining industry standards for infrared communication. They provide specifications, like IrOBEX for free download, feature products using IrDA and post information about industry events.

Lotus Corporation

http://www.lotus.com

Lotus Domino and Notes Lotus include are variety of mobile enhancements, which allow to access enterprise data from pervasive devices.

Puma Technology

http://www.pumatech.com

Puma Technology provides synchronization solutions. Their main product is called IntelliSync.

Starfish Software

http://www.starfish.com

Starfish Software is a leading supplier for wireless synchronization solutions.

SyncML

http://www.syncml.org

Several world leaders in mobile computing and synchronization formed the SyncML Initiative to define and promote a new synchronization protocol that enables ubiquitous access to updated information with any application. SyncML is working on specifications, a reference toolkit, and demonstrations, which will be publicly available from their homepage. This site offers interested parties the possibility to get involved by joining as a SyncML supporter at no cost.

Part V
New Services

Services are an important counterpart to technology. They are crucial for the benefit and value of an application. For example an ordinary credit card uses only a very simple magnetic stripe plastic card. But the infrastructure of the credit card organizations establish a versatile payment method, which is accepted worldwide and used self-evidently by masses. Credit cards are applied for retail, Internet shopping, and mail ordering. All involved parties benefit:

- The customer has a means of paying almost everywhere in a very convenient way.
- The vendor tries to attract more clients, which spend more money.
- The credit card companies base their revenue on offering this type of payment service.

Pervasive Computing devices like PDAs, mobile phones, or set-top boxes function as the delivery point of service offerings. With new devices and applications new kinds of services arise, helping us to increase the value for the user. While prices for computer chips and software decline, subscriptions, service contracts, and licenses are the profitable gatekeepers of information. The telecommunication industry was among the first to apply a service based business model: cellular phones are deployed for free but the connection fees make that deal profitable.

Services increase the value for the user

Such services can be distinguished in three categories (Figure V.1):

- Content services provide information. They comprise web shops, databases, or virtual libraries which are published somewhere in a network. Commerce, retail, and entertainment industry are the major contributors of this kind of services. Yet, only first beginnings of this service offerings and infrastructure are visible.
- Communication services transmit information. They are the domain of telecommunication companies and network operators. They maintain and exploit worldwide networks, such as GSM. Email, paging, and instant messaging are the typical kind of services offered here.
- Access services provide access to networks. Typical access services are dial-in points of Internet Service Providers (ISP) like AOL or the wireless Palm.Net. They allow users to connect their systems physically to some

network in order to access information. Another kind of access service is getting more and more important: Internet portals, such as Yahoo, offer logical entry-points through which selected content can be accessed.

*Figure V.1:
Services for
Pervasive
Computing*

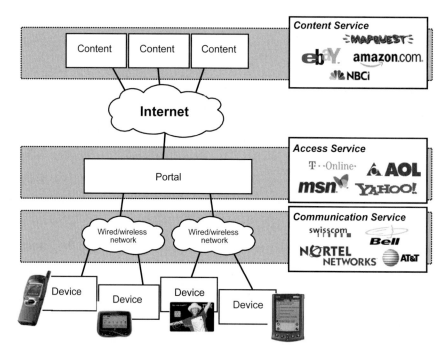

*Horizontal and
vertical services*

Horizontal services address the needs of a vast of consumers. SMS, interactive television offerings, and services for home appliances are typical examples. Vertical services target dedicated user groups or focus on specific branches, like retail or finance. Other examples are Business to Business offerings, or company internal portal services for accessing corporate information. Such services can be part of private intranets and protected by secure authentication mechanisms.

21 Home Services

More and more households already have more than one computer. A networked home will not just be these computers connected with each other using a Local Area Network (LAN), for example.

Pervasive Computing is going far beyond this. It will enable a whole set of new services as well as improve existing services. In this chapter, you will learn about the new services and business models that this will bring to your home.

21.1
The System View

A networked home will be connected to an external network, like the Internet, which connects the devices on your local home network with different service providers. Between the local and the external network a service gateway acts as a gateway or firewall. Your service providers will be able to download services to this gateway as well as to administer them.

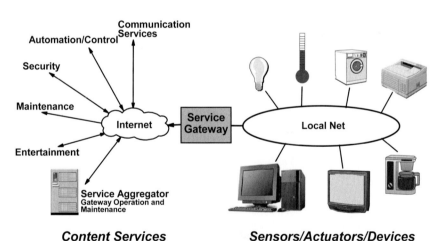

Figure 21.1:
Networked Home
System View

Content Services   **Sensors/Actuators/Devices**

The service gateway supports many wide-area networks and local network interfaces as well as local devices. It will act as the application server for many value-add services, like energy metering and management, remote home health care or home automation. Having a central point of control allows trusted service providers to aggregate and deliver services to a client's household. The client gets the possibility to choose services from different services providers that run together in that service gateway.

The service gateway is an embedded, zero-admin, server that will allow managing and integrating the existing devices, like PCs, set-top boxes, or cable modems, as well as new devices like intelligent washing machines or refrigerators.

It will enable the rapid mass deploying of these services to an increased number of households.

A networked home is of course not the only place a service gateway is required and will be used. In automobiles, escalators, trains or any other intelligent devices the service gateway will ease the remote maintenance and control.

21.2
Communication Services

The increasing number of Internet enabled devices and the availability of broadband Internet access, like DSL or Cable modem, will drive the demand for interconnecting these devices at home. One broadband access to the Internet could then be used from every Internet-enabled device in the home. Appliances would be able to interact with each other and specific resource, like printers or back-up tapes could be shared.

Home networks
Some recent developments in wireless, phoneline, and powerline networking do not require the installation of new wires in a household to create a network. Some of the evolving standards for a network at home are:

- Bluetooth – a standard for a short-range radio frequency network between devices.
- PLC – Power Line Carrier, it uses the existing electrical wiring to build the network.
- HomePNA – The Home Phoneline Networking Alliance standard uses the existing telephone cables.
- IEEE 802.11 – it uses radio frequency to connect to a device or network.
- HomeRF – creates a wireless LAN using radio frequency.
- HAVi – Home Audio Video interoperability, is focusing on in-home entertainment networking.

Using a service gateway, a home network will be connected to the outside world, most probably to the Internet using one of the following Internet access technologies:

Internet access technologies
- Analog modem or ISDN – this is already available for a long time, but the transmission rates are not high enough for current and future applications.
- DSL – Digital Subscriber Line – a high-speed way of accessing the Internet using today's telephone cables.

- Cable modem services – this uses broadband cables, as they are used for television today.
- Satellite link – this ensures high-speed wireless access to the Internet.

The communication services segment will most likely be the one that is driving the home services market. The primary consumer applications are shared Internet access, both for PC and intelligent appliances, the ability to interconnect multiple devices at home as well as value-added telephone and video-conference services, such as voice over IP and multiple phone connections over one line.

This provides the service providers or carriers the opportunity to offer a full range of communication products to the consumer. In a deregulated environment, as it is already in Europe and in the US, the operators can increase consumer loyalty and differentiating themselves from their competitors by offering these additional services.

Service providers and operators have a twofold benefit from a networked home. The service provider will be able to generate additional revenue from these services and the operator will get additional revenue from customers who are using these services and are creating additional switching costs.

21.3
Home Automation

The essential drivers of this segment are ease of use, simplified maintenance, and the penetration of networked devices. The usually long lifetime of household devices will limit the rapid adoption of these new devices in a large number of households.

It is relatively inexpensive to enable these types of devices for network connectivity during manufacturing, but it is relatively expensive to retrofit intelligence at a later point in time. Therefore these new devices will probably need at least one decade to be in almost every household, which would mean that it will take more than a decade.

The networked device will automatically handle many tasks, like setting the appliance clock, registering new devices for warranty purposes, providing product information for services, ordering appliance supplies, and displaying product user manuals.

Many appliance manufacturers are already enabling their devices for network access to allow download of the services into the devices later. One of the most compelling cases is remote diagnostics and remote maintenance. If this just enables a service engineer to find out which parts are broken or defect in a device, before driving to the consumer's household, this would save a trip back to the warehouse in case the engineer does not have the right parts with him.

The product could be improved and enhanced via network connection, video clips could be delivered over the Internet, like making recipes, removing stains, storing food, changing settings, adding suppliers, or making repairs. In addition, the controls could be adjusted to the needs of the person using it; children could have simpler choice than adults.

Mainly the device manufacturers will offer new and more convenient services to get a closer relation with their customers and to have them buy the next devices from them again.

21.4
Energy Services

The energy market is currently changing dramatically in Europe and USA, due to deregulation and the end of the monopoly of the power companies.

Automatic meter reading will most likely be one of the first services. Today meter reading is usually done once a year in Europe and monthly in the US. After the deregulation, the consumers now have the chance to switch the power supplier monthly, or maybe even in shorter terms. In some pilot installations, consumers can already buy electricity in the supermarket. The utility company sells smart cards, which contain the amount of electricity the user bought. At home, the customer inserts the smart card in a slot in his meter, in the same way a phone card is used to make calls at a public phone. With all these developments, the times a meter must be read will increase dramatically, compared to once a year.

Consumers will be switching electricity suppliers frequently, based for example on the lowest price or the best service. Additional value-added services, like energy management will enable power companies and service providers to generate additional revenue as well as to increase consumer convenience and loyalty.

Having the lights in a house automated and connected to a service gateway would enable a customer to turn on the lights or check if all of them are turned off from remote using the Internet. There are many more devices at home that could benefit from being connected, like thermostats or sensors.

Not all of these need to be separated services; they could exchange information and work together. A security service could perhaps turn on the lights in case it detects a theft or it could open the garage, turn on the lights, and switch on the radio at the time it detects the homeowner coming home.

21.5
Security Services

There is already an existing market for security services today. However, the emerging industry trends for integrated service bundles as well as new entrants in the market from the telecommunications and energy industry are changing the current business landscape.

The number of monitored residential security systems is increasing consistently; especially in the US over 17 million homes already have security systems installed. The expectation is that this number will grow with about 650,000 new subscribers each year.

The networked home will give consumers a more flexible platform and enable them to easily subscribe to these services without the need to install an extra set of wires in their homes. The monitoring could then also be extend to other systems in the household, like checking if the light is turned and all cooking devices in the kitchen are switched off.

The homeowner with a networked home is now able to check from remotely, if the back door is locked at home, maybe even by using his WAP-enabled mobile phone. There will be no need anymore to return home, just because one wasn't sure if all windows and doors were closed and locked.

Integrating these new security services with existing core services to offer a brand range of services to homeowners is the primary business opportunity for service providers. Players from other industries, like telecommunication or utility companies, will integrate these services with their current communication and energy management services.

All these services will use the same infrastructure, which will make it more efficient and simpler to install and manage.

21.6
Remote Home Healthcare Services

The aging population in developed countries together with the breakdown of the extended families and the increase mobility of the workforce creates the demand for new and innovative remote home healthcare services. Of course, these services can't replace the human interaction and visits, but are a useful addition to make the life of these people more convenient.

The initial applications in this segment are likely to provide "peace of mind" to relatives of elderly and disabled people through security and monitoring systems, and to offer easy-to-use communication systems such as a user friendly home shopping and video telephone.

There are two target customer groups in this market. First, the elderly themselves with their primary need for easy-to-use and fail-safe devices as well as reminders. These new services give them an increased independence, more convenience, and a feeling of keeping in touch.

The second target group is the relatives of the elderly, with their primary need to be sure that everything is OK between visits and to keep in touch with the elderly. The most likely applications in this area are monitoring and sensing systems, like motion detectors as well as personal alarms, assistance tools, and communication.

21.7
Further Readings

Gatespace

http://www.gatespace.com

Gatespace is a provider of service gateways. They also provide useful information related to networked homes on their homepage.

E2Home

http://www.e2-home.com

E2Home is a joint venture between Ericsson and Electrolux providing home services and devices for a networked home. See which appliances and services are already available today.

IBM Pervasive Computing

http://www.ibm.com/pvc

IBM is a leading technology provider. Their web pages provide detailed information about the future of networked homes. Even a toolkit to develop for an OSGi compliant service gateway is available for download as part of the WebSphere Everyplace Embedded Software.

Merloni Elettrodomestici

http://www.merloni.com

Merloni provides the latest information about their Internet enabled kitchen appliances, like Leon@rdo, a kitchen monitor or Ariston Digital, a new generation of kitchen appliances.

Open Services Gateway Initiative

http://www.osgi.org

The Open Services Gateway initiative is an open group of industry leaders, which define the specification for a service gateway. Without their work, no interoperability would be possible between different service providers. The benefit for the consumer is that there will be the need for only one service gateway in a household, because it is able to host all different kinds of services from different service providers.

The OSGi homepage provides details about the OSGi specification, a list of members with links, and recent press releases.

Panja Home Systems

http://www.panja.com

This site features a photo tour through a networked home installation.

Parks Associates

http://www.parksassociates.com

Parks Associates is the organizer of the Connection series of conferences, as well as doing research in the area of networked home. On their homepage are presentations from recent Connection conferences, as well as research reports.

Point Clark

http://www.pointclark.net

A commercial provider for residential gateway software and services.

Residential Area Highway Project

http://rah.ele.tue.nl

This page includes interesting links to standardization activities, documents, and projects related to in-house networks and residential gateways.

22 Travel and Business Services

22.1
Travel Services

The travel industry evolved from centralized reservation systems, which were only accessed by travel agencies, to direct customer interaction using the Internet. The next step is extending the availability of travel agency services to ubiquity, allowing travelers to take advantage of these services with the help of mobile devices, even at the time these travelers are already on the road. It seams to be self-evident that many early adopters of Pervasive Computing devices were mobile professionals, traveling frequently on business.

Pervasive travel applications are more than just porting an online booking and reservation system from a PC to a mobile phone or handheld computer. There are new kinds of services and applications, which will arise with the new technological possibilities. For the travel industry, these new value added services are a way to establish customer loyalty. The traveler is informed and can take the required actions anywhere in the world. He can be integrated into business processes, for example when using his WAP phone to trigger a flight check-in or he can carry along an electronic tickets stored on a smart card chip. *Value-add for customer loyalty*

A traveler has basically three different kinds of entry points for accessing information or requesting services (see Figure 22.1): *Accessing services*

- At home, he can visit web sites from his regular PC or from immobile pervasive devices, such as an interactive television set or a screenphone. Time tables, hotel information, and multimedia travel guides can easily be retrieved from the web and are already commonly used. The Internet gives quick access to ticket services, reservations, and fares. Individual user profiles on the travel agency's server store preferred seats, discount rates, and applicable bonus programs.
- On the road, the traveler can access driving directions, download maps, check for latest updates of schedules, or change his itinerary using personal mobile devices. Weather, events, or traffic information are easily accessible. A wide range of communication services help to stay in contact and be reachable. These devices can either carry around useful content, which has been provided earlier, or they can function as an access media to relevant online information, provided by a network. Depending on the capabilities

of the device, such information can be retrieved ubiquitously using wireless communication services like WAP. Other devices, which require a wired connection, can be attached to a PC or phone-line, in order to download the content of interest.

- Finally, various kinds of intelligent ticket vending machines, multipurpose kiosks, and check-in terminals are present in hotel lobbies, airport terminals, and railway stations. They give access to the systems in the back, answer inquiries, find vacancies, determine lowest fares, compose quickest connections, and print out tickets and boarding cards. Smart cards authenticate the traveler, when accessing electronic tickets or frequent flyer accounts stored on the back-end system of a carrier. Smart cards can even be used to store an electronic ticket directly on the card itself, in a secure and fraud proofed manner. Hotel vouchers, room keys, and electronic purses are other objects, which could be carried around on a smart card and used to access travel related services.

Why pervasive travel services?

All these kinds of services, applications, and information offerings are supplied by the travel industry for several reasons:

First of all, this is a strategic part of customer relationship management (CRM) and is often integrated in loyalty programs. Mobile travel services are a value add for the client, and help companies to differentiate from competitors. Airlines, hotels, and rental car companies can consult their clients better and can offer individual care, based on customer profiles.

Another reason is the reduction of the transactions costs to a minimum using low-cost distribution channels like the Internet and establishing business processes, which directly involve the customer.

Finally, providing interesting content, such as maps, weather forecasts or local tourists information is a way to advertise and attract new customers.

Figure 22.1: Travel Services

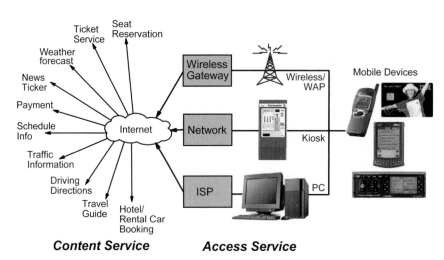

22.2
On the Ground ...

Handheld devices are helpful tools for navigation. With an attached GPS system and a speech output module, driving directions can be given from the backseat. For example, a Dutch brewery supplies an application running on a Palm device, which is equipped with a GPS system and points the user to the closest pub, selling beer of that brand.

The Scandic Hotels offer a WAP based booking service. Room reservations can be made and changed with a mobile phone.

The German railway company offers travelers the opportunity to buy a ticket via Internet. The ticket is printed out at home. The web server sends a receipt directly to the conductor of the selected train, allowing him to validate the self-printed ticket.

Network connected cars allow very sophisticated travel applications scenarios:

Automotive Services

- Pervasive devices inside the car access real-time information about weather, traffic, shopping opportunities, and tourist facilities. PIM applications running on a board computer system synchronize with the desktop computer at home. The engine's diagnostic controller can notify the driver's PDA about maintenance schedules, while customer appointments and address book entries are used to program the navigation system. The optimized itinerary is displayed on the dashboard.
- During the trip, latest traffic information is received from a service provider. The travel route will be will updated automatically, in order to avoid traffic jams. According to the driver's personal profile, a notification will occur, when he passes points of interest, like shopping malls, restaurants, or tourists sights.
- Beside navigation and travel information, other entertainment, computing, and communication capabilities, like in-car TV, games, and email are gaining importance. Multimedia capabilities and wireless connectivity make vehicles an important platform to deliver Internet based services (see Figure 22.2).
- Other valuable offerings from automotive industry are: Diagnostic data is sent from the car to the closest repair shop as soon as a problem is detected. This allows ordering the necessary parts before the car arrives at the service station. Further, the system can arrange an appointment for the repair, based on severity, and personal calendar entries.
- An online connection between the car's GPS system and an administration system could be used for fleet management. For instance, the headquarter of an express delivery company could always track the current position of fleet members and would be able to generate the most efficient pick-up plan. Information downloaded to each driver could depend on his current location.

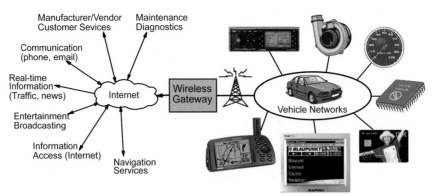

Figure 22.2: Automotive Services

Manufacturer/Vendor Customer Sevices

Maintenance Diagnostics

Communication (phone, email)

Real-time Information (Traffic, news)

Internet

Wireless Gateway

Vehicle Networks

Entertainment Broadcasting

Information Access (Internet)

Navigation Services

Content Services   Access Service   Sensors/Actuators/Devices

22.3
... And in the Air

Airlines are currently very busy in preparing their systems and infrastructure for Pervasive Computing. In early pilots, they provide a set of new services to their passengers:

- In December 1999, Swissair introduced a system that enables selected customers to check in from Web-enabled mobile phones. Instead of queuing in front of the airline counter, passengers dial the WAP gateway and execute an online check-in dialog on their mobile phone. They can select seats and access latest flight information, such as the gate number and departure time. A formal boarding card is printed out automatically and can be picked up, when entering the plane.
- Delta Air Lines supports Palm VII devices as well as mobile phones to access flight schedules and up-to date departure information like gate, time, or delays.

22.4
Business Services

Pervasive devices are changing from more or less private gadgets to serious business tools offering key vertical applications for specific industries. PDAs, phones, and smart cards extend the reach of enterprise networks. Sales forces can access up-to-date product catalogues and enter orders immediately at the customers site. Documents and data are no longer bound to dedicated locations, but always accessible. Workflow processes can include mobile workers, which can initiate transactions from remote, receive event notifications, or task assignments. Incoming communication is automatically redirected to the current location of the recipient. Inside offices, an entirely new service infrastructure grows:

devices like printers, fax machines, handheld computers, smart cards, phones, and door openers interconnect with each other and collaborate. All these new connectivity and access services improve productivity.

For mobile devices, there are three typical entry points to corporate networks:

- The dominant gateway to the network is still the Personal Computer, residing on the personal desk. Pervasive devices like phones, handheld companions, and other peripherals connect and interact with the PC.
- Wireless services comprise the mobile access from remote, for example while on a business trip, as well as the cordless access within the office building. The latter provides convenient and flexible connectivity without having to care about network plugs when visiting another room or making a roundtrip through the building.
- Public business centers and Internet cafés provide communication services to mobile professionals in traveler lounges, hotels, or other even in the seat of a transportation carrier. They offer network adapters to connect a personal mobile device to the Internet. With Virtual Private Network (VPN) technology a mobile device can tunnel a secure connection through the public web to a corporate Intranet. Office services like fax machines and printers can be used to route personal information to the current location.

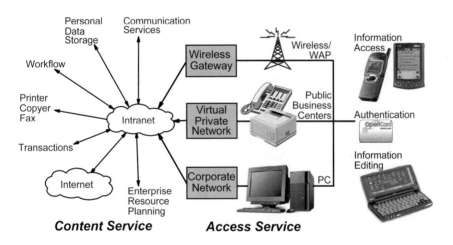

Figure 22.3: Business Services

22.5
Field Support

A lot of time and money was invested by companies to develop applications, which provide real-time information to their employees. For the mobile workforce it gets increasingly important to immediately access information from a variety of corporate data repositories. Gathering information and presenting reports and proposals are major tasks for mobile workers. They need billing, contract, technical, and demographic data stored in corporate databases and legacy systems.

In order to generate customer invoices, bills, and reports financial personnel needs up-to-date information. Delayed data entry into corporate databases hampers an organization's ability to access current information about customers, inventory status, work orders, work schedules, billing history, and other important data.

One part of a real-time mobile solution is the mobile device itself. The other part is the application, which is especially designed for meeting the needs of mobile professionals. The application should be able to gather information, execute business logic, and to access and synchronize with corporate data.

22.6
Further Readings

Delta Airlines

http://www.delta-air.com

Delta Airlines offers travelers the possibilities to check the worldwide flight schedule, as well as up-to-date departure and arrival information using mobile phone or wireless handhelds.

Heiniken BarTrek

http://www.heiniken.com/bartrek

BarTrek provides travelers the way to the closest pub selling Heiniken beer. As soon as the users downloaded the BarTrek application to their handhelds, they are able to connect their devices to a mobile GPS system. The application displays the current position and the route to the closest Heiniken pub.

OffRamp

http://offramp.co.uk

OffRamp delivers a list of nearby restaurants, hotels and shops to mobile devices via Web, WAP or WebTV.

OnStar

http://www.onstar.com

This site offers a variety of vehicle related services, including travel information, traffic news, diagnostics, and other kinds of assistance.

Swissair

http://www.swissair.com

Swissair offers their customers at Zurich airport the possibility to check-in from WAP-enabled mobile phones, instead of waiting in a line at the ticket counter.

Wap.com

http://www.wap.com

This site provides a long list of links to WAP-enabled services and related topics.

YourWAP

http://www.yourWAP.com

YourWAP provides different services for user of mobile devices, like access to emails and contacts. It always could also send birthday reminders to your mobile device.

23 Consumer Services

Similar to the travel industry, retailers use Pervasive Computing for Customer Relation Management and as an additional distribution channel for their products. Mobile Commerce (M-Commerce) extends today's PC based e-Commerce to the variety of pervasive gadgets. A new revenue stream is expected, when reaching the customer everywhere at any time. In order to attract and retain customers, retailers offer manifold loyalty programs based on mobile devices. Personalized customer services evaluate purchase history, individual profiles, and maybe even the client's current location.

The main change in commerce is a significant shift from a product "push" (for example in a brick-and-mortar store) to a customer "pull" (for example when retrieving product information from the Internet). That impacts advertisement, loyalty programs as well as the shopping itself.

23.1
Interactive Advertisement

The retail and commerce industry will take advantage of the new possibilities of interactive television and will update their way of advertising. Interactive advertisement allows vendors and customers to get in touch already during the TV spot. Graphical elements such as icons and banners known from the Web may be placed anywhere. The customer can jump into requested information pages to obtain more information. Or he can purchase the product immediately, using online form sheets. Additionally, the vendor is enabled to trigger individual local data from remote. For example, he can display the location of the shop that is closest to the viewer. Maybe a vendor wants to distribute loyalty points to those who watch his commercial spots. Or the viewer can click during an action movie on the car that the hero just has crashed into a canyon in order to find out its product specification. Smart agents will explore the Internet and find the cheapest offering for that car and create an online order for you. With interactive TV, vendors hope to reach their customers more efficiently. They can use a television program not only for their commercials but as a new market place. Additionally, vendors will enable convenient online shopping even with the familiar television set. Those living without a PC might accept a set-top box to connect to the Internet. Opening the Internet for TV viewers will trigger the next wave of e-Business.

23.2
Loyalty

Retailers are seeking to exploit different technologies and improve their loyalty programs. Smart cards offer the necessary security characteristics to store bonus points, vouchers, coupons, and awards securely, without requiring an online authentication at the point of sale. This simplifies the infrastructure when implementing cooperations and alliances between different stores.

Loyalty and payment functionality integrated in one card and interacting dynamically enhances the value for all stakeholders. Partnerships between a credit card company and a supermarket chain are one typical example for such a card.

23.3
Shopping

The list of services leverage shopping is endless:

- Some shops offer mobile devices, which allow the consumer to scan-in a barcode himself of the items he is putting in his shopping cart. The list of items bought is then transferred to the cashier who is charging the customer for the products.
- Smart labels attached to products make the check-out obsolete.
- A soft drink vending machine at the airport in Helsinki accepts payment per mobile phone. Each cleaning program has its own phone number. A short message text is used to transmit a payment order to the telecommunication provider, which charges the amount due to the telephone bill.

The "Shop & Go" application of the British supermarket chain Safeway Stores is a good example how technology and service offerings can be combined: Safeway has distributed a modified version of Palm devices to their customers, which can be used to submit grocery orders. From a list of available goods an individual shopping list can be composed. This is simplified by an attached barcode reader: Products can be scanned at home or in the store. Past purchases can be used as a template for the shopping list. Via phone line new special offerings and latest price updates are downloaded to the device and completed orders are transmitted to Safeway. The ordered goods can be picked up at an agreed-on time.

The philosophy of this application is to provide a simple and convenient front-end to the customer and to use sophisticated systems in the background. An IBM System/390 runs a DB2 database and data mining software (IBM Intelligent Miner) to analyze the customer's past three months shopping history and to provide customized special offers. New products are promoted based on prior purchases and purchases of other customers with similar profiles. To prevent unauthorized access to Safeway's host system, each Palm Pilot is initially setup with a customer identification. Thus, any tampering is flagged and the originator can be identified.

*Figure 23.2:
Safeway's
"Shop & Go"
Application*

23.4
Payment Services

The incredible growth of the Internet forces a rapid movement of financial institutions into the electronic arena. Banks are enabled to use Internet technologies to provide a more flexible way to deliver financial services, such as bill display and payment, personalization, brokerage, insurance, etc. on ATMs and point of sale kiosks.

Banks like the Handelsbanken in Sweden or the Spanish Banesto bank extended their Internet banking services to mobile devices. Thus, customers are able to access certain Internet banking systems using their WAP phones.

The Paybox System

However, these new services are not restricted to the banking sector. For instance, a system called Paybox was recently introduced in Germany. Paybox allows the customers to perform Internet payments via their cellular phones – not even WAP phones are required. Customers and merchants have to register with the Paybox. system. When a customer wants to purchase goods in a Web shop, the registered mobile phone number has to be entered. The trader sends the transaction to the Paybox system which calls the customer back who has to confirm the transaction with a PIN.

Electronic Purse Card

Other payment systems which became popular in the last few years are smart card based electronic purse systems like the German "Geldkarte" or the French "Moneo" system. The transfer of money onto the purse card takes place via a reference bank account or via terminals accepting cash. The PIN protected chip card is used to authenticate the card owner and to load money onto the electronic purse card.

When the card owner purchases goods using this card, the money is not transferred between the accounts of the card owner and the merchant, but between the card itself and the reference account of the merchant.

23.5
Further Readings

Geldkarte

http://www.heise.de/ct/99/13/031

An article of the computer magazine C't on the Geldkarte payment system (in German language).

Paybox

http://www.paybox.de/international/english.html

The homepage of the paybox system. Here you get detailed information about the paybox system features and about the partners of the paybox organization.

Safeway Demo

http://www-3.ibm.com/pvc/tech/safeway.shtml

Here you can get information about the Safeway Easi-Order system. A demo of the application can be executed online.

Part VI
Appendices

A   Bibliography

[COU02]   Courtois, Pieprzyk,,
            http://eprint.iacr.org/2002/044

[DAY99]   Day, B., *Program Java Devices*, Java World, July 1999,
            http://www.javaworld.com

[DTV00]   *Digital TV: The Future of E-Commerce. E-Commerce Times*, 20.3.00,
               http://www.ecommercetimes.com/news/articles2000/000320-1.shtml

[ECMA262]   ECMA, *Standard ECMA-262, ECMAScript Language Specification*, 1997

[EMV96]   *EMV '96 - Integrated Circuit Card Specification for Payment Systems*,
            Europay, Mastercard, Visa, 1996

[ERI00]   *Ericsson R 320 White Paper*, Ericsson, 2000

[FOX99]   Fox, A., *Information Delivery Infrastructure in the Mobile, Wireless Age*,
            ProxiNet, 1999, http://www.proxinet.com/technology

[FLA99]   Flanagan, D., *Java in a Nutshell - A Desktop Quick Reference*,
            O'Reilly & Associates, 1999

[HAN98]   Han, R., Bhagwat, P., et. al., *Dynamic Adaptation in an Image Transcoding
            Proxy for Mobile Web Browsing*. IEEE Personal Communications Maga-
            zine, December 1998.

[HAN02]   Hansmann, U., Nicklous, S., Schaeck, T., Seliger, F., *Smart Card Applica-
            tion Development Using Java, Second Edition*, Springer, 2002

[HAN03]   Hansmann, U., Purakayastha, A., Mettalä, R., Thompson, P., *SyncML – Syn-
            chronizing and Managing Your Mobile Data*, Prentice-Hall, 2003

[HEN00]   Henry, E., *Sun's Blueprint*, JavaWorld, February 2000,
            http://www.javaworld.com

[IDC00]   Hause, K., et al., *Review and Forecast of the Worldwide Information*

[ISO7816]   ISO 7816, *Identification Cards – Integrated Circuit Cards with Contacts*,
               ISO

[IrMC99]   Infrared Data Association, *Specifications for Ir Mobile Communications,
               Version 1.1*, 1999, http://www.irda.org

[ISO99]   International Organization for Standards, 1999, http://www.iso.ch

[JC00]   *Java Card 2.1.1 Platform Specification*, http://www.javasoft.com/javacard

[JCF00]   Java Card Forum, http://www.javacardforum.org

[KOB87]   Koblitz, N., *Elliptic Curve Cryptosystems, Mathematics of Computation*, v.
               48, n. 177, 1987, pp. 203-209

[KUM00]   Kumar, D., et al., *End-to-End Management with Tivoli - Managing PDAs*,
               IBM, 2000, http://www.redbooks.ibm.com

[MIL86]   Miller, V.S., *Use of Elliptic Curves in Cryptography, Advances in Cryptology
               – Crypto '85 Proceedings*, Springer-Verlag, 1986, pp. 417-426

[NOK99]   *The Demand for Mobile Value-Added Services*, Nokia, 1999,
               http://www.nokia.com/press/background/index.html

[NOK00]   *Multimedia Messaging White Paper*, Nokia 2000

[PALMVII]   *Palm VII – Wireless Internet Access*, Palm: 1998,
               http://www.palm.com/pr/palmvii/7whitepaper.pdf

[RUL00]   Ruley, J., *Microsoft debuts Pocket PC*, Byte, 20.3.2000

[RSA00]   RSA Laboratories, *Frequently asked questions about today's cryptography
               Version 4.1,* 2000, http://www.rsa.com

[SlotTrot]   Slot-Trot Software Oy AB, *WinWAP Browser*, 2000,
               http://www.slottrot.com

[SWE99]   Swedlow, T., *Enhanced Television*, American Film Institute, 1999,
               http://www.itvt.com/etvwhitepaper.html

[TIV00]   *Pervasive Management - Expanding the Reach of IT Management to Perva-sive Devices*, Tivoli Systems, 2000
              http://www.tivoli.com/products/solutions/pervasive

[VCAL]   The Internet Mail Consortium (IMC), *vCalendar – The Electronic Calendar and Scheduling Exchange Format Version 1.0*, 1996,
              http://www.imc.org/pdi/vcal-10.doc

[VCARD]   The Internet Mail Consortium (IMC*), vCard – The Electronic Business Card Version 2.1*, 1996, http://www.imc.org/pdi/vcard-21.doc

[WAESpec]   WAP Forum, *Wireless Application Environment Specification 19990524*, 1999, http://www.wapforum.org/http://www.wapforum.org/
              http://www.wapforum.org

[WBXML]   WAP Forum, *WAP Binary XML Content Format 19990616*, 1999,
              http://www.wapforum.org

[WEBB99]   Webb, W., *Embedded Technology transforms the Automobile*, EDN, August 1999, http://www.ednmag.com/ednmag/reg/1999/081999/17df1.htm

[WMLScript]   WAP Forum*, WMLScript Specification 19990617*, 1999,
              http://www.wapforum.org

[WMLSLib]   WAP Forum, *WMLScript Standard Libraries Specification 19990617*, 1999,
              http://www.wapforum.org

[WRI99]   Wright, M., *Prep for a multimedia Future*, EDN, August, 1999,
              http://www.ednmag.com/ednmag/reg/1999/081999/17cs.htm

[WSPSpec]   WAP Forum, *Wireless Session Protocol Specification 19999528*, 1999,
              http://www.wapforum.org

[YAN00]   Yang, D. J., *A boob tube with brains. Business & Technology*, 13.3.00,
              http://www.usnews.com/usnews/issue/000313/webtv.htm

B Glossary

- **3G** – Third generation global mobile radio standard. Operates in the 2 GHz band and offers data rates of up to 2 Mbps.
- **ACL** – Access Control List includes information, which restrict access to files and directories in a file system.
- **AMPS** – Advanced Mobile Phone System. Developed by Bell Labs in the 1970s and first used commercially in the United States in 1983.
- **APDU** – Application Protocol Data Unit. In context with smart cards, APDUs are the formatted messages exchanged between card readers and smart cards
- **Applet** – A small program that can be downloaded to and executed on a client. In this book, the term applet is used both for programs running in web browsers as well as for programs running on Java Cards.
- **Asymmetric Cryptographic Algorithm** – A cryptographic algorithm that uses different keys for encryption and decryption. Examples are RSA and ECC.
- **Authentication** – A sequence of actions proving the *authenticity* of an entity to a second entity.
- **Byte Code** – Machine-independent code, e.g. generated by a Java or Visual Basic compiler.
- **CAN** – Controller Area Network. A bus system typically applied within in-vehicle networks.
- **Card Applet** – A Java application that is stored and executed on a Java Card. A card applets inherits from the class Applet of the Java Card Framework. Card applets have no dependencies on Java's AWT.
- **CDMA (1)** – Code Division Multiple Access. A method for allowing multiple users to share a frequency band by spreading the information across the frequency spectrum according to a specific code. Employs spread-spectrum technology.
- **CDMA (2)** – Often used to refer to the IS-95 CDMA mobile phone system deployed in the U.S.
- **cdma2000** – A third generation technology proposal based on technological evolution from the North American IS-95 CMDA system. This 3G proposal optimizes compatibility with existing IS-95 CDMA networks.
- **cdmaOne™** – IS-95 CMDA system. This term designates the complete end-to-end wireless system calling for standard interfaces between network components. Trademark of the CDMA Development Group.

- **Chip** – An information particle sent out on a single frequency by a CMDA system. Many chips combine to form symbols, which can represent binary ones and zeros.
- **Chip rate** – A significant parameter of a CMDA system that helps determine the raw bit rate capacity. Specified as chips per second.
- **Conduit** – Applications, which typically run on desktop computers and are used to synchronize data of corresponding mobile devices.
- **Cryptographic Protocol** – Protocol that employs a sequence of cryptographic operations to authenticate entities or transmit information.
- **DECT** – Digital Enhanced Cordless Telecommunications developed by ETSI. Can be used for data as well as voice applications.
- **DES** – Digital Encryption Standard, invented by IBM and standardized by NIST, is the most well known widely used symmetric cryptographic algorithm.
- **Digital Signature** – Encrypted digital fingerprints of data, ensuring data integrity and authenticity
- **Downlink** – The radio link transmitting from the base station transceiver to the mobile device.
- **DSI** – Device Support Infrastructure is a device management architecture for pervasive Devices.
- **DSL** – Digital Subscriber Line. Modem telecommunications technology that enables broadband data transmission over regular telephone line.
- **DTV** – Digital Signal Television. A digital broadcasting standard, for two-way and high-bandwidth data transmission allowing interactive television.
- **Dynamic Memory** – Volatile memory used during application execution.
- **ECC** – Elliptic Curve Cryptography. An asymmetric, or public key, cryptographic algorithm. Requires a shorter key length than other algorithms for a given level of security.
- **EDGE** – Enhanced Data rate for GSM Evolution. Uses advanced modulation techniques to maximum speeds of 384 kbps.
- **ERP** – Enterprise Resource Planning System.
- **FAT** – File Allocation Table. The file system known from DOS.
- **FDD** – Frequency Domain Duplex. A method for two-way communication that works by using separate frequencies for transmission and reception.
- **FDMA** – Frequency Division Multiple Access. A method for allowing multiple users to share a frequency band by assigning a separate carrier frequency to each user. Often used by analog cellular phone systems.
- **Forward link** – downlink.
- **GPRS** – General Packet Radio Service is a successor of GSM and enhances the bandwidth of mobile data transmission up to 115 kbps.
- **GPS** – Global Positioning System. A satellite based system to determine one's current location.
- **GSM** – Global System for Mobile Communications. The first European digital standard, developed to establish cellular system compatibility throughout Europe.

- **Handoff** – The U.S. term for the process which takes place when control of a mobile station is passed from one base station to another.
- **Handover** – The European term for handoff.
- **Hash** – A digital fingerprint of data. Hash algorithms are often used in conjunction with *Digital Signatures*.
- **HDTV** – High Definition Television is a standard for digital broadcasting which provides about five times more resolution than conventional television.
- **Heap** – Memory that is dynamically allocated by the system. Unlike stack memory, heap memory can be allocated or deallocated at any point during program execution.
- **IDB** – Intelligent Transportation Systems Data Bus. A network technology for vehicles, which allows to connect consumer devices to on-board systems.
- **IMT-2000** – A set of specifications published by the International Telecommunications Union for third generation mobile radio communication employing satellite as well as terrestrial components. Calls for global standardization of high-speed (144 kbps – 2 Mbps) network access. Takes its name from the group goal of completing specification work in 2000 as well as from the proposed operating frequency band of 2000 MHz.
- **IrDA** – Infrared Data Association.
- **IS-136** – An extended TDMA system based on IS-54 and incorporating elements from GSM.
- **IS-54** – North American TDMA system compatible with AMPS.
- **IS-95** – North American CDMA system compatible with AMPS.
- **ISDN** – Integrated Services Digital Network.
- **ISP** – Internet Service Provider. Companies offering Internet access either using a dial-up or broadband connection.
- **Java Card** – A *smart card* that has the capability of running *Java* programs using a restricted command set and library on-card.
- **JVM** – Java Virtual Machine. The execution environment for Java programs. The JVM interprets the Java byte code. The JVM also enforces security restrictions for the executed code.
- **LDAP** – Lightweight Directory Access Protocol. A protocol for accessing online directory services.
- **LIN** – Local Interconnect Network. A low-cost in-vehicle network.
- **MP3** – abbreviation for MPEG-1, audio layer 3. A standardized algorithm that compresses audio tracks without a significant reduction of sound quality.
- **MPEG-2** – standard for digital video and audio compression for moving images defined by the Motion Picture Experts Group.
- **PCS** – Personal Communications Service.
- **PDA** – Personal Digital Assistant, such as handheld computers or organizers.

- **PDC** – Personal Digital Cellular. This is a TDMA-based Japanese standard operating in the 800 and 1500 MHz bands.
- **PIM** – Personal Information Management typically comprises applications like calendar, address book, to do lists, mail, and memos.
- **PKCS** – The Public-Key Cryptography Standards are a series of standards initiated by RSA to foster interoperability of cryptographic systems. Especially PKCS#11 and PKCS#15 are relevant for smart cards.
- **POTS** – Plain Old Telephone System. Standard wireline telephone system.
- **Private Key** – Private Keys are used in public key cryptosystems to generate digital signatures or decrypt messages. Private keys must be kept in secret to assure that only the owner can use them.
- **PSTN** – Public Switched Telephone System. Standard wireline telephone system.
- **Public Key Algorithm** – Asymmetric cryptographic algorithm, where one key is revealed to the public and one key is kept in private. *Public keys* are used for encryption of data or validation of digital signatures while *private keys* are used for decryption of data or generation of digital signatures.
- **Public Key** – Public Keys are used in public key cryptosystems to verify digital signatures or to encrypt messages. Public keys are usually published so that anybody can use them.
- **RAM** – Random Access Memory
- **RDS** – Radio Data System. It is used to transmit data on regular FM audio signals.
- **Reverse link** – uplink.
- **ROM** – Read Only Memory.
- **RSA**– Most important *asymmetric cryptographic algorithm*. The acronym stands for the inventors Rivest, Shamir and Adleman.
- **RTOS** – Real-time Operating System..
- **Screenphone** – Phone supplied with a screen, which can be used to surf the Internet.
- **SDMA** – Space Division Multiple Access, a possible component of 3G Digital Cellular.
- **Secret Key** – A key for a symmetric algorithm, that needs to be kept in secret to assure security of the system in which it is used.
- **Set-top box** – An electronic device that sits on top of a TV set and allows it to connect to the Internet, game consoles, or cable systems.
- **SIM** – Subscriber Identity Module. A smart card chip used in GSM mobile phones to securely hold subscriber-specific information.
- **Smart Card** – A credit card sized plastic card with a computer chip.
- **Smart Phone** – Intelligent mobile phone offering handheld computer like capabilities, such as Personal Information Management.
- **SMS** – Short Message Service is used to transmit text to and from a mobile phone.

- **Stack** – An area of memory where storage for variables is allocated when a subprogram or procedure begins execution and is released when execution completes..
- **Storage Memory** – Persistent storage memory used to store applications as well as data.
- **Symmetric Cryptographic Algorithm** – Cryptographic algorithm, where the same key is used for encryption and decryption. Examples are *DES* and IDEA.
- **Tamper Proof** – Tamper proof devices are built so that they loose all stored information when somebody tries to tamper with the device.
- **TCP/IP** – Transmission Control Protocol based on Internet Protocol. This protocol allows reliable delivery of streams of data from one host to another.
- **TDD** – Time Division Duplex. A method for two-way communication that works by allocating separate time slots for transmission and reception.
- **TDMA** – Time Division Multiple Access. A method for allowing multiple users to share a frequency band by assigning time slots for data transmission to each user.
- **TD-SCDMA** – Time-Division Synchronous CDMA. A third generation technology proposal developed by the Chinese Academy of Telecommunication technology (CATT). Compatible with existing GSM and TDMA networks. Allows higher data rates by adding a special synchronization signal to the data frames.
- **Thread** – The basic unit of program execution. An application can have several threads running concurrently, each performing a different job.
- **TMC** – Traffic Message Channel. It is used to transmit up-to-date traffic information.
- **Transcoding** – Content adaptation to meet the specific capabilities of the targeting device.
- **UMTS** – Universal Mobile Telephone Standard. A third generation cellular technology proposal developed by ETSI. This specification includes satellite as well as terrestrial network communication.
- **Uplink** – The radio link transmitting from the mobile device to the base station transceiver.
- **UTRA** – UMTS Terrestrial Radio Access. A third generation cellular technology proposal using W-CMDA developed by ETSI.
- **VBI** – Vertical Blanking Interval. A free line within the analog TV signal, which can be used to transmit closed captioning or other data.
- **WAP** – Wireless Application Protocol allows accessing Internet based content using mobile devices.
- **W-CDMA** – Wideband CDMA. Defines the air interface in the UMTS proposal. This air interface is optimized for compatibility with existing GSM networks.
- **X.25** – A ITU recommendation specifying packet switching technology for connecting remote terminals with host computers.

C Index

hardware abstraction layer (HAL) 138
hardware security 53
hash 197
hashing 180
HAVi 406
HDML 218
HDTV 102, 111
heating, ventilation, and air conditioning (HVAC) 73
Hellman, Martin 194
Hewlett-Packard 179
High Definition Television (HDTV) 102, 111
High-Speed Circuit Switched Data (HSCSD) 278, 279
Hitachi 106
home automation 408
home banking 50
Home Location Register 272
home network 70, 303
Home Services 409
HomePlug 304
homePNA 304, 406
homeRF 304, 406
horizontal service 404
HotSync 32, 145, 148
HTML 119, 174, 175, 366
HTML BODY 208
HTML HEAD 208
HTTP 101, 136, 173, 175, 204, 205, 216, 329, 331, 334, 394, 396
HVAC 73
hybrid smart card 51
hyper video 100
Hypertext Markup Language (HTML) 206

I

IBM 29, 109, 126, 176, 249
IBM Portal Zone 368
IBM Smart Card for e-business 160
IBM Tivoli Access manager 366
IBM Voice Application Access 373
IBM WebSphere Everyplace Access 370
IBM WebSphere® Portal 363
ICC 179, 180
ICC Service Provider 180

IDB Forum 93
identification card 50
IEEE 1394 104, 107
IEEE 802.11 406
IETF 175
IFD Handler 180
IMC 220
i-mode 232
imprint 50
IMT-2000 283, 292
information access devices 25
infotainment 86
infrared 42, 131, 136, 145, 389
Infrared Data Association (IrDA) 21, 300, 392
Infrared Exchange Manager API 145
infrared low level API 146
Infrared Mobile Communication (IrMC) 392
initialization 55
INS 55
install(...) 159, 162
instruction byte (INS) 55, 161
integrated circuit 50
integrity 190, 198
Intel 179
intelligent appliances 26
intelligent control 82
Intelligent Notification Services 372
interactive advertisement 421
interactive services 103
interactive television 17, 100, 124
interactive video 100
International Mobile Equipment Identity (IMEI) 271
International Subscriber Mobile Identity (ISMI) 271
Internet access 33
Internet Explorer 37
Internet Mail Consortium (IMC) 220
Internet TV 100
inventory control 61
iPic Web server 27
IPSec 195
IPv6 279
IrDA 21, 300, 392
IrMC 392
IrOBEX 303

Organization for the Advancement of
  Structured Information Standards 252
OSEK 82

P

packet data 217
packet switched data 268
Palm 29
Palm Computing 29
palm computing device 145
Palm OS 29, 151, 182, 186
Palm OS communication standards 145
Palm OS devices 17, 33, 141, 145
Palm VII 328
passivation layers 53
passthrough synchronization 389
payment 50
Pay-TV 100
PC 377
PC/SC 179
PDA 25, 124
PDC 270, 277
Personal Digital Assistant (PDA) 25, 87,
  94, 124
Personal Digital Cellular 270
Personal Information Management (PIM)
  25, 131
personal video recording 100
Personal Wireless Telecommunications
  289
personality protocol 320
personalization 55
PersonalJava 124, 171
Pervasive Computing 14
Philips 59
phone card 49
Phone line networking 310
Phone.com 215
phones 26
physical examination 53
physical security 52
Pico-cell 283
piconet 295
PIM 25, 35, 88, 131, 415
PIM application 388
PKCS 195
Plain Old Telephone System 259

Planetweb 105
Plug and Play 316
Pocket PC 131
point-of-sale terminals 26, 124
polygon 105
Portal 355
portal gateway 324
portlet 253, 356
portlet API 370
POS 379
Power Line Carrier 406
Power Line Networking 70, 305
power management 74
Power Management API 170
Power Monitoring API 170
powertrain controller 82
privacy 190
private key 51, 198
process(APDU) 159
production 52
profile 123
program memory 131
programming languages 115
Provisioning 229
proxy 318
Pseudo-random Noise code 262
Psion 37, 151, 170
PSTN 259, 272, 293
public key 198
public key algorithm 194
public network 50
Public Switched Telephone Network 259
public transport 51
push service 288
PWT 289

Q

QNX 88, 125
queue manager 182

R

Radio Data System 85
radio frequency antenna coil 58
radio wave 58, 59
RC2 193

RC3 193
record protocol 200
register() 159
remote diagnostics 407
remote home healthcare 409
remote maintenance 407
Remote synchronization 388
residential gateway 72, 335
Response APDU 55
retail 16, 423
reverse channel 263
Rijndael 193
Rivest, Dr. Ronald L. 193
RMI 318
RSA 195, 196
RSA Data Security 193, 196

S

S/MIME 195
Salutation 298, 315, 320
Salutation Manager 320
Samsung 29
San Francisco Framework 123
sandbox 118
SAP 249
satellite 101, 103
scatternet 295
Schlumberger 179
screenphone 19, 26, 40, 44, 170, 413
secret key 60
secret key algorithm 190
secure authentication 189
Secure Hash Algorithm (SHA) 196
Secure Socket Layer (SSL) 171, 199, 217
Secure Socket Layers API 170
secure token 50
security 60, 116, 200, 251, 379
security services 409
SecurityManager class 159
Sega Dreamcast 106
SELECT APDU 160
select() 159, 161, 173
Self-registration 359
Semantic Interpretation 241
sensor 82
serial communication 146
Service and Selection API 173

service broker 247
service discovery 320
service gateway 405
Service layer 177
service provider 215, 247
service provider portal 358
service requestor 247
Service Session 320
servlet 119
servlet interface methods 342
servlet model 325
Servlets 364
session beans 345
setOutgoingAndSend(...) 162
set-top box 26, 102, 124, 378
SHA 196
Short Message Service (SMS) 41, 85,
    170, 217, 263
Siemens 40, 179
Signaling System Number 7 273
signature 51
SignatureCardService 177
signing data 195
Simple Object Access Protocol 251
Simple Service Discovery Protocol
    (SSDP). 317
simplicity 21
Skins 367
smart actuator 67
smart agents 421
smart antenna 286
smart appliance 72
smart card 26, 57, 103, 408, 414
smart clock 72
smart control 26, 67
smart identification 66
smart label 26, 66
smart phones 43
smart sensor 67
smart tag 58
SmartCard 178
SMS 217, 383
SOAP 247, 251
software distribution 378
software radio 286
software vendor 215
Solaris 182
Sony 29, 104

X

Printing (Computer to Plate): Saladruck Berlin
Binding: Stürtz AG, Würzburg